T0122586

Progress in IS

More information about this series at http://www.springer.com/series/10440

Klaus North · Ronald Maier
Oliver Haas
Editors

Knowledge Management in Digital Change

New Findings and Practical Cases

Editors
Klaus North
Wiesbaden Business School
RheinMain University of Applied Sciences
Wiesbaden
Germany

Oliver Haas
Deutsche Gesellschaft für Internationale Zusammenarbeit (GIZ)
Bonn
Germany

Ronald Maier
School of Management
University of Innsbruck
Innsbruck
Austria

ISSN 2196-8705 ISSN 2196-8713 (electronic)
Progress in IS
ISBN 978-3-030-08809-5 ISBN 978-3-319-73546-7 (eBook)
https://doi.org/10.1007/978-3-319-73546-7

Printed on acid-free paper

This Springer imprint is published by the registered company Springer International Publishing AG part of Springer Nature
The registered company address is: Gewerbestrasse 11, 6330 Cham, Switzerland

Preface

Digital is all about flows; the information flows, knowledge flows and minds flow.[1]

The disruptive power of digital change is a major challenge for knowledge-based value creation worldwide. The transformation toward a digitized economy and society deeply changes how we manage information and knowledge, how we connect, collaborate, learn, and decide within and across organizations. While digitalization offers new opportunities for disruptive renewal, knowledge workers, managers, and organizations will have to recreate their governance, leadership, innovation, knowledge, and learning processes and practices as well as their work organization. New business models and digitally enabled co-creation emerge, requiring new ways of managing knowledge. The "Knowledge ladder 4.0" is the guiding conceptual model of this publication.

Given the complexity of digital transformation at different levels, this book will not cover all aspects related to the subject. In particular, legal and governance issues are not covered by the contributions.

This book focusses on digitally enabled knowledge-intensive value creation. We offer cutting-edge contributions including case studies from practitioners and academics working on managing knowledge in transformational contexts, divided into the following four sections:

(1) Digital enrichment of resources to leverage human performance,
(2) Collaboration and networking,
(3) Leading and learning and, finally,
(4) New forms of digitally enabled knowledge-intensive value creation.

A glossary of key terms enriches the book.

This publication provides guidance to academics, managers, consultants, trainers, coaches, and those interested to learn about transforming organizations in a knowledge economy 4.0.

[1] http://futureofcio.blogspot.de/2014/11/knowledge-management-best-quotes.html.

Our thanks go to the contributors who furthered our insights into the topic and to Christian Rauscher; Executive Editor at Springer for actively supporting the book project.

We thank in particular Christopher Drodge for language editing and proof-reading of the book chapters and Christina Sarigianni for merging individual contributions into one consistent book format as well as compiling the glossary.

We wish inspiring reading and look forward to feedback.

Wiesbaden, Germany — Klaus North
Innsbruck, Austria — Ronald Maier
Bonn, Germany — Oliver Haas
Spring 2018

Contents

Editors and Contributors

About the Editors

Klaus North, Dr.-Ing. is Professor of International Management at Wiesbaden Business School, Germany. His research covers knowledge and innovation management. He was Founding President of the German Knowledge Management Association (GfWM) and was Scientific Director of the German Knowledge Management Award. His textbook "Wissensorientierte Unternehmensführung" ("Knowledge Management", Springer Texts in Business and Economics) available in many languages has become a reference work on the subject.

Ronald Maier, Dr. is a Professor of Information Systems at the School of Management at the University of Innsbruck, Austria. He received his Ph.D. in Management Information Systems from WHU Otto Beisheim School of Management in Koblenz, Germany and a habilitation degree from University of Regensburg, Germany. His research interests include collaboration engineering, connectivity, crowdsourcing, and knowledge management. His research has appeared in journals such as Journal of Strategic Information Systems, Business & Information Systems Engineering, Computers in Human Behavior, IEEE Transactions on Learning Technologies, and Journal of Knowledge Management.

Oliver Haas is Head of the Global Division "Health, Education, Social Inclusion" at the Deutsche Gesellschaft für Internationale Zusammenarbeit (GIZ). He has been working in global transformation for more than 15 years. This led to various postings (Malaysia, South Africa, and Vietnam), where he managed projects in Vocational Education and Training. Prior to his current position, he worked at the World Bank in Washington D.C., USA as Senior Operations Officer. He holds a Masters in Sociology (Free University Berlin) and is a certified Organizational Developer. He has lectured at various European universities and institutes (Barcelona, Vienna, and Darmstadt) and is an editor of and a regular contributor to Germany's leading magazine on organizational development "OrganisationsEntwicklung."

Contributors

Maribel Acosta Karlsruhe Institute of Technology (KIT), Karlsruhe, Germany

Rebecca Arnold Johns Hopkins Center for Communication Programs (CCP), Baltimore, USA

Dominik Aronsky Semedy AG, Zug, Switzerland

Vanessa Bachmaier Innsbruck, Austria

Andréa Belliger IKF, Lucerne, Switzerland

Per Bergamin Institute for Research in Open, Distance and eLearning (IFeL), Swiss Distance University of Applied Sciences (FFHS), Brig, Switzerland

Piers J. W. Bocock Johns Hopkins Bloomberg School of Public Health, Baltimore, USA

Sabrina Bresciani University of St. Gallen, St. Gallen, Switzerland

Angela Fessl Know-Center, Graz, Austria

Fabricio Foresti Universidade Federal de Santa Catarina, Florianopolis, Brazil

Oliver Haas Deutsche Gesellschaft für Internationale Zusammenarbeit (GIZ) GmbH, Bonn, Germany

Peter A. Henning Karlsruhe University of Applied Sciences, Karlsruhe, Germany

Franziska S. Hirt Institute for Research in Open, Distance and eLearning (IFeL), Swiss Distance University of Applied Sciences (FFHS), Brig, Switzerland

Patrick Hofer Media Interface GmbH, Zurich, Switzerland

Ilona Ilvonen NOVI Research Center, Tampere University of Technology, Tampere, Finland

Jörgen Jaanus Tallinn University, Tallinn, Estonia

Sebastian Kernbach University of St. Gallen, St. Gallen, Switzerland

Michael Kohlegger University of Applied Sciences Kufstein, Kufstein, Austria

Christian Kreutz University of Applied Sciences, Darmstadt, Germany

David J. Krieger IKF, Lucerne, Switzerland

Tobias Ley Tallinn University, Tallinn, Estonia

Rupali J. Limaye Johns Hopkins Bloomberg School of Public Health, Baltimore, USA

Edith Maier University of Applied Sciences, St. Gallen, Switzerland

Ronald Maier Dept. of Information Systems, Production and Logistics Management, University of Innsbruck, Innsbruck, Austria

Angelika Mittelmann GfWM, Linz, Austria

Everton R. Nascimento Federal University of Santa Catarina, Florianópolis, Brazil

Klaus North Wiesbaden Business School, RheinMain University of Applied Sciences, Wiesbaden, Germany

Roberto C. S. Pacheco Federal University of Santa Catarina, Florianópolis, Brazil

Viktoria Pammer-Schindler Institute for Interactive Systems and Data Science, Graz University of Technology, Graz, Austria

René Peinl Hochschule Hof, Hof, Germany

Thorsten Petry RheinMain University of Applied Sciences, Wiesbaden, Germany

Christian Ploder Management Center Innsbruck, Innsbruck, Austria

Ulrich Reimer University of Applied Sciences, St. Gallen, Switzerland

Achim Rettinger Karlsruhe Institute of Technology (KIT), Karlsruhe, Germany

Anja Richert RWTH Aachen University, Aachen, Germany

Hans-Peter Schnurr Semedy AG, Zug, Switzerland

Stephan Schäper Julius Blum GmbH, Höchst, Austria

Isabella Seeber University of Innsbruck, Innsbruck, Austria

Tara M. Sullivan Johns Hopkins Bloomberg School of Public Health, Baltimore, USA; Johns Hopkins Center for Communication Programs (CCP), Baltimore, USA

Nina Suomi Tallinn University, Tallinn, Estonia

York Sure-Vetter Karlsruhe Institute of Technology (KIT), Karlsruhe, Germany

Stefan Thalmann Pro2Future GmbH and Know-Center GmbH, Graz, Austria; Institute for Interactive Systems and Data Science, Graz University of Technology, Graz, Austria

Gregorio Varvakis Universidade Federal de Santa Catarina, Florianopolis, Brazil

Rosina O. Weber Drexel University, Philadelphia, USA

Daniel Weihs Israel Institute of Technology, Haifa, Israel

Dirk Wenke Semedy AG, Zug, Switzerland

Gudrun Wesiak Know-Center, Graz, Austria

Stefan Zander Darmstadt University of Applied Sciences, Darmstadt, Germany

Value Creation in the Digitally Enabled Knowledge Economy

Klaus North, Ronald Maier and Oliver Haas

Abstract This chapter discusses the critical question of how to manage knowledge for value creation in digitally enabled economies. We introduce the concept of "Knowledge 4.0" to set the developments of how companies and organisations use digital technologies for knowledge creation and sharing into a historic perspective. We explain the chain of activities that create value in the digitally enabled knowledge economy following the model of the "knowledge ladder 4.0". The model helps to relate enabling technologies to changes and new forms of managing knowledge and knowledge work. In addition, this introductory chapter summarises the key findings of the contributions presented in the subsequent chapters that we group into the four topic areas: (1) digital enrichment of resources to leverage human performance, (2) collaboration and networking, (3) leading and learning and, finally, (4) new forms of digitally enabled knowledge intensive value creation.

1 Towards Digitised Knowledge Societies

The move towards an increasingly digital world is rapidly changing the ways in which people and organisations create, use & share data, information and knowledge. A common definition of 'digital transformation' is the one coined by

K. North (✉)
Wiesbaden Business School, RheinMain University of Applied Sciences, Wiesbaden, Germany
e-mail: Klaus.North@gmail.com

R. Maier
Dept. of Information Systems, Production and Logistics Management, University of Innsbruck, Innsbruck, Austria
e-mail: ronald.maier@uibk.ac.at

O. Haas
Deutsche Gesellschaft für Internationale Zusammenarbeit (GIZ) GmbH, Bonn, Germany
e-mail: oliver.haas@giz.de

K. North et al. (eds.), *Knowledge Management in Digital Change*, Progress in IS,
https://doi.org/10.1007/978-3-319-73546-7_1

Bounfour (2016), namely 'the change associated with the application of digital technology in all aspects of human society'. The corresponding digitisation of previously analogue operations, tasks and managerial processes profoundly impacts companies and organisations (Iansiti and Lakhani 2014).

We are witnessing a development towards digitised knowledge societies on a global scale. What does this mean? Knowledge societies are dominated by professional experts and their scientific methods. Knowledge economies are marked by the expansion of knowledge-producing or knowledge-disseminating occupations (Burke 2000; see also Adolf and Stehr 2017). "Knowledge 4.0" refers to a societal stage where applications of digital technologies are pervasive in everyday life, leading to a "digital ubiquity" (Iansiti and Lakhani 2014), and also contribute a significant share to value creation. Researchers find that smart, connected products with their four capabilities of monitoring, control, optimisation and autonomy transform competition in the digitally-enabled knowledge economy (Porter and Heppelmann 2014). Thus, professional expertise is increasingly leveraged or "augmented" Davenport and Kirby (2016) by cognitive and networked systems. For example, McKinsey forecasts a potential economic impact of five to seven trillion US$ through the automation of knowledge work by 2025 (Manyika et al. 2013).

Figure 1 shows this development in a historic perspective (cf. Van Doren 1991; Burke 2000) starting with the *"Age of Reason" (Knowledge 1.0)*. Even though in ancient times there have been schools of philosophers reflecting about knowledge, at least in Europe, the sixteenth century is considered as the start of a systematic scientific exploration of nature and the development of a more widely accepted scientific method. From about 1700 it became possible to pursue an intellectual career not only as a teacher or writer but also as a salaried member of certain organisations dedicated to the accumulation of knowledge, notably the academies of science (Van Doren 1991, p. 27).

16th – 17th Century	18th-19th Century	20th Century	21st Century
"Age of reason"	Industrial Society	Information and Knowledge Society	Digitized Knowledge Society
• Scientific penetration of nature (Rousseau, Galiliei, Newton ...) • Development of a "Scientific Method": systematic-methodical appropriation of new knowledge • Interaction between scholars and craftsmen, Emergence of "knowledge instiutions" (universities)	• Knowledge production permeates all areas of life • Industrial Revolution Separation of knowledge (planning / design) and execution (knowledge embedded in machines) • Professionalization of knowledge producers (engineers, doctors)	• Knowledge becomes the dominant production factor • Emergence of Computer, Internet Artificial Intelligence; Algorithms for routines • Dominance of professional experts and their scientific methods	• Digitization of everyday life and value creation • Cognitive, social, collaborative and networked systems, Augmented Intelligence • Digital penetration of professions and education
Knowledge **1.0**	Knowledge **2.0**	Knowledge **3.0**	Knowledge **4.0**

K.North 2017

Fig. 1 Phases of knowledge production and dissemination

The insights gained in the "Age of Reason" enabled the development of an *"Industrial Society" (Knowledge 2.0)* in the eighteenth century. Knowledge was increasingly embedded in machines and production systems. Knowledge creation had been professionalised.

The twentieth century witnessed the upcoming of an *"Information and Knowledge Society" (Knowledge 3.0)*. Information and knowledge became dominant production factors. From an organisational perspective, researchers saw the way knowledge is handled as a source for competitive advantage advocated by the resource-based view (Grant 1991) and the knowledge-based theory of the firm (Kogut and Zander 1992; Spender 1996). Organisations address the need for constant communication and acquisition of knowledge dispersed among employees (Hayek 1945) by applying organisational and IT mechanisms to establish an environment supportive of knowledge work (Davis 2002), also called knowledge management systems (Alavi and Leidner 2001; Maier 2007; North and Kumta 2018). Professional expertise and scientific methods are pervasive in this "Knowledge 3.0" stage.

In the *"digitised knowledge society" (Knowledge 4.0)*, digital transformation strategies take on a different perspective and pursue different goals. From a business-centric perspective, they focus on the transformation of products, processes, business models and organisational aspects owing to new technologies (Manyika et al. 2013) such as big data (Mayer-Schönberger and Cukier 2013), business analytics (Chen et al. 2012), cloud computing (Martens et al. 2011), cognitive systems (Samulowitz et al. 2014), robots (Brynjolfsson and McAfee 2014), social software (Kaplan and Haenlein 2010) and the Internet of Things (Porter and Heppelmann 2014). From a human-centred perspective, knowledge management's focus on collections of (documented) knowledge has been extended to comprise connections between people (Kaschig et al. 2016) and to embrace social relations with their corresponding technology support (Von Krogh 2012), also called social knowledge environments (Pawlowski et al. 2014).

Be it in business or in everyday life, digital transformation strategies have certain elements in common. These elements can be ascribed to four dimensions: *use of technologies*, *changes in value creation*, *structural changes*, and *financial aspects* (cf. Matt et al. 2015). The transformation of analogous assets into electronic representations is associated with new forms of cognition.

2 Understanding Value Creation: The Knowledge Ladder 4.0

Let us now have a closer look at how digital technologies enable value creation based on data, information and knowledge. We will explain the relationships following the model of the "knowledge ladder" (North 2005; North and Kumta 2018).

Value creation in a knowledge economy is a step by step process including many learning loops in which resources are enriched. The organisation of symbols into data represents the first step in the creation of value, which, in a next step, are given meaning to become information. Information serves as input for decision-making and actions, which requires the capability of selection, sensemaking and interpretation. From this perspective, knowledge is the result of information processed by the conscious mind. While information is organised data, knowledge refers to the tacit or explicit understanding about relationships among phenomena. It is embodied in routines or algorithms to perform activities, in organisational structures and processes. Knowledge is embedded in believes and behaviours, a large part of it is tacit. The value of knowledge becomes evident only if the "know-what" is converted into "know-how" which manifests as actions. The ability or capacity to act appropriately in a specific situation is known as competence. von Krogh and Roos (1996, p. 45) clarify the dynamics of competent acting: "*... we view competence as an event, rather than asset. This simply means that competencies do not exist in the way a car does; they exist only when the knowledge (and skill) meet the task*". This capacity to make an appropriate choice of actions depends upon a wide repertoire of action potentials which is based on experiences and expertise developed over time. Value is the result of the interplay between multiple competencies of a person, a group, a network, an intelligent system or an institution based on its unique information and knowledge resources (North and Gueldenberg 2011). From this perspective, competitiveness is the result of the capability to bundle competencies uniquely and to renew them to create a unique customer value (cf. Hamel and Prahalad 1994; Teece 2009).

How do digital technologies enable and change these value creation processes? To explore this, we have created a "knowledge ladder 4.0" shown in Fig. 2. It relates the steps of knowledge-based value creation to selected enabling technologies (found below the knowledge ladder in the lower part of Fig. 2) and to their effects on the digitally enabled enrichment of resources (displayed above the knowledge ladder in the upper part of Fig. 2).

We will explore the technological developments and the effects they have on knowledge-based value creation in two steps: Firstly we will move up the knowledge ladder linking technologies and repercussions on managing data, information and knowledge. Secondly, we will look into four application areas (digitally enabled enrichment of resources to leverage of human performance, collaboration and networking, leading and learning, digitally enabled value creation, see Sects. 3, 4, 5 and 6).

Let us now move up the digitally enabled knowledge ladder and look into some critical steps.

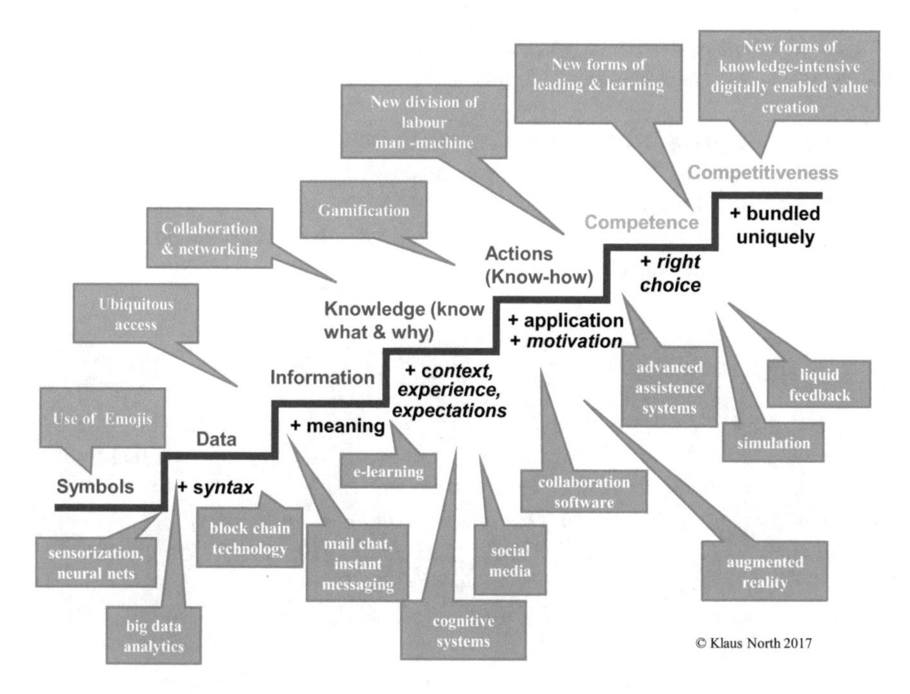

Fig. 2 Knowledge ladder 4.0: digital technologies for knowledge-based value creation

2.1 From Data to Information—Data & Surveillance Capitalism

Increasingly high-performance data analytics (HPDA) enables the acquisition and analysis of huge volumes of data and its subsequent transformation into information as a basis for actionable insights. Algorithms such as neural networks are able to interpret sensory data, recognise patterns, cluster and classify enormous amounts of data (e.g. face recognition of thousands of people).

> Researchers from Google have used a deep-learning network to find and read the house numbers on many millions of Google Street View shots, even if they were rotated, tilted or uncommon. This served to locate the houses exactly on Google Maps. A team of people would have been involved with such a task for many years. The computer managed it in less than an hour. Source: Eberl (2016)

Such systems create actionable information but require humans with the knowledge to be able to act on the basis of that information. This means that the analytic capabilities of systems and the sensemaking capacity of humans and organisations have to match. Van der Aalst and Damiani (2015) argue that a major challenge is to relate massive amounts of event data to processes that are highly dynamic.

Researchers have associated the capabilities of big data analytics to a "data capitalism" which is "cashing in on our privacy" (Thornhill 2017). In this view, data has become an important source of monetisation as it enables the analysis of customer preferences and provide user-optimised advertising, products and services, and to further develop them.

Surveillance capitalism

Zuboff (2016) argues that we are entering a "surveillance capitalism" where the game is selling access to the real-time flow of our daily life –our reality— in order to directly influence and modify people's behaviour for profit: *"This is the gateway to a new universe of monetisation opportunities: restaurants who want to be your destination. Service vendors who want to fix your brake pads. Shops who will lure you like the fabled Sirens. The "various people" are anyone, and everyone who wants a piece of your behaviour for profit. Small wonder, then, that Google recently announced that its maps will not only provide the route you search but will also suggest a destination".*

As data and its presentation are a source of revenues they are increasingly "manipulated".

"Data curation" includes processes to create, maintain, and validate data to ensure the value of the data and present it under the perspective of generating revenues.

Hofmann (2017) reports on recent studies examining the data policies of digital platform providers. Although platform members increasingly produce, evaluate and circulate content, they rarely control the flow of information. The rise of new media channels also increases the power of the algorithms. Facebook, for example, currently categorises, filters and hierarchises approximately 500,000 comments per minute. This is done according to rules, which are not disclosed, but in fact decide about light and shadow in the communication flow. Digital platforms primarily "reward" those contributions with visibility in the news stream, which have the greatest prospects for further spread and thus promise not only attention, but also advertising revenues. This radical decoupling of quality and popularity of content, for example political news, explains why targeted misreporting (fake news) enjoys often the largest spread in the social networks. The algorithmically curated information flows (cf. Domingos 2015) or "newsfeeds" do not address people as political citizens, but as a data source whose presence should be held on the platform to gain continuous up-to-date information on their interaction behaviour (Urbinati 2014; Hofmann 2017). Summing up, there is a lot of debate about the major governance issue of how to deal with ownership and control of consumer data which manifests, among other forms, in the initiative towards a European Charter of Digital Fundamental Rights.[1]

Apart from the use of data for monetisation, there is a contrary movement towards open data which builds on the foundations laid by the highly visible and sustainable open source software initiatives and covers fields such as open content, open data, open government (OECD 2016), open innovation (Chesbrough 2006), open science (Le Dinh et al. 2015) or citizen science (Newman et al. 2012). These "open" movements advocate accessibility, collaboration and therefore the power of free or "democratised" innovation for digitised knowledge societies (Von Hippel and Von Krogh 2003; Von Hippel 2005, see also Pacheco et al.'s contribution on digital science in this book).

2.2 *From Knowledge to Competence—The (R)evolution of Knowledge Work*

At the centre of the knowledge ladder is the issue of how knowledge is put into action to create business value. *Enabling technologies* provide tools for agile communication and collaboration as well as intelligent systems leveraging human performance. Concepts such as "Augmented Intelligence" or "Advanced Artificial Intelligence" or "Cognitive Computing" describe systems that learn at scale, reason with purpose and interact with humans naturally (Kelly III 2015). "Cognitive" refers to the properties that the system integrates knowledge from diverse sources including current state and past experiences made by the system, "naturally" interacts with the user plus that the system generates and evaluates new hypotheses and capabilities (Samulowitz et al. 2014). What are the limits of these systems? (Davenport and Kirby 2016) argue that people are better able to interpret unstructured data, have the cognitive breadth to simultaneously act on different tasks as well as the judgment and flexibility that come with these basic advantages. Bostrom (2014) raises the questions what happens when machines surpass humans in general intelligence? Will there remain distinctive capabilities of humans if machine brains surpassed human brains in general intelligence?

The difference between artificial intelligence and cognitive computing
If your smartwatch had machine learning algorithms 'fed' inside it such that it can predict your health diagnosis by measuring your heart pulse: it might be a good example of AI—but a bad fit for cognitive computing, as it is still not interacting 'naturally' to humans. A Cognitive Computing system would rather have:

[1]https://digitalcharta.eu/wp-content/uploads/2016/12/Digital-Charta-EN.pdf.

1. taken your verbal command of 'Hey AI Doctor, please tell me what is wrong with my health';
2. and would have 'arrived upon a plan' to check your pulse. You as user did not ask to 'check your pulse'—you only told that you were not feeling good—The agent arrived upon the plan to check your heart pulse itself using its intelligence. It could have arrived upon the plan to check your temperature using a thermometer as well.
3. and deduced the repercussions of what a low heart pulse would do once it was detected on your wrist. And would have explained the repercussions to you in 'natural language' much like how a human doctor would do.
4. A Cognitive Computing smartwatch would have memorised your health records from a time period and would have recorded the latest state or environment you are in, and would have given personalised recommendation based on that. So from its memory it would have memorised that you are a diabetic patient and you recently attended your son's wedding and ended up eating lots of sweets—and its recommendation will take these two facts into account.

Source: adapted from[2]

While these technologies change everyday life, they have a particular influence on managing knowledge work. In the following we summarise major trends that will affect knowledge work in the future, as have been identified by a number of studies (cf. Intel 2014; BMAS 2015; Lehtiniemi et al. 2015; Telekom 2015). The issues summarised here will be discussed in further detail in Sects. 3, 4, 5 and 6.

New forms of interaction between humans and machines: Smart systems will emerge and collaborate with humans, changing the nature of work, and driving a re-imagination of work content and work process. Various forms will coexist in the future; from people who control machines, machines as people's peers, to the merging of machine and human or the complete takeover of activities by intelligent systems. This will lead to a redefinition of expertise. If in the future expertise will be defined as human (expert) plus intelligent system, a major issue will be how people and machines will learn together? How will systems develop common sense and ethics (tacit knowledge).[3] Who will evaluate potential courses of action? How will systems weigh chances and risks? How will we appropriate the created value? Will humans remain capable of action if the assistance systems fail? Creative activities, for now, remain a domain for humans. Will intelligent systems become creative in the future? How will we escape the implied competition between

[2]https://www.quora.com/What-is-the-major-difference-between-Cognitive-Computing-and-Artificial-Intelligence.

[3]Compare: http://www.huffingtonpost.com/entry/ethics-and-creativity-in-artificial-intelligence-an_us_593047b4e4b09e93d7964848.

humans and machines? Concerning the latter, (Davenport and Kirby 2016) suggest to view the rising capabilities of machines not as threat and replacement for humans, but with a mindset of augmentation, that is to employ machines as partners and collaborators in creative problem solving.

Distributed value generation: The new world of work is characterised by networks. Standardised back-end processes are shared between companies, without being visible to customers or employees. This creates jobs without a clear organisational membership and products without a clear sender. Boundaries within and between organisations fade as work is organised in temporary projects done by people with temporary affiliations.

Work without borders: Highly qualified specialists work around the world as part of project work. Qualifications are globally transparent and comparable. The spatial location of the service provider is no longer relevant. For the first time, labour thus acquires the same mobility as capital. The traditional places and times of work dissolve. For employees, this results in new options, for example to improve the compatibility of family and work life, but also new burdens ("always on") (Mazmanian et al. 2013; Waizenegger et al. 2016). Offices will serve as temporary anchor points for human interaction rather than daily travel destinations. Office as a Service (OaaS) will become a strategic tool to connect employees in the right place, at the right time.

Crowdworking: Companies are increasingly focusing on customers instead of employees. Many (digitisable) services are offered by volunteers and free of charge, for example in open innovation contests a crowd submits ideas for innovations to a contest sponsor, usually hosted by an open innovation platform (Adamczyk et al. 2012; Boudreau and Lakhani 2013). Prosumerism blurs the boundaries between producers and consumers. Volunteered digital work complements or replaces professional employment. In addition, digital services are divided into ever smaller parts and delegated to "virtual labourers". Big data analytics can be used to assign value contributions to specific individual workers. Cloud- or clickworkers offer online services on a crowdsourcing platform such as Amazon MechanicalTurk, usually paid on a per-task basis for, for example, web research, text creation, tagging, categorisation and translation. In the foreseeable future, many of these activities will be fully digitised. While such developments certainly offer enormous opportunities for individuals, organisations and the society at large, some researchers have also described risks involved in this new form of global organisation of labour (e.g., Ettlinger 2016).

Self-management as a core competence: As traditional work relationships and processes are dissolved, knowledge workers have to learn self-management including self-organisation and self-control. Self-management means, amongst others, to organise work, to define or redefine work objectives, to choose adequate means and methods, to organise one´s own competence development as well as to find a sustainable work-life-balance.

Digital Leadership: The distribution of work in different locations is accompanied by a shift from a "presence culture" to a "result culture" ("do your work wherever you are"). Leaders need to learn to align individual interests of these dispersed workforces with organisational goals. The art is to build and maintain personal ties through impersonal channels enabled by information and communication technologies.

2.3 *From Competence to Competitiveness*

Moving up the knowledge ladder, the ultimate objective is to create unique customer value based on the capabilities of an organization. Unique knowledge in the form of a superior proprietary technology can lead to long-term market dominance, as we have seen in the case of Google's superior PageRank algorithms. Such proprietary technology can be the base for competitive advantage in a "winner takes all" manner due to the network effects created. However, as a result of disruptive technologies, the traditional boundaries for industries are being blurred and barriers for new entrants are lowered due to the pervasiveness of digitally enabled knowledge sharing and low equipment costs.

Switching costs as source of maintaining competitive advantage Switching costs might prevent customers from exchanging one service for another that might be functionally better, which might be more than offset by the cost of transfer to the new service or might (initially) offer inferior network advantages. Examples for such effects are social networks. "Switching cost is also essentially what makes Facebook so difficult to defeat: For a user, to move into another similar social medial platform implies the cost of building up a new «friends» base without any guarantees that his or her friends will do the same. The Google+ debacle is a powerful reminder of how resilient switching cost may be…."[4]

To retain and rebuild competitive advantage organisations need to develop capabilities for digital renewal and learn how to create and implement digital business strategies. The literature increasingly links digital transformation to the development of "dynamic capabilities" (Teece et al. 1997; Eisenhardt and Martin 2000; Yoo et al. 2012; Bharadwaj et al. 2013; Teece 2017). Dynamic capabilities have been defined by Teece et al. (1997) as an organisation's ability to integrate, create and reconfigure both internal and external competences to address changing environments. Karimi and Walter (2015) ascertain the role of dynamic capabilities in response to digital disruption. Their empirical results on the digital transformation of the newspaper industry suggest that dynamic capabilities are positively

[4](https://salvadorbaille.com/2017/02/07/so-you-think-you-have-a-competitive-advantage-i/).

associated with building digital platform capabilities, and that these capabilities impact the performance of a company's response to digital disruption.

Dynamic Capabilities and (Digital) Platform Lifecycles
To adopt a longer-term perspective on the competitive requirements of their platform-based business, managers should understand the dynamics of (digital) platforms: Managers should reflect on the four-stages—Birth, Expansion, Leadership, and Self-Renewal—of the platform lifecycle in terms of its dependence on the dynamic capability categories of sensing, seizing, and transforming. The requirements evolve from a heavy emphasis on generative sensing and planning-stage seizing in the birth phase, through greater emphasis on "seizing" activities and minor transformations as the platform, ideally, grows and stabilises. When platform renewal is called for, the emphasis returns to sensing future possibilities and generating new ideas for a platform and business model, developing them alongside the existing business, and eventually undertaking a major transformation to restart the platform lifecycle.
Source: Teece (2017)

The development of dynamic capabilities is closely linked to learning and managing knowledge acquisition, creation and sharing within and across organisations. Following Pavlou and El Sawy (2011)'s argument that dynamic capabilities are based on sensing, learning, integration and coordination, we will explore how these capabilities are related to managing knowledge in digital transformation and what are the specific challenges of coping with such turbulent and disruptive environments (North and Varvakis 2016). (Chap. 12)

Sensing capability: turbulent and disruptive environments require (1) receptiveness to weak signals, (2) a constant gathering of information on the business environment, market and technology trends, plus customer needs, followed by (3) the interpretation of this information with the available knowledge and (4) to draw conclusions. The challenge here is to effectively communicate internally across units and fields of knowledge what is changing and create a shared understanding of what this means for the organisation.

Learning capability: new business opportunities and threats to existing business arise from digitisation, which require new knowledge and skills to offer new or revised products, services or to change business models. The challenge here is to integrate quick learning loops into daily operations and business development.

Integration capability: integration focuses on overall sense-making and on building of a shared understanding throughout the organisation. Shared tacit knowledge is at the core of an integration capability. New or changed ways of doing business require the ability to combine individual knowledge into new operational processes and practices of a team or a business unit. The challenge here is to ensure that everybody understands and shares what digitisation means for the business and is enabled to assume their new role in the "digital game".

Coordination capability: coordination focuses on orchestrating individual tasks and activities. Organisations need to maintain an attitude that accepts change, establish monitoring systems and ensure the availability of financial and human resources. The challenge lies in empowering employees who need to develop the knowledge, skills and attitude needed to decide, monitor and act in an entrepreneurial spirit in a "work 4.0" environment.

The above described capabilities are a basis for **developing and implementing digital business strategies** (Mithas et al. 2013). Competence development needs to be aligned with a digital business strategy in order to create business value that differentiates a company from its competitors. (Bharadwaj et al. 2013) argue that a digital business strategy is different from traditional IT strategy in the sense that it is much more than a cross-functional strategy, and it transcends traditional functional areas and various IT-enabled business processes. Therefore, digital business strategy can be viewed as being inherently "transfunctional" (see also Koch and Windsperger 2017).

Yoo et al. (2010, p. 724) argue that pervasive digitisation gives birth to a new type of product architecture: "*The layered modular architecture extends the modular architecture of physical products by incorporating four loosely coupled layers of devices, networks, services, and contents created by digital technology.*" For example, as most subsystems of an automobile are becoming digitised and connected through vehicle-based software architectures, an automobile has become a computing platform on which other firms outside the automotive industry can develop and integrate new devices, networks, services, and content (Henfridsson and Lindgren 2010).

Along similar lines, Koch and Windsperger (2017, p. 2) propose a network-centric view, where firms may achieve competitive advantage by co-creating value with interconnected firms in the digital environment. They refer to a digital ecosystem as a network of companies and other institutions that is inter-linked by complementary interests to create and sustain value around a common digital platform. Therefore a digital business strategy extends the scope beyond firm boundaries and supply chains to dynamic ecosystems that can even cross traditional industry boundaries (Bharadwaj et al. 2013).

2.4 Implications for Managing Knowledge

In the past, organisations primarily engaged in knowledge management (KM) practices that focused on managing current knowledge and past experiences with a strong emphasis on documentation (Pawlowsky et al. 2011). KM has y been acknowledged as a factor that impacts on an organisation's performance (Zack et al. 2009) in an environment characterised by complexity and turbulence. A hypercompetitive "VUCA" environment (volatile, uncertain, complex, ambiguous), changed communication behaviours and the evolution towards knowledge work 4.0 set the scene for managing knowledge within and across organisations in the digitised society.

In analogy to the concept of "ambidexterity" (Tushman and O'Reilly 1996), KM has to support a number of conflicting knowledge activities such as "exploitation" and "exploration" (March 1991) or "sharing" and "protection" (Manhart et al. 2015; Loebbecke et al. 2016) at the same time in such VUCA settings. In the light of the ensuing conflict between stability and flexibility, KM stabilises the organisation's capabilities in a mode of protection and exploitation on the one hand and concurrently supports dynamic capabilities in a mode of exploration and sharing to enhance agility and renewal. An organisation's ability to manage such seemingly contradictory processes and practices increasingly gains importance with digital transformation. Let us look in more detail into these two functions of KM (North and Haas 2014).

Operational KM as stabiliser

Also in the future, operational KM will continue to aim at making the right knowledge available at the right time and place to support the employees of an organisation, plus the relevant stakeholders in the organisation's environment for day-to-day operations. The means and ways of how to achieve this ambitious objective, however, will change under a KM 4.0 perspective. Organisations can engage in the following activities to stabilise the portfolio of competencies in an organisation:

(1) **Facilitate ubiquitous and curated knowledge flows**: Quick, easy and ubiquitous access to the knowledge base of the organisation and across organisations gains importance and can be characterised by decentralized, and increasingly peer-networked repositories augmented by rapidly evolving machine intelligence. Murray and Wheaton (2016) argue that there is a need for "knowledge curation" as even advanced technologies such as machine-readable ontologies have not yet come close to being able to extract deep meaning or accurately organize content into proper contextual categories. Curation establishes, maintains and adds value to repositories of knowledge and helps to keep them relevant and up-to-date. In practice, curation could mean that an expert compiles a selection of links and shares them, adding a clear explanation of the selection criteria used to compile the list as well as brief introductions explaining why each link is relevant (Spiro 2017). However, the decisions necessary in such a process might also be augmented by machine intelligence, by a team or crowd who are engaged in the domain that is curated by the expert.

(2) **Enable collaboration**: The emphasis of KM has shifted from the support for collecting to connecting knowledge activities (Kaschig et al. 2016) that help to make collaboration work. Connecting knowledge activities are viewed comprehensively to comprise connections between people, that is joint knowledge creation, sharing and acquisition, and connections of knowledge both in an abstract and a manifest form—the integration of knowledge from diverse sources be it people,

documents or algorithms. KM needs to help people to develop the competencies needed for work 4.0, amongst which competencies for technology-mediated collaboration and collaboration with machines as "team mates" (Seeber et al. 2018) stand out.

(3) **Monitor and control augmented learning and decision-making**: As organisations increasingly develop and deploy algorithms to automate routine knowledge tasks and decisions plus provide decision support in known situations, such automated knowledge behaviour needs to be monitored and controlled to be not only efficient, but also compliant with an organisation's internal and external regulatory system. The corresponding experiences made need to be systematically reflected and interpreted in this respect, KM will have to ensure transparency of cognitive technologies, so that users will always be aware of how cognitive systems "think" and act. A particular challenge here is to identify and leverage the tacit knowledge of subject matter experts or communities and to provide the means for humans to keep up to date with the exponential growth of opportunities created by self-learning systems.

Strategic KM as catalyst

In an increasingly turbulent and complex environment, it is the responsibility of KM to critically examine knowledge and competencies of the organisation, a network or business ecosystem and identify its "blind spots". Here, KM takes on the role of an innovator and "irritates the system" by questioning past learning, established behaviours and practices (North and Haas 2014). KM must succeed in supporting the development of "dynamic capabilities" of organisations to reconfigure, realign and integrate core competencies with the help of external resources. Organisations can engage in the following activities to productively foster the growth of capabilities for improved organisational performance under shifting environmental conditions:

(1) **Identify critical knowledge**: KM needs to provide deep insight into the critical knowledge assets required to embark on the learning journey involved in the activities to pursue future organisational goals. Therefore, KM also questions current core competencies, intellectual property rights, market and industry comprehension, and customer understanding and expectations (MacMillan et al. 2017). KM should identify the pockets and islands of knowledge creation within and beyond the organisational boundaries that can be connected to acquire new core competencies that can be appropriated by the organisation. Hence, organisations need to integrate isolated knowledge on and views of the environment to make sense of information as a basis for seizing new opportunities and transforming the organisation. Strategic knowledge mapping helps to uncover and take an integral view on critical knowledge assets, providing the context for discovering the most promising digitalization strategies (MacMillan et al. 2017).

(2) **Facilitate sensemaking and shared understanding** as a basis to act: Klein et al. (2006) describe sensemaking as a way of understanding connections between people, places and events that occur now or occurred in the past, in order to

anticipate future trajectories and act accordingly. The ability to frame (set in context) and reframe problems and observations is particularly important when big data analytics seem to provide answers without adequate context knowledge (Madsbjerg 2017). Deep insights and shared understandings emerge through multiple discourses of people (Kurtz and Snowden 2003; Kolko 2010). The underlying mechanisms of meaning making can be seen as the essence of collaboration (Stahl et al. 2006) and highlight that negotiation processes are interactive, reciprocal and that meaning resides in the social realm and can be manifest in socio-technical systems (Dennerlein et al. 2016). Sensemaking is a shared and communal activity that produces knowledge appropriate for action, but biased heavily based on the individuals doing the sensemaking—that is, each group of people who have the various sensemaking conversations will "talk into existence" a very different set of situations, organisations, and environments (Weick et al. 2005). In this view sensemaking is a process that is highly collaborative, effective for organisational growth and planning in both the short and long-term, and highly dependent on interpretation.

The increasing complexity of work tasks intensifies the demand for collaboration, which in turn requires KM to support the creation of shared understanding among work groups (Bittner and Leimeister 2014). On the organisational level, shared understanding among organisations that collaborate in business ecosystems is vital for efficient knowledge creation in such ecosystems. Researchers found that at the beginning of business ecosystem formation, organisations need to share their capabilities, expertise, and knowledge and in particular make the tacit knowledge explicit in order to boost integration (Annanperä et al. 2016).

(3) **Encourage renewal, agile learning and reflection**: To ensure renewal in an ever changing and often disruptive environment, firms have to learn how to systematically develop new business models and non-profit organisations need to be capable of redesigning their missions in an accelerated manner (cf. Kotter 2014). KM can play a key role in these above described issues related to render organisations more dynamic in the future. In an environment that is characterised by unpredictability and various unanticipated crises, KM must support quick problem-solving, encourage constant experimenting, foster collaborative learning and facilitate professional reflection to learn from mistakes. For example, KM can be responsible for developing a "next practices" process in an organisation. Future developments in a business or technology area, or in a business model can be explored in cross-departmental workshops which include a range of stakeholders such as customers and the scientific community.

(4) **Build platforms for engagement**: In an era of information overload, human attention is a scarce resource. In order to attract heterogeneous and unexpected knowledge it is of strategic importance to build platforms that engage members in and beyond the organisational boundaries. Ghazawneh and Henfridsson (2010) point to the importance of governing third-party development through specific knowledge which they call "platform boundary resources". These include the design of technical boundary resources such as software development kits and application programming interfaces and social boundary resources such as

incentives, intellectual property rights, and control systems. KM's role is to build platforms that attract engagement of a wider community for the strategic development of organisational competencies, products and services.

After having clarified how digitalisation interrelates with managing knowledge in general we will now look into four application areas (digitally enabled enrichment of resources to leverage of human performance, collaboration and networking, leading and learning, digitally enabled value creation, see Sects. 3, 4, 5 and 6) and summarize the contributions of this book.

3 Digitally Enabled Enrichment of Resources to Leverage Human Performance

As we have explained above, the model of the knowledge ladder symbolises how resources such as data or information are connected, given meaning, related to contexts and thus enriched to enable value creation. The contributions which are grouped in this section have in common that they explore how this enrichment works, what are limitations and future perspectives. Particular emphasis is put on the interplay between smart systems and knowledge workers.

A key enabler for the enrichment of resources is the area of **Semantic Technologies**. While most semantic technologies originate from the vision of representing the existing Web in a machine-processable format, it's most notable success so far are large cross-domain "knowledge graphs". They are created by collaborative human modelling and linking of structured and semi-structured data. Rettinger et al. introduce the latest innovations in modelling knowledge using knowledge graphs and how those knowledge graphs enable value creation by making unstructured content, like text documents accessible by machines and humans, and finally how semantic technologies help to make hard- and software components in cyber physical systems interoperable.

An application of semantic technologies can be found in **clinical decision support systems (CDS)**. Healthcare professionals often make clinical decisions under time constraints within a highly complex patient situation. The aim of CDS, therefore, includes the improvement of clinical decisions by providing and applying evidence-based medical information at the right time of decision making. Amongst others, intelligent algorithms can detect specific patterns that are indicative of clinical conditions or diseases. Schnurr and colleagues explain the interplay of such "intelligent systems" and healthcare professionals and how knowledge is created and maintained through a collaborative process between knowledge engineers and clinicians. A major issue is how intelligent systems and users learn together, or from each other. The authors argue, that in future, we have to train and teach our computers. Maybe computers will have to pass exams and need to be certified to support humans in critical application domains.

Smart systems do not only provide guidance but increasingly support or interact with humans in physical tasks such as care robots in smart homes or smart robots within smart factories. Humans and machines will work side by side in so-called "hybrid teams." The success of future production or assistance concepts will strongly depend on the successful implementation of direct cooperation between humans and robots. As a step further, robots should be able to identify and adapt to individual strengths and weaknesses and take over the role of a workmate, helping to construct knowledge in social, teamwork-oriented processes.

In her contribution Anja Richert explores the interaction of **hybrid teams of humans and robots**. The empirical part researches if the appearance of the robot and its behaviour influence the perception of the robot as a partner as well as the human cooperation behaviour.

Kohlegger and Ploder take a further look into the **interplay between digital assistance systems and knowledge workers** to allow new, deep insights into phenomena and support business value creation. A model of data driven knowledge discovery is presented that describes how this interplay could look like and is critically discussed using real-world cases. The main conclusions are that it is crucial to (1) separate data-driven and expert-based analysis in knowledge discovery, (2) clearly describe the problem that should be solved by the analysis, (3) understand the particular domain that analysis is applied to, (4) complement data-driven with expert-based analysis, and (5) understand the relation of analysis and action implementation.

Digital change and Industry 4.0 concepts do not erase the need for human insight or experience. Experience plays an eminent role particularly in highly complex and automated digitised work environments. At the same time, digital transformation opens up new opportunities for implementing solutions for **advanced experience management** by automatically capturing, exchanging and preserving lessons learned and offer support that is both context-aware and situation-specific. These are based on key technologies such as information extraction from texts, process mining and text mining. Maier and Reimer discuss in their contribution various technological solutions for automating (parts of) capturing and providing experience-based knowledge:

- integrating knowledge provision into the work processes in a way that is both context-aware and specific to the situation
- using process mining to predict an employee's next activities and provide relevant knowledge
- extracting information from texts and text mining to identify good practices e.g. from discussions on social media.

The authors argue that the suggested approaches help to solve the dilemma that on the one hand companies deem experience and its transfer and exchange very important, while on the other hand well-known methods for capturing and preserving valuable experience within the company are rarely used due to the effort and time they require.

4 Collaboration and Networking

The digital transformation supports a move towards more flexible, collaborative and agile approaches to doing business. This requires the development of mindsets and instruments for digital and social collaboration and networking. The contributions which we have grouped under this topic exemplify how technologies and their applications can foster networking and collaboration in different contexts.

In his contribution Peinl provides an overview over the development of **digital collaboration solutions** and shows how social software, and machine-understandability have changed to better support knowledge processes. To enhance flexibility and agility Active Case Management (ACM) is increasingly applied. ACM is characterised by goal- and data-orientation, transparency, runtime flexibility, continuous improvement and integration of information systems. It looks at tasks from a case perspective, which is familiar to doctors or lawyers. Within a case, tasks can be arranged in sequence, but will most of the time have no strict order. A case therefore provides context for tasks as well as data and collaborating people.

Digitised visualisations are gaining importance in collaboration and networking. "Picture" based social networks are becoming increasingly popular in private and business contexts. Visual digital knowledge sharing tools, for example, can provide guidance to a meeting by collecting input from participants, keeping a record of participants' contributions (that can easily be shared online or by email), assessing options with voting systems or mapping different opinions. In this context, visualisation emerges as a powerful way to support knowledge work. Visualisations are also well suited to externalise tacit knowledge. Kernbach and Bresciani argue that **visual knowledge mapping** is a very effective way of sharing, integrating and creating knowledge and value for collaborative work in organisations. The authors present ten visual tools, evaluate and classify them. As one of the conclusions a role of a *Visual tool curator* is proposed: Have someone in your organisation with good knowledge about the problem and the visual tools to test different tools and check requirements of the internal IT department for the implementation.

Organisations rely on employees to externalise their tacit knowledge in order to effectively conduct knowledge-intensive work processes. Tacit knowledge externalisation is particularly important in times of digital transformation where product and service innovation cycles become shorter and require creative, quick decision-making. Hence, it is essential for businesses to provide employees with rich **social interaction opportunities** so that they can articulate their personal tacit knowledge in order to create understandable, usable knowledge that can be stored and made available for organisational members for increased sustained competitive advantage. Tacit knowledge externalisation (TKE) represents one of those knowledge-intensive processes that is fostered by corporate social media (CSM). In their contribution Bachmeier and Seeber investigate which TKE mechanisms have the strongest effects on knowledge utilisation when supported by CSM. Results of a survey from 381 employees from the hospitality industry using the CSM platform *hotelkit* are reported. The main findings are that TKE on CSM has strong positive effects on knowledge utilisation and that storytelling and practice demonstration are the most powerful mechanisms for that.

The creation of digital innovations requires active participation and knowledge sharing on behalf of all collaboration partners in inter-organisational settings. However, while the participants collaborate, they also have their own interests and as they are competitors in many cases, they have to protect their competitive knowledge. Collaboration thus requires **balancing of knowledge sharing and protection** on both the organisational and individual level. Thalmann and Ilvonen review literature from several domains from the point of view of how the balancing act is scoped and what kind of measures prior research on the area identifies. The balancing act is examined on the channel, partner and artefact levels.

Based on the literature review the authors conclude that the balancing happens on three levels of detail: (1) the decision about using certain communication channels, (2) the decision to share/collaborate with certain partners and (3) the decision to share a certain knowledge artefacts. Thereby, it turned out that balancing on all three levels require a careful consideration of the benefits and risks of sharing or not sharing, and the norms and values of the knowledge sharing community.

Collaboration and the **development of networks** requires a shared repertoire of understanding based on **social, geographical and professional "proximities"** of the members. In their contribution Thalmann and Schäper, therefore, investigate how networks of organisations can make use of proximities to enhance the localisation of knowledge. This is a particularly relevant question for SMEs in need to access and absorb new knowledge. The contributors found that social proximity is perceived as very important while evaluating a problem, creating shared understanding, and finding new solutions. It seems that a high social proximity leads to a more open communication by lowering the barriers for contribution and active engagement in the previous activities of the localised learning process. It seems that the geographic proximity plays a complementary role in building and strengthening social, organisational, professional, and cognitive proximity. Professional proximity seems to be an important dimension for localising knowledge in networks. A high professional proximity can speed up the localisation of external knowledge by making use of the professional expertise. Therefore, the members' professional identity plays an important role for facilitating the collaboration and finding solutions.

5 Leading and Learning 4.0

With digital transformation's move towards more flexible, collaborative and agile approaches, also demands and learnings have changed for leadership and management to doing business. Referring to the knowledge ladder, leading and learning go hand in hand in order to acquire, develop and apply competencies.

Such individual, team and organisational competencies are needed to create digitally enabled knowledge-based products and services as well as business models.

The substantial technological changes challenge organisations because they find themselves in a volatile, uncertain, complex and ambiguous (VUCA) environment that demands changes, or rather, extensions of traditional leadership approaches. In his chapter, Thorsten Petry suggests that managers need to adopt digital leadership, an adaption of their leadership style to the challenges of a VUCA environment. Petry conceptualises digital leadership as agile, participative, networking, open and trust-based. The chapter describes a number of instruments to support such digital leadership. Petry calls for ambidextrous leadership so that the instruments of digital leadership should be seen as a complement to traditional management instruments, rather than their replacement.

Within the same realm, how digitisation affects managers, Daniel Weihs' chapter offers a scenario of management set in 2035 where so-called autosomes, reasoning autonomous systems employed as knowledge workers. Weihs imagines autosomes as self-motivated, self-guided machines that are capable of holding management and even executive positions, and interact with other autosomes and humans to work towards their goals. The scenario describes how such hybrid teams of humans and autosomes would collaborate, what issues would arise, how humans would accept the change in their work environments, even being subordinated to a machine, how autosomes would deal with the limitations of humans such as human information processing capabilities or the need for time off-duty. The chapter raises a number of important questions about issues such as responsibility, that is who is held responsible in case something goes wrong, motivation meaning what ambitions would one design such autosomes to have, reaction to failure, how should an autosome react to the autosome's own faults or to external, uncontrollable issues or the advancement of knowledge, meaning how to deal with the autosome's self-guided learning on the job.

The question of how decision power should be distributed between humans and machines is also at the core of another chapter written by Per Bergamin and Franziska Hirt. The chapter focusses on technology-enhanced learning and discusses non-traditional settings such as open, distance, online or informal workplace learning. Such settings on the one hand demand and promote self-regulatory learning strategies by the users of digital learning environments, but on the other hand also offer external regulation to guide users through the plethora of materials and choices available. Bergamin and Hirt suggest that designers of digital learning environments face a so-called self-regulation dilemma and need to strike a balance between self-regulation and external regulation. The chapter then discusses the three basic loci of control, that are learners, other persons or learning systems and describe concepts of adaptive learning systems for the latter. The chapter finally discusses options of shared control between learner and system plus the concepts of scaffolding and fading to address the self-regulation dilemma and gives an outlook on advancements in artificial intelligence and sensor technology and their impact on the design of digital learning environments.

The design of digital learning technology is also the focus of the next chapter written by Jörgen Jaanus, Nina Suomi and Tobias Ley. The chapter proposes a learning oriented architecture of knowledge management technology for business organisations. The aim of this architecture is to address the gap between the increased speed with which business demands are changed in a digitally enabled economy and their transformation into learning needs. The learning architecture is intended to guide the design of digitally enabled knowledge workplaces that boost just in time learning, meaning informal, collaborative and socially embedded learning at the workplace. The architecture builds on four cases of companies in the professional service sector which, according to the authors, jointly represent the core challenges of digitisation that the sector faces. Jaanus et al. summarise those challenges as a disconnect between knowledge organisation systems and value creation activities in knowledge work and postulate that learning gets a more prominent position in knowledge management platforms. The chapter discusses how their learning oriented architecture connects IT-supported development of knowledge workers' competencies with the three core business goals of efficiency, quality and sustainability.

The subsequent chapter authored by Angelika Mittelmann approaches the development of competencies for the digitally enabled economy from yet another direction. Instead of asking how we might change leadership styles or should design technology, Mittelmann focuses on what personal competencies knowledge workers and managers need to develop in order to succeed in what she calls "Work 4.0", meaning digitised work places in a digitally transformed collaborative and organisational environment. Mittelmann decomposes the challenges involved in Work 4.0 and describes 13 competencies structured along the three categories intrapersonal, interpersonal and information and communication technology-related competencies. She further explains how her Fitness Circuit for Personal Knowledge Management takes the concept of circuit training from sports and turns it into a multi-stage program to develop such Work 4.0 competencies. The chapter is rounded up by a discussion of the Agile Competency Development Cycle which shows how such competency development can be implemented in a blended learning approach at the organisational level.

The changes in competence development due to digital transformation also challenge learning models and didactical methods which need to be adapted or replaced by new models. Under the label of Learning 4.0, Peter Henning observes substantial changes in learning behaviour in education, industry and society. The chapter discusses how the four learning paradigms, behaviourist, cognitivist and constructivist learning, plus connectivism fit to describe, explain and facilitate learning (behaviour) in the digital age. Henning describes advances in learning models such as a hypercube model of learning where learning is seen as movements of the learner along learning pathways in a hypercube of knowledge objects. The chapter critically reflects on what pedagogical reactions can (or cannot) be taken on learning analytics data and what chances and risks are involved in learners increasingly relying on the plethora of material available in the (mobile) internet. Henning finally provides an optimistic outlook to future learning and characterises

Learning 4.0 as largely digital, network-oriented, diverse, constructive, individualised and adaptive learning based on semantically enhanced material.

Reflective learning represents one such approach for future learning, which aims to enhance knowledge transfer to professionals at their workplaces. Angela Fessl, Gudrun Wesiak and Viktoria Pammer-Schindler combine elements of gamification and reflective learning into a medical quiz. The quiz contains content and reflective questions and was implemented in the learning platform Moodle as an instrument for playful reflective learning, that aims to stimulate the transformation of theoretical knowledge into daily work practice of healthcare professionals. The authors report the findings of a pilot evaluation study in which the quiz was administered to nurses working in German stroke units as a complement to their formal professional education. The exploratory evaluation revealed a positive association between the usage of the quiz and the perceived learning effects while reflection was found to be only of secondary importance under the conditions of this study. The chapter discusses lessons learned from the study and concludes that reflective learning approaches can provide a motivating and efficient way to transfer knowledge at the workplace.

6 New Forms of Knowledge-Intensive Digitally Enabled Value Creation

Today the half-life of knowledge decreases as we speak while at the same time the level of complexity of value creation increases. As a result, it becomes evident that paradigms and tools in how to manage work can't just be adjusted, they need to be completely revisited and challenged as a whole. Not surprisingly, we cling on to those mechanisms that have proven to be successful in the past and that we have become used to. Only slowly we realise that turbulence, volatility and unpredictability in collective value creation become the new default mode in organisational management.

How grave these changes are, becomes evident when we look at the roles organisations have played in the past compared to how they will most likely operate in the future. In the past organisations were anchors of stability that provided a "home" for staff over a significant amount of time, if not for life. Change and disruption were episodic occurrences that either happened when the top level was replaced by someone new or in times of severe crisis. Consequently, transformational change was seen as rare, unusual and above all, very painful. The way to address organisational renewal was to plan, to execute the plan and to communicate the results. What we witness today in organisational value creation is more or less the complete opposite, where value creation is organised in loosely coupled networks where collaboration is an inherent part. Resources are allocated internally according to market dynamics and much less due to internal power dynamics. Ultimately, change is considered as a permanent opportunity that goes as deep as rethinking rules of the game and much less adjustments of business models. The planning paradigm will be replaced by improvisation, trial-and-error and fast paced learning. Transformational results are communicated on the fly and not at the end. Campaigning and communication is part and parcel of change processes, irrespective the size or sector of the organisation.

It should have become evident by now, that digitally enabled value creation is not a question of organisational design alone, but goes much deeper. It goes deeply into the DNA of the organisation and touches upon what drives an organisation in its inner core, its culture. No doubt, startups with lean structures, a minimum of overhead costs and a "permission to fail" might have it easier to operate as an agile player. But don't be fooled: To get rid of hierarchies and top-down decisions is and cannot be the solution to the organisational challenges of the twenty first Century. On the contrary, any organisation that consists of more than 30 people experiences the need for order, which often leads to exactly that: hierarchy and top-down decision making. That's why it would not only be unwise but also dangerous to force multinationals to give up hierarchies as it would create uncontrollable chaos. The challenge however is and must be, to create agile, knowledge-intensive and digitally enabled value creation as part of existing hierarchical structures. You guessed it, this is nothing that can be done via structures, processes and procedures. This requires leadership that is competent and able to manage day to day business (exploit) while thinking out of the box and existing routines (explore).

In their chapter, Bellinger and Krieger illustrate vividly that in all areas of society a paradigm shift from thinking in terms of closed systems to thinking in terms of open networks is taking place. Living in a connected world includes online and offline networks alike. Although the term "network" is part and parcel of management and business talk, let's take a closer look: Networks are non-hierarchical, inclusive, connected, complex, and open. They consist of both humans and nonhumans. In today's world of deconstructed value chains, networks have become some sort of blueprint for the way in which societies are organised. A sector that shows exactly that is healthcare. Healthcare is no longer primarily something that takes place in the intimacy and confines of the doctor-patient relationship. Instead, health care is distributed throughout a complex network of both human and nonhuman actors such as databases, hospital information systems, digital health records, electronic health cards, online patient communities, health related apps, smart homes with ambient assisted living technologies, etc. Networks operate most efficiently when they conform to norms such as connectivity, flow of information, communication, participation, transparency, and authenticity. It's these norms that guide the production and utilisation of health related information and knowledge. They condition how health related knowledge can create value both with regard to efficiency and quality of care. Bellinger and Krieger take a closer look at how these norms of digital transformation have changed managing knowledge in health care networks and through that provide a first-hand look at how the disruptive force of digitization has had an impact on health care.

Digitally enabled value creation is a global phenomenon that has the potential to support eradicating poverty and help boost economic prosperity. However, until recently, digitally enabled Knowledge Management in developing countries has more often than not been dismissed as unrealistic given challenges with access to electricity and the internet. However, a number of recent examples of holistic activities in Knowledge Management, including digital elements, have demonstrated a measurable contribution to improved outcomes with some of the world's poorest people. In their chapter Bocock, Sullivan, Arnold and Limaye focus on such a case. They look at how a digitally enabled Knowledge Management program was designed, piloted, and measured, in two districts in Bangladesh. Ultimately, the program aimed to help rural community-based health workers be more informed about, and helpful in, providing health and nutrition guidance to some of the world's poorest people.

To Foresti and Varvakis the "industry 4.0" is a new productive paradigm that is rooted in digitisation. To the authors the phenomenon is based on so-called "cyber-physical systems". These systems allow its users absolute transparency and manageability of an entire value chain. Consequently, the awareness of the production process increases for all stakeholders involved. The authors feel that such a phenomenon can best be characterised as "ubiquity", that is, virtual presence in various places, at all times. Over the past 5 years the authors have conducted extensive bibliographic research and found out that the new emerging business models with 4.0 Industry are essentially based on the ubiquity of information, things, products, and consumers. Ultimately, ubiquity allows for new ways of interaction between customers and suppliers, as well as innovative ways of producing and managing organisations. The question is, are managers and leaders prepared to manage ubiquity and permanent access to information and knowledge at all times? Or put differently, are they ready to manage the unmanageable?

Have you ever lost your key chain? It's a drag to remember all the keys that were attached to it and if worst comes to worst you might even have to replace door locks. Now can you imagine how it must feel like to lose 50 million USD at one go? This is exactly what happened to the DAO, the world's first "distributed autonomous organisation". It was founded on May 15, 2016 and in just a few weeks, the investment fund gathered 119.5 million USD from more than 50,000 investors. What is a DAO? A decentralised autonomous organisation (DAO), sometimes labelled a decentralised autonomous corporation (DAC), is an organisation that is run through rules encoded as computer programs called smart contracts. The DAO was not only the biggest crowdfunding campaign of all time, it was also ground zero for the biggest cybercrime in IT history. Just a month after its launch, hackers succeeded in siphoning 50 million USD out of the fund. Patrick Hofer tells with much detail how the DAO came to life and what can happen when an organisation feels too big to fail. What reads like a fictional cybercrime story is in fact brutal reality and can severely hinder global and digitally enabled financial transactions.

How open source innovation sets the path for future business models is discussed in Christian Kreutz' contribution. Start-ups are innovation leaders thanks to leveraging open source network innovation.

Every day, more and more promising new start-ups with competing business models are founded; something that would be impossible without the internet. For most companies, the new product-and-service-development process is changing drastically in terms of speed, little location relevance and sophisticated international collaboration opportunities. Thanks to their openness and flexible organisation models, start-ups are being able to include faster digital technologies to challenge large companies. On the other hand, you have established companies, with clearly defined rules of the game, policies on everything that needs to be managed and a vertical command chain. These organisations heavily rely on their internal innovating capacity and in the light of disruptive technological breakthroughs and ever-changing market dynamics, face various challenges in sustaining their products and market shares. "Unless you have the capacity to innovate alternative business models, someone else will likely get a crack at it first.", as said by Henry Chesbrough nicely illustrates exactly that. Christian Kreutz believes that to achieve this in times of ongoing innovation, companies have to transform themselves from different angles. They have to absorb digital changes, and find new ways to be able to participate in a global and highly dynamic open innovation eco-system. In his chapter he shows how this can be done.

Digital change and scientific development have mutual implications. On one hand, science and technology development has been a major factor to digital change. On the other hand, the digital era has brought major changes to the production of scientific knowledge. First, there is a cyberinfrastructure—not only infrastructure for computing, but a major virtual lab where all professionals in science and technology (e.g., researchers, engineers, technicians) can collaborate and exchange data, information, and knowledge. In Europe, this new infrastructure is referred to as e-science. Second, the digital era has increased the co-production beyond frontiers of traditional players, inviting new players to contribute to the scientific debate. Pacheco, Nascimento and Weber express that this kind of co-work is central to both citizen science and transdisciplinary knowledge coproduction, where non-academic players engage in activities such as planning, data gathering, and impact assessment of science. In their chapter, the authors define digital science as a convergent phenomenon of cyberinfrastructure, e-science, citizen science and transdisciplinarity. They examine how digital science has been as disruptive factor to traditional scientific development, changing productivity, expanding frontiers and challenging traditional process in science, such as planning and assessment.

References

Adamczyk, S., Bullinger, A. C., & Möslein, K. M. (2012). Innovation contests: A review, classification and outlook: Innovation contests. *Creativity and Innovation Management, 21*(4), 335–360.

Adolf, M., & Stehr, N. (2017). *Knowledge*. London: Routledge.

Alavi, M., & Leidner, D. E. (2001). Review: Knowledge management and knowledge management systems: Conceptual foundations and research issues. *MIS Quarterly, 25*(1), 107.

Annanperä, E., Liukkunen, K., & Markkula, J. (2016). Managing emerging business ecosystems– A knowledge management viewpoint.

Bharadwaj, A., El Sawy, O. A., Pavlou, P. A., & Venkatraman, N. V. (2013). Digital business strategy: Toward a next generation of insights.

Bittner, E. A. C., & Leimeister, J. M. (2014). Creating shared understanding in heterogeneous work groups: Why it matters and how to achieve it. *Journal of Management Information Systems, 31*(1), 111–144.

Bmas, G. F. M. O. L. S. A. (2015). Re-imagining work green paper: Work 4.0.

Bostrom, N. (2014). *Superintelligence: Paths, dangers, strategies*. Oxford: OUP.

Boudreau, K. J., & Lakhani, K. R. (2013). Using the crowd as an innovation partner. *Harvard Business Review, 91*(4), 60–69, 140.

Bounfour, A. (2016). *Digital futures, digital transformation*. Berlin: Springer.

Brynjolfsson, E., & Mcafee, A. (2014). The second machine age: An industrial revolution powered by digital technologies. *Digital Transformation Review*, (5), 12–17.

Burke, P. (2000). A social history of knowledge.

Chen, H., Chiang, R. H., & Storey, V. C. (2012). Business intelligence and analytics: From big data to big impact. *MIS Quarterly, 36*(4).

Chesbrough, H. W. (2006). The era of open innovation. *Managing Innovation and Change, 127*(3), 34–41.

Davenport, T. H., & Kirby, J. (2016). *Only humans need apply: Winners and losers in the age of smart machines*. New York: Harper Business.

Davis, G. B. (2002). Anytime/Anyplace computing and the future of knowledge work. *Communications of the ACM, 45*(12), 67–73.

Dennerlein, S., Seitlinger, P., Lex, E., & Ley, T. (2016). Take up my tags: Exploring benefits of meaning making in a collaborative learning task at the workplace. In *European Conference on Technology Enhanced Learning*. Berlin: Springer.

Domingos, P. (2015). *The master algorithm: How the quest for the ultimate learning machine will remake our world*. New York: Basic Books.

Eberl, U. (2016). *Smarte Maschinen. Wie Künstliche Intelligenz unser Leben verändert*. München: Hanser.

Eisenhardt, K. M., & Martin, J. A. (2000). Dynamic capabilities: What are they? *Strategic Management Journal, 21*(10–11), 1105–1121.

Ettlinger, N. (2016). The governance of crowdsourcing: Rationalities of the new exploitation. *Environment and Planning A, 48*(11), 2162–2180.

Ghazawneh, A., & Henfridsson, O. (2010). Governing third-party development through platform boundary resources. In *The International Conference on Information Systems (ICIS), AIS Electronic Library (AISeL)*.

Grant, R. M. (1991). The resource-based theory of competitive advantage: Implications for strategy formulation. *California Management Review, 33*(3), 114–135.

Hamel, G., & Prahalad, C. (1994). *Competing for the future*. Boston, MA: Harvard Business School Press.

Hayek, F. A. (1945). The use of knowledge in society. *The American Economic Review,* 519–530.

Henfridsson, O., & Lindgren, R. (2010). User involvement in developing mobile and temporarily interconnected systems. *Information Systems Journal, 20*(2), 119–135.

Hofmann, J. (2017). Demokratie im Datenkapitalismus. *WZB Mitteilungen Heft, 155,* 14–17.

Iansiti, M., & Lakhani, K. R. (2014). Digital ubiquity: How connections, sensors, and data are revolutionizing business. *Harvard Business Review, 92*(11), 90–99.

Intel. (2014). The future of knowledge work. White paper workplace transformation.

Kaplan, A. M., & Haenlein, M. (2010). Users of the world, unite! The challenges and opportunities of Social Media. *Business Horizons, 53*(1), 59–68.

Karimi, J., & Walter, Z. (2015). The role of dynamic capabilities in responding to digital disruption: A factor-based study of the newspaper industry. *Journal of Management Information Systems, 32*(1), 39–81.

Kaschig, A., Maier, R., & Sandow, A. (2016). The effects of collecting and connecting activities on knowledge creation in organizations. *The Journal of Strategic Information Systems, 25*(4), 243–258.

Kelly III, J. E. (2015). Computing, cognition and the future of knowing: How humans and machines are forging a new age of understanding. *IBM Research.*

Klein, G., Moon, B., & Hoffman, R. R. (2006). Making sense of sensemaking 1: Alternative perspectives. *IEEE Intelligent Systems, 21*(4), 70–73.

Koch, T., & Windsperger, J. (2017). Seeing through the network: Competitive advantage in the digital economy. *Journal of Organization Design, 6*(1), 6.

Kogut, B., & Zander, U. (1992). Knowledge of the firm, combinative capabilities, and the replication of technology. *Organization Science, 3*(3), 383–397.

Kolko, J. (2010). Sensemaking and framing: A theoretical reflection on perspective in design synthesis. *Design Research Society.*

Kotter, J. P. (2014). *Accelerate: Building strategic agility for a faster-moving world.* Boston, MA: Harvard Business Review Press.

Kurtz, C. F., & Snowden, D. J. (2003). The new dynamics of strategy: Sense-making in a complex and complicated world. *IBM Systems Journal, 42*(3), 462–483.

Le Dinh, T., Nomo, T. S., & Ayayi, A. (2015). Towards a cyberinfrastructure for social science research collaboration: The service science approach. In *International Conference on Exploring Services Science*. Berlin: Springer.

Lehtiniemi, T., Kuikkaniemi, K., Poikola, A., Nelimarkka, M., Valtonen, T., Floréen, P., & Turpeinen, M. (2015) Trends of knowledge work and needs for knowledge work tools. Re: Know White Paper.

Loebbecke, C., Van Fenema, P. C., & Powell, P. (2016). Managing inter-organizational knowledge sharing. *The Journal of Strategic Information Systems, 25*(1), 4–14.

Macmillan, I., Ihrig, M., & Steinhour, J. (2017). Mapping critical knowledge for digital transformation. From http://knowledge.wharton.upenn.edu/article/management-knowledge-assets/.

Madsbjerg, C. (2017). Sensemaking: The power of the humanities in the age of the algorithm.

Maier, R. (2007). *Knowledge management systems: Information and communication technologies for knowledge management.* Berlin: Springer Science & Business Media.

Manhart, M., Thalmann, S., & Maier, R. (2015). The ends of knowledge sharing in networks: Using information technology to start knowledge protection. In *ECIS.*

Manyika, J., Chui, M., Bughin, J., Dobbs, R., Bisson, P., & Marrs, A. (2013). *Disruptive technologies: Advances that will transform life, business, and the global economy.* San Francisco, CA: McKinsey Global Institute.

March, J. G. (1991). Exploration and exploitation in organizational learning. *Organization Science, 2*(1), 71–87.

Martens, B., Poeppelbuss, J., & Teuteberg, F. (2011). Understanding the cloud computing ecosystem: Results from a quantitative content analysis. *Wirtschaftsinformatik, 2011,* 16.

Matt, C., Hess, T., & Benlian, A. (2015). Digital transformation strategies. *Business & Information Systems Engineering, 57*(5), 339–343.

Mayer-Schönberger, V., & Cukier, K. (2013). *Big data: A revolution that will transform how we live, work, and think.* New York: Houghton Mifflin Harcourt.

Mazmanian, M., Orlikowski, W. J., & Yates, J. (2013). The autonomy paradox: The implications of mobile email devices for knowledge professionals. *Organization Science, 24*(5), 1337–1357.

Mithas, S., Tafti, A., & Mitchell, W. (2013). How a firm's competitive environment and digital strategic posture influence digital business strategy. *MIS Quarterly, 37*(2), 511.

Murray, A., Wheaton, K. (2016). Welcome to curation 2.0. *KM World January 2016, 25*(1), 1–2. http://www.kmworld.com/Articles/Column/The-Future-of-the-Future/Welcome-to-Curation-2.0-108249.aspx.

Newman, G., Wiggins, A., Crall, A., Graham, E., Newman, S., & Crowston, K. (2012). The future of citizen science: Emerging technologies and shifting paradigms. *Frontiers in Ecology and the Environment, 10*(6), 298–304.

North, K. (2005). *Wissensorientierte Unternehmensführung* (6th edn 2016). Wiesbaden: Gabler Verlag.

North, K., & Gueldenberg, S. (2011). *Effective knowledge work: Answers to the management challenges of the 21st century*. Bingley: Emerald Group Publishing.

North, K., & Haas, O. (2014). Zwischen Experiment und Routine: Wie wird Wissensmanagement erwachsen? *OrganisationsEntwicklung, 3*(2014), 50–56.

North, K., & Kumta, G. (2018). *Knowledge management: Value creation through organizational learning*. Berlin: Springer.

North, K., & Varvakis, G. (2016). Competitive strategies for small and medium enterprises. In *Increasing Crisis Resilience, Agility and Innovation in Turbulent Times*. Cham: Springer.

OECD. (2016). New skills for the digital economy. OECD Digital Economy Papers. Paris. **258**.

Pavlou, P. A., & El Sawy, O. A. (2011). Understanding the elusive black box of dynamic capabilities: The elusive black box of dynamic capabilities. *Decision Sciences, 42*(1), 239–273.

Pawlowski, J. M., Bick, M., Peinl, R., Thalmann, S., Maier, R., Hetmank, L., et al. (2014). Social knowledge environments. *Business & Information Systems Engineering, 6*(2), 81–88.

Pawlowsky, P., Gözalan, A., & Schmid, S. (2011). Wettbewerbsfaktor Wissen: Managementpraxis von Wissen und Intellectual Capital in Deutschland-Eine repräsentative Unternehmensbefragung zum Status quo.

Porter, M. E., & Heppelmann, J. E. (2014). How smart, connected products are transforming competition. *Harvard Business Review, 92*(11), 64–88.

Samulowitz, H., Sabharwal, A., & Reddy, C. (2014). Cognitive automation of data science. In *ICML AutoML Workshop*.

Seeber, I., Bittner, E., Briggs, R. O., De Vreede, G. –J., De Vreede, T., Druckenmiller, D., Maier, R., Merz, A., Oeste-Reiß, S., Randrup, N., Schwabe, G., Söllner, M. (2018). Machines as teammates: A collaboration research agenda. In *Hawaii International Conference on System Sciences (HICSS)*, 3–6.1.2018, Waikoloa, HI, USA.

Spender, J. C. (1996). Making knowledge the basis of a dynamic theory of the firm: Making Knowledge. *Strategic Management Journal, 17*(S2), 45–62.

Spiro, K. (2017). *70:20:10 challenges: Turning curation into knowledge sharing*. https://elearningindustry.com/turning-curation-into-knowledge-sharing-70-20-10-challenges.

Stahl, G., Anderson, T., & Suthers, D. (2006). Computer supported collaborative learning: An historical perspective, 2006. *Cambridge Handbook of the Learning Sciences,* 409–426.

Teece, D. J. (2009). *Dynamic capabilities and strategic management: Organizing for innovation and growth*. Oxford University Press on Demand.

Teece, D. J. (2017). Dynamic capabilities and (digital) platform lifecycles.

Teece, D. J., Pisano, G., & Shuen, A. (1997). Dynamic capabilities and strategic management. *Strategic Management Journal, 18*(7), 509–533.

Telekom. (2015). Arbeit 4.0. from https://www.telekom.com/medien/konzern/285970.

Thornhill, J. (2017). *Data capitalism is cashing in on our privacy…for now*. Financial Times.

Tushman, M. L., & O'reilly, C. A. (1996). Ambidextrous organizations: Managing evolutionary and revolutionary change. *California Management Review, 38*(4), 8–29.

Urbinati, N. (2014). *Democracy disfigured: Opinion, truth, and the people*. Cambridge, MA: Harvard University Press.

Van Doren, C. L. (1991). *A history of knowledge: Past, present, and future*. New York: Random House Digital, Inc.

Van Der Aalst, W., & Damiani, E. (2015). Processes meet big data: Connecting data science with process science. *IEEE Transactions on Services Computing, 8*(6), 810–819.

Von Hippel, E. (2005). Democratizing innovation: The evolving phenomenon of user innovation. *Journal für Betriebswirtschaft, 55*(1), 63–78.

Von Hippel, E., & Von Krogh, G. (2003). Open source software and the "private-collective" innovation model: Issues for organization science. *Organization Science, 14*(2), 209–223.

Von Krogh, G. (2012). How does social software change knowledge management? Toward a strategic research agenda. *The Journal of Strategic Information Systems, 21*(2), 154–164.

Von Krogh, G., & Roos, J. (1996). Five claims on knowing. *European Management Journal, 14*(4), 423–426.

Waizenegger, L., Remus, U., & Maier, R. (2016). The social media trap—How knowledge workers learn to deal with constant social connectivity. IEEE.

Weick, K. E., Sutcliffe, K. M., & Obstfeld, D. (2005). Organizing and the process of sensemaking. *Organization Science, 16*(4), 409–421.

Yoo, Y., Boland, R. J., Lyytinen, K., & Majchrzak, A. (2012). Organizing for innovation in the digitized world. *Organization Science, 23*(5), 1398–1408.

Yoo, Y., Henfridsson, O., & Lyytinen, K. (2010). Research commentary—The new organizing logic of digital innovation: an agenda for information systems research. *Information Systems Research, 21*(4), 724–735.

Zack, M., Mckeen, J., & Singh, S. (2009). Knowledge management and organizational performance: An exploratory analysis. *Journal of Knowledge Management, 13*(6), 392–409.

Zuboff, S. (2016). *The secrets of surveillance capitalism.* Frankfurt: Frankfurter Allgemeine Zeitung.

Author Biographies

Klaus North, Dr.-Ing. is Professor of International Management at Wiesbaden Business School, Germany. His research covers knowledge and innovation management. He was founding president of the German Knowledge Management Association (GfWM) and was scientific director of the German Knowledge Management Award. His textbook "Wissensorientierte Unternehmensführung" ("Knowledge Management", Springer Texts in Business and Economics) available in many languages has become a reference work on the subject.

Ronald Maier, Dr. is a Professor of Information Systems at the School of Management at the University of Innsbruck, Austria. He received his Ph.D. in Management Information Systems from WHU Otto Beisheim School of Management in Koblenz, Germany and a habilitation degree from University of Regensburg, Germany. His research interests include collaboration engineering, connectivity, crowdsourcing and knowledge management. His research has appeared in journals such as Journal of Strategic Information Systems, Business & Information Systems Engineering, Computers in Human Behavior, IEEE Transactions on Learning Technologies and Journal of Knowledge Management.

Oliver Haas is Head of the Global Division "Health, Education, Social Inclusion" at the Deutsche Gesellschaft für Internationale Zusammenarbeit (GIZ). He has been working in global transformation for more than 15 years. This led to various postings (Malaysia, South Africa, and Vietnam), where he managed projects in Vocational Education and Training. Prior to his current position, he worked at the World Bank in Washington D.C., USA as Senior Operations Officer. He holds a Masters in Sociology (Free University Berlin) and is a certified Organizational Developer. He has lectured at various European universities and institutes (Barcelona, Vienna, and Darmstadt) and is an editor of and a regular contributor to Germany's leading magazine on organizational development "OrganisationsEntwicklung."

Part I
Digitally Enabled Enrichment of Resources to Leverage Human Performance

Semantic Technologies: Enabler for Knowledge 4.0

Achim Rettinger, Stefan Zander, Maribel Acosta and York Sure-Vetter

Abstract Semantic technologies are a key enabler for Knowledge 4.0. Specifically, knowledge graphs have caused significant practical implications for managing knowledge in the digital economy. While most semantic technologies originate from the vision of representing the existing Web in a machine-processable format, it's most notable success so far are large cross-domain knowledge graphs. They are created by collaborative human modelling and linking of structured and semi-structured data. So far, they exhibit only little but still very powerful semantics, which have shown benefits for numerous applications. This chapter introduces the latest innovations in modelling knowledge using knowledge graphs and explains how those knowledge graphs enable value creation by making unstructured content, like text documents accessible by machines and humans. Finally, we show how semantic technologies help to make hard- and software components in cyber physical systems interoperable.

1 Introduction

Semantic technologies intend to bridge the gap between human knowledge and computational knowledge. They try to capture knowledge in an explicit computational knowledge representation that is both, accessible by humans and processable by computers in a meaningful manner. This is a feature that other approaches lack, such as representations created by the nowadays popular deep learning approaches.

Knowledge graphs (KGs) are currently seen as one of the most advanced components to realize the vision of explicit and computational knowledge representations. The term "knowledge graph" was reintroduced by Google in 2012 (Singhal 2012) and is now being used for any graph-based knowledge repository. In

A. Rettinger · M. Acosta · Y. Sure-Vetter (✉)
Karlsruhe Institute of Technology (KIT), Karlsruhe, Germany
e-mail: york.sure-vetter@kit.edu

S. Zander
Darmstadt University of Applied Sciences, Darmstadt, Germany

K. North et al. (eds.), *Knowledge Management in Digital Change*, Progress in IS,
https://doi.org/10.1007/978-3-319-73546-7_2

semantic technologies, the recommended data model to publish graph-based data on the Web is defined in the Resource Description Framework (RDF). Therefore, in the remainder we use the term knowledge graph for any RDF graph. An RDF graph consists of a finite set of triples where each triple (s, p, o) is an ordered set of the following terms: a subject s, a predicate p that associates the subject and the object, and an object o. An RDF term is either a URI, a blank node, or a literal. A triple allows to express a statement about a real-world fact. With these basic building blocks, knowledge graphs allow the representation of objects, their abstract relations and classes (groups) of objects, as well as their instantiations as real-world objects, called entities, and their concrete relations and class memberships (Färber et al. 2016).

Based on those basic technological building blocks, in recent years, several noteworthy large, cross-domain, and openly available KGs have been created. These include DBpedia, Freebase, OpenCyc, Wikidata, and YAGO. They have grown to an impressive number of triples, like Freebase which is the largest KG with over 3.1B triples.

In the following sections, we will first outline how human knowledge is being captured in KGs, then explain how they can help to access the content expressed in unstructured sources like text documents and finally demonstrate that semantic technologies can be used to facilitate the interoperability gap between different cyber physical systems' components.

2 Semantic Technologies for Knowledge Engineering

The application of semantic technologies in knowledge engineering allows for the creation and management of knowledge-based systems. Systems that exploit knowledge and data semantics enable advanced capabilities in all the tasks of knowledge engineering processes.

Nowadays, the largest amounts of knowledge collected with the help of semantic technologies are the result of combining data harnessed from a wide range of sources—including humans and the Web—and representing it as semantic interconnected entities in knowledge graphs (KGs). KGs allow a wide range of analytical tasks including query processing through declarative languages, question answering, visualizations, and further data analysis including statistical analysis, data mining, etc. (see Fig. 1). Semantic technologies support the construction of KGs through semantic enrichment of non-semantic sources, data integration of several sources, and data curation. As depicted in Fig. 1, knowledge captured from humans can be stored in Knowledge Bases (KBs) and used across all the processes of knowledge engineering. Semantic technologies exploit human knowledge from KBs and computation knowledge encoded in data semantics to construct KGs.

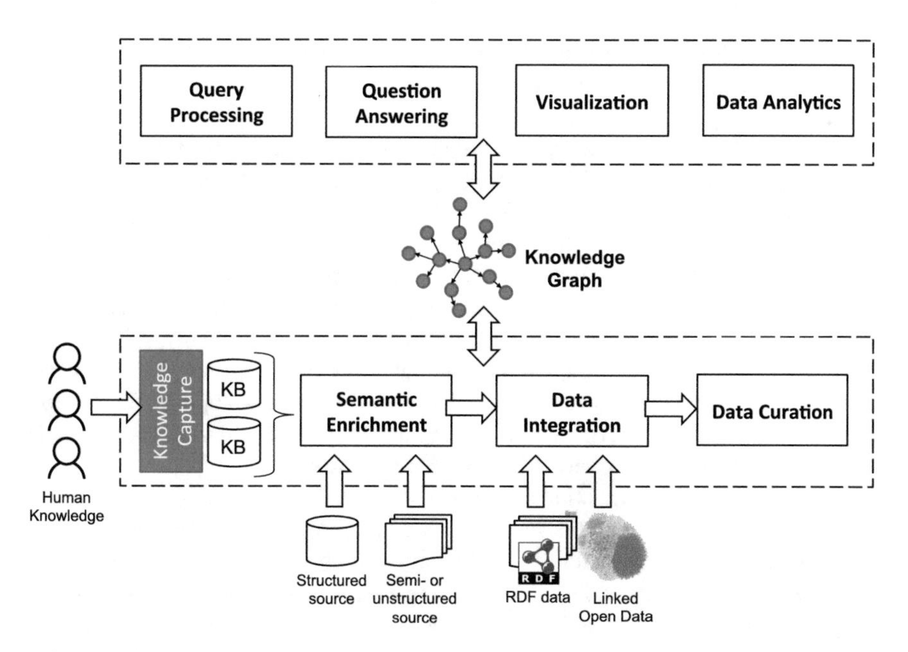

Fig. 1 Overview of the application of semantic technologies for knowledge engineering processes based on knowledge graphs

2.1 *Knowledge Capture*

Knowledge capture is the process of extracting and representing knowledge from reference sources directly or indirectly provided by humans. In this context, knowledge is modelled as artefacts including rules, ontological terms, and conceptual objects that symbolize physical objects or phenomena of a domain.

Typically, in Enterprise Systems, knowledge is harnessed through knowledge acquisition tasks that involve systematic interviews with domain experts: knowledge is collected through observation of experts including the usage of questionnaires. Nowadays, the rise of the Web and semantic technologies have allowed for capturing knowledge from experts as well as lay contributors to create semantic KBs in a distributed and collaborative way. Exemplary for successful knowledge capture frameworks that exploit Semantic Web technologies are Semantic MediaWiki (Krötzsch et al. 2006), DBpedia (Auer et al. 2007; Lehmann et al. 2015), and Wikidata (Vrandečić and Krötzsch 2014).

Semantic MediaWiki (SMW, https://www.semantic-mediawiki.org) is an extension to a wiki implementation that integrates semantic capabilities into the process of collaborative knowledge management through wikis. SMW enables human contributors to capture knowledge by semantically enriching and reusing content in the wiki. KBs created with SMW combine unstructured content and semantic (semi-) structures. KBs acquired with SMW are modelled using RDF, represented with terms from the Web Ontology Language (OWL) (Bao 2012), and can be queried using declarative languages, for example, the SPARQL query language.[1]

DBpedia (http://dbpedia.org) is a project that implements an extraction framework to gather data from different language versions of Wikipedia. The result of this framework is the DBpedia KB which is created by extracting data via declarative mappings or wrappers. In this context, wrappers are components that parse the data embedded in Wikipedia articles and translate it into concepts defined in the DBpedia ontology. The DBpedia wrappers are the result of a collaborative effort to the DBpedia project, where human contributors manually specified the wrappers to semantically enrich Wikipedia data. The resulting wrappers are stored in a knowledge base and currently available at the DBpedia Mappings Wiki (http://mappings.dbpedia.org). DBpedia data is published following the RDF data model and can be accessed and queried through web services.

Wikidata (http://www.wikidata.org) is a Wikimedia project that manages facts mainly from Wikipedia but also from its sister projects like Wikivoyage and Wikisource. Wikidata constitutes a collaborative multi-lingual KB that serves as a centralized source to provide unified and consistent facts across the multiple language versions of Wikipedia. Facts in the Wikidata KB are annotated with a list of reference sources that support the veracity of the facts. Wikidata contributors include humans and machines (bots) that create and maintain data in the KB. The content of the Wikidata KB is exported in different data models including RDF and can be also accessed through web services.

2.2 *Semantic Enrichment*

In semantic enrichment, the data of non-semantic sources is annotated with semantic descriptions from vocabularies or ontologies. The challenges and current solutions for performing semantic enrichment of sources highly depend on the characteristics of the underlying data.

Data accessed in knowledge-based systems may have different data models and structures. In the case of structured (e.g., relational databases) or semi-structured (e.g., XML files, CSV files, etc.) data sources, data can be semantically annotated using rules specified by experts in a knowledge capture process and stored in KBs. These rules translate elements from the sources into ontological concepts, properties, and instances. Particularly, in the case of relational databases, these rules or mappings can be specified using the W3C recommended languages Direct Mappings (DM) (Arenas et al. 2012) or R2RML (Souripriya et al. 2012). DM or R2RML mappings are executed by processors able to generate semantic graph-based data from relational databases following the RDF data model.

[1]SPARQL 1.1 Query Language. Technical report, W3C (2013)

Unstructured sources that provide natural language documents or visual information can also be semantically annotated. Depending on the nature of the unstructured data, different semantic enrichment approaches are applied (see Section "Semantic Technologies for Understanding Unstructured Context").

Depending on the data available at the sources, the process of semantically enriching data might be expensive in terms of time and computational resources. In particular, when dealing with large volumes of data, it is not practical to enrich and materialize the entire content of a source. To tackle this problem, current Big Data architectures (Auer et al. 2017) have focused on solutions that semantically enrich data on-demand, following the paradigm of schema or data on-read.

2.3 *Data Integration*

Data integration is the process of consolidating data from heterogeneous sources. Heterogeneity may occur at different levels: physical infrastructure (hardware and location), network protocols, data models, and data representation. As explained in the section "Semantic Enrichment", the integration of data sources can also be performed offline, where all the content of the sources is integrated in a pre-processing step, or on-demand based on users' queries as performed by traditional data integration systems (Lenzerini 2002).

The integration of semantic data allows for the construction of knowledge graphs from different sources. Data that has been created following the Linked Data principles assumes a common network protocol to access the data (HTTP or SPARQL) and a common graph-based data model (RDF). In cases where a common data representation cannot be assumed, Linked Data integration approaches may exploit rules and other knowledge artefacts (e.g., ontological definitions) specified by users and maintained in KBs and apply reasoning over this knowledge in order to consolidate semantically heterogeneous sources. Nonetheless, when entities, classes, and properties in a Linked Data source are linked or aligned to other sources, it is possible to assume that the sources are providing a common data representation. In line with this assumption, current federated SPARQL engines—e.g., ANAPSID (Acosta et al. 2011), FedX (Schwarte et al. 2011), SPLENDID (Görlitz and Staab 2011), and SemaGrow (Charalambidis et al. 2015)—are able to on-the-fly integrate RDF data from distributed and autonomous sources that are semantically homogeneous during query processing.

2.4 *Data Curation*

Data curation includes processes to create, maintain, and validate data to ensure the value of the data, in this case, of the KG that represents semantically annotated and integrated data. To perform data curation, knowledge-based systems may exploit the semantics encoded in the data as well as knowledge captured in KBs.

In the context of knowledge engineering, one of the key tasks in data curation is the creation or completion of data. State-of-the-art solutions have investigated different reference sources or oracles to complete web data and knowledge graphs by, for example, automatically extracting data from web tables (Dong et al. 2014) and NLP graphs (Welty et al. 2012), respectively. Besides automatic approaches, a branch of state-of-the-art solutions resort to crowdsourcing, where humans act as oracles to complete databases or knowledge graphs (Franklin et al. 2011; Marcus et al. 2011; Park and Widom 2013; Acosta et al. 2015, 2017).

One instance of a system that applies crowdsourcing for knowledge graph completion is HARE (Acosta et al. 2015, 2017). HARE is a query engine able to enhance the completeness of knowledge graphs on-demand based on queries posed by users. HARE relies on the topology of the knowledge graph to identify potential missing values. To resolve missing values, HARE exploits the semantics of the data encoded in the knowledge graph to generate human-readable questions to be answered by a crowd composed of experts or lay users (contacted via crowdsourcing platforms). HARE stores the answers collected from the crowd in KBs. In this way, the knowledge captured from the crowd can be reused in subsequent queries. Empirical results evidence that HARE can reliably augment the completeness of knowledge graphs from different domains including Life Sciences. Furthermore, the results show that non-expert crowds can produce high quality answers achieving accuracy values from 0.84 to 0.96. Furthermore, the human-readable questions produced by HARE by exploiting the semantic description of entities in knowledge graphs are able to provide assistance to the crowd to produce high quality answers and to speed up the process of KG completion.

3 Semantic Technologies for Understanding Unstructured Content

The amount of entities in large knowledge graphs (KGs) has been increasing rapidly, enabling new ways of semantic information access, like keyword and semantic queries over entities and concepts mentioned in unstructured content, like text documents and videos. While entity search has become a standard feature (most prominent is the Google Knowledge panels shown next to the search results when searching for named entities), major Web search engines are still limited in their semantic processing capabilities: it is not possible to disambiguate search terms manually, search terms in one query can't be in different languages, the retrieved content items have to be in the same language as the search terms and search results are not gathered across heterogeneous content representations; like natural language and visual information. Most importantly, they don't allow to ask complex queries that spans information across multiple content sources. Recently developed systems have shown that it is possible to overcome those issues by using semantic technologies.

3.1 *Annotation*

Semantic technologies enable computers to access the knowledge that is captured in unstructured documents like text or images. The key to semantic processing of unstructured content is annotating it with unique identifiers as provided by KGs. Since KGs have grown considerably over the last years they reached a size that makes high-quality annotations possible for general domains as covered by Wikipedia. The task of identifying mentions of entities in text documents and disambiguating them with their corresponding unique identifier in the KG has be termed entity linking.

One instance of an entity-linking system is X-LiSA (Zhang and Rettinger 2014) which is an infrastructure for cross-lingual semantic annotation. It allows to bridge the ambiguity of unstructured data and its formal semantics as well as to transform such data in different languages into a unified representation. The architecture of X-LiSA is shown in Fig. 2, where cross-lingual groundings extraction is performed offline to generate the indexes used by the online cross-lingual semantic annotation component.

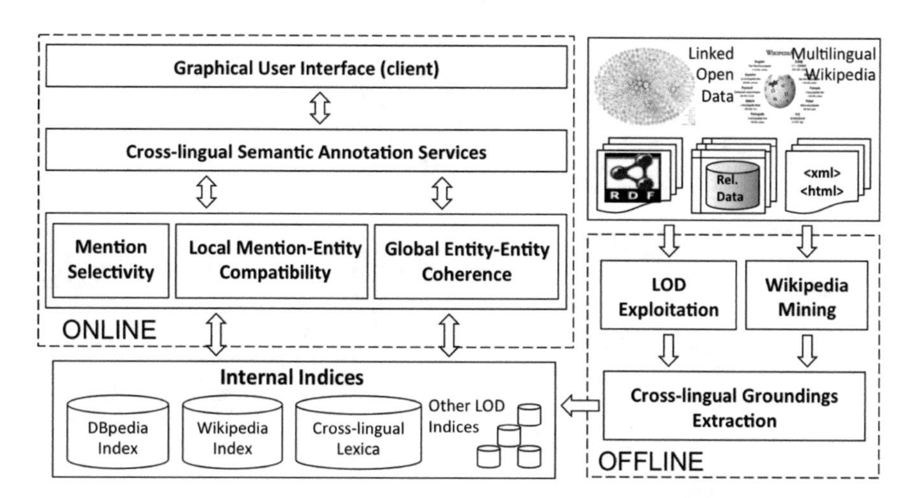

Fig. 2 Technical Components of the X-LiSA cross-lingual semantic annotation system. In an offline pre-processing step, large knowledge sources like Linked Open Data and Wikipedia are exploited and indexed in large repositories. This way, a cross-lingual lexicon of linguistic phrases that refer to entities in the knowledge graph is created. In the online step, text documents obtained for annotation are then split into potential phrases to be annotated, the phrases are scored and linked by the Mention Selectivity, Local Mention-Entity Compatibility and Global Entity-Entity Coherence modules. The output is a text document where all entities mentioned are annotated and linked to the corresponding KG entity

When such a system is used to annotate natural language text that comes from text documents in potentially different languages, text that was extracted from speech in videos and text from social media (again in different languages), a common data model is needed to make the content semantically accessible (see Fig. 3). To allow for semantic interoperability an RDF vocabulary is defined and tailored specifically to the different modalities: text, audio and video. It extends other vocabularies, such as the Dublin Core7, SIOC8 and KDO9. For each entity annotation, the predicates that define the start and end positions of the entity mention are used in a flexible manner and may define character positions, in the case of text, or milliseconds/frame numbers in case of audio/video. Each category annotation captures one topic of the media content. In any case, each entity mentioned in or each topic covered by any content item should relate to a resource in the knowledge base, namely an entity or a category in DBpedia.

Once the content is annotated in this way and if combined with semantic (keyword) query interpretation this allows for a semantic access to cross-modal cross-lingual content.

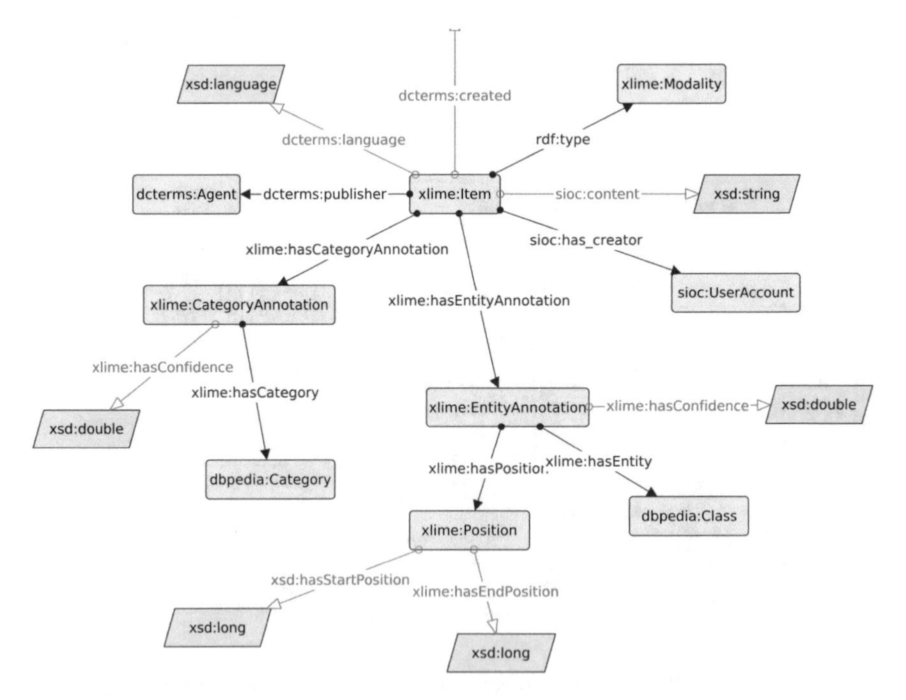

Fig. 3 To capture the meta information about annotations of different media items, a common data model is needed. The figure shows a model used in the xLiMe project to annotated text, audio and video content, denoted as "item". Meta information includes: entity mentions and the position of the mention (phrase in the text, bounding box in the image, time in the video), category of the item is being annotated, as well as the creator, publisher, the modality (text, audio or video) and the language (in case of text)

- **Document retrieval**: The search process starts with a keyword query in any language, which can even contain keywords in multiple languages. Instead of retrieving media items directly by keywords, a semantic search approach first finds the query entity vector (QEV), which represents a subgraph of the semantic graph of the knowledge base with nodes representing entities and edges describing their relations such that for each query keyword there is at least one entity in the subgraphs matching it. For each content item, we construct the data entity vector (DEV), where the entries contain the confidence scores of the annotations (i.e., the linked entities), which are generated by the cross- lingual semantic annotation. The semantic similarity between the QEV and each DEV can be calculated based on standard similarity measures, such as cosine similarity, which is then used for ranking of retrieved media items.
- **Document classification, clustering, recommendation**: The goal of a cross-lingual and cross-media semantic recommendation system is to find the similar content items posted across languages, modalities and channels. Here, we focus on a knowledge-centric approach to semantic recommendation using explicit semantics. This allows the semantic annotations to be further used for finding similar items. Once the entities inside documents are extracted, they can be used to calculate document similarity. Firstly, a subgraph is constructed from the entities identified for each document. As shown (Paul et al. 2016), the subgraphs of both documents are used to find the bipartite graph and graph-based similarity is then applied by computing the pairwise entity similarities based on the hierarchical and traversal scores.
- **Content analytics and complex queries**: Advanced data analytics across unstructured content has become a major necessity, which currently cannot be supported by non-semantic search engines. Using the knowledge extracted by semantic technologies from different media and languages in combination with additional background knowledge in DBpedia, a semantic approach allows to ask complex questions, such as "Which politicians from the Conservative Party of UK were most present in social media in the last two weeks before the Brexit referendum in different languages?". Such questions are formulated as a SPARQL query and are answered by aggregating, counting and averaging knowledge across several media items. The semantic integration provides the ability to study differences and commonalities among media channels and languages.

The just outlined technologies have shown benefits in several business cases from media monitoring to product recommendation and have been used by social, media and political scientists to analyse content. See e.g. (Zhang et al. 2017) for further details.

4 Semantic Technologies for Cyber-Physical Systems

In the last part of this chapter, we outline some of the benefits semantic technologies in general, and ontology-based description frameworks in particular can provide to cyber-physical systems (CPS) and smart factories. We discuss that along the two aspects: *utilization* and *interoperability* of hard- and software components that constitute such systems. An integral aspect is the usage of ontology languages (Motik et al. 2009) and description logics (Baader 2003; Krötzsch et al. 2006; Rudolph 2011) as knowledge representation frameworks that allow for the axiomatic description of component metadata models (see Zander and Hua 2016). Detailed information together with extensive formal specifications of description logic based formalisms can be found in the previous references. Those axiomatic descriptions can be processed by reasoners, i.e., software programs that are able to understand the formal model-theoretic semantics injected by an ontology-based description framework into the component metadata models. This process is called *reasoning* and allows machines to infer new, implicitly contained knowledge or to check the consistency of a model or knowledge base, particularly when new data are added to it. More details are given in the paragraph about machine processability.

We discuss utilization and interoperability of CPS components simultaneously, as interoperability requires the utilization of a component's metadata model. Throughout this chapter, we understand the term 'metadata model' as a technical description of a component's characteristic features using a specific representation framework and format. In many technical specifications, a component's metadata model is also called *information model.*

The unobtrusive collaboration of a multitude of different and hard- and software components is a central aspect in smart factories that employ intelligent, self-regulatory production lines. Together with the Internet of Things (IoT), cognitive and cloud computing, they are the main pillars of the recently emerging Industry 4.0 paradigm. The multitude of different hardware and software components like services, tools, software agents etc., raises new challenges in addressing the structural, schema and semantic heterogeneity introduced by new technologies, protocols, description frameworks, interfaces, data structures and formats. The importance of addressing these challenges is amplified by the increasing complexity of tools, systems, and other software components embedded in business processes and demand the continuous integration of technical data and expert knowledge throughout the entire value-creation network. The interplay of autonomously operating hardware- and software components is one piece of the puzzle towards the broad realization of smart factories.

In order to reconcile different efforts and emerging formats, leading industrial players such as ABB, Daimler, Huawei, Kuka, Siemens etc. (to name just a few) recently started standardization initiatives for Industry 4.0 specific description frameworks. One such initiative is AutomationML, a well-adopted and fast growing XML-based description standard that already caught the attention of Industry 4.0

communities.[2] It covers engineering aspects including topology, geometry, kinematics, logic and communication (Drath et al. 2008) that can be used for describing properties and functionalities of a CPS component. Data contained in the AML description of one component can be exposed to the communication network of a CPS system and consumed by other components (Schleipen et al. 2014).

Unfortunately, the expressive power and flexibility of most CPS-related description frameworks are limited (Zander and Hua 2016). From AutomationML alone, for instance, it is not able to determine whether two components are compatible and able to work together based on the interface descriptions they exhibit. As a consequence, many of such description frameworks are not able to provide that form of understanding, in particular not in a machine-processable fashion.

This is the starting point where semantic technologies and ontology-based description frameworks in particular will help complementing such standardized industrial-driven description frameworks in meaningful ways. Thus making a contribution towards the realization of a technical interoperability infrastructure where hard- and software components are able to autonomously collaborate and exchange information together with their intended semantics. Semantic technologies help in doing that through the following aspects:

(a) **Logical Model**: The description framework of Web ontologies is RDF, the logical model of which is built upon a graph-based data structure rather than a tree-based structure, which is the case for XML and XML-based formats such as AutomationML. Graph-based data structures not only provide a greater flexibility in representing information, they also circumvent the one-to-many mapping problem between graph- and tree-based representation formats and thus mitigating the problem of *structural heterogeneity* and *modelling ambiguity*.
(b) **Expressivity**: The graph-based representation model of Web ontologies allows to treat relationships (so-called *properties* in terms of the Web Ontology Language (OWL) or *roles* in description logic terms) as *first-class citizens* and explicitly specify their characteristics and semantics. This is one of the main distinguishing features of Web ontologies compared to the object-oriented paradigm in which the creation of a relation is bound to the existence of a class or object. In Web ontologies, relations can be used independently of classes or resources (member of classes) and exist in a self-contained manner. Moreover, their semantics can be defined upon the formal, model-theoretic semantics of the ontology language used to describe and represent an ontology.
(c) **Machine processability**: The logical theory upon which an ontology language and subsequently the ontology it describes is defined determines the expressive power and accuracy through which elements of a domain of interest (the so-called *universe of discourse* in ontology terms) can be described. Since the formal semantics of an ontology and its language elements (terms and concepts) are defined upon a logical theory, they can be processed automatically by

[2]Reference architecture model Industrie 4.0. Technical report, ZVEI. (2015).

machines in the form of reasoning engines. Reasoning describes the process of deducing logical entailments from the axioms constituting an ontology, the so-called *knowledge base*. Reasoning allows the determination of the consistency of a knowledge base, i.e., it checks whether an axiom (statement in ontology terms) is satisfiable for the given knowledge base or whether the introduction of a new fact to a knowledge base violates its consistency. Reasoning also allows for the deduction of new facts based on implicitly contained knowledge (as we will demonstrate in the following example).

These are some of the main features that distinguish ontologies and ontology languages from well-established representation frameworks introduced to computer science such as the Unified Modelling Language (UML) (Krötzsch et al. 2014).

In the following paragraph, we demonstrate how ontologies can serve as a semantic shell for enhancing the information models of cyber physical systems' components and allow for the deduction of new knowledge that fosters interoperability and data exchange between collaborating components.

In a first step, an AutomationML description needs to be analysed and transformed, i.e., uplifted into a compliant semantic graph represented as RDF description using transformation rules and domain heuristics (Björkelund et al. 2011; Kovalenko et al. 2015; Zander and Hua 2016). Such an uplifted semantic graph can then be processed by a reasoner in order to automatically classify a component with respect to specific classification systems and complemented its information model with domain knowledge derived from domain ontology axioms (some of them have been developed by several initiatives and projects such as KNOWROB, NIST Robot Ontology, OMRKF, ORA WG, and ReApp) (Schlenoff et al. 2012; Tenorth and Beetz 2013). Figure 4 illustrates an RDF representation of an excerpt of an uplifted AML description describing some technical aspects of a Sick S30B Laser Scanner. For reasons of readability and comprehensibility, ontology namespaces have been omitted throughout this chapter.

The concrete instance of the laser scanner is identified and represented via its UUID (line 1). Several technical parameters are then added in the form of RDF triples such as manufacturer and model information (line 2), starting and end angles (line 3), the maximal measurement range of the laser scanner in meter (line 4) and so on.

```
1 <urn:uuid:f81d4fae-7dec-11d0-765-00a0c91e6bf6>
2   :hasManufaturer "Sick" ; :hasModelName "S30B-2011GA" ;
3   :startAngle "135"^^xsd:integer ; :endAngle "135"^^xsd:Integer ;
4   :maxMeasurementRangeInMeter "40"^^xsd:integer ;
5   :maxProtectiveFieldRangeInMM "2000"^^xsd:integer ;
6   :maxWarningFieldRange "8000"^^xsd:integer ;
7   :maxSimultaneousFieldEvaluations "0"^^xsd:integer .
```

Fig. 4 Excerpt of an uplifted AutomationML description of a Sick S30B Laser Scanner. The code represents an excerpt of the corresponding RDF graph and is serialized in the RDF N3 notation (also known as Notation3)

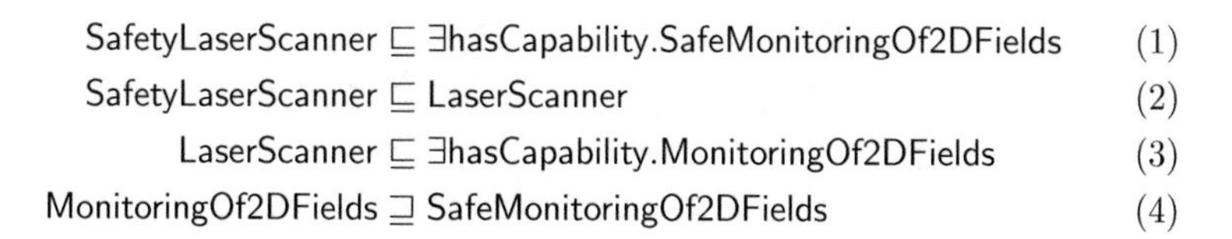

Fig. 5 Excerpt of a set of description logic axioms that represent a classification hierarchy for components (Axiom 2 and 3) and capabilities (Axiom 4) and link concepts of a domain ontology to complex concept expression (right part of Axiom 1) in order to represent capability information that a reasoner can infer

Those technical assertions are then processed against the domain knowledge encoded in several hardware, software and capability ontologies that contain axioms as illustrated in Fig. 4. These axioms allow for the classification of a component, i.e., based on the uplifted information, the component represented by its UUID can be classified as SafetyLaserScanner. This process is automatically conducted by a reasoner in order to infer additional information and use it to complement the component's information model.

Axiom 1 states that components classified as safety laser scanners have the default capability of safe monitoring of 2D fields (see Fig. 5). The concept representing 'SafeMonitoringOf2DFields' is defined in an external capability ontology and linked to the 'SafetyLaserScanner' concept through the property 'hasCapability'. Such information is encoded as TBox knowledge, i.e., as schema knowledge. Such schema knowledge together with classification information, i.e., asserting that the Sick S30B is a safety laser scanner (usually this can be inferred through the technical information provided by the AutomationML description), allows a reasoner to infer that a Sick S30B laser scanner has the default capability of safe monitoring of 2D fields. When such information is inferred from the constituting knowledge base axioms and additional domain knowledge such as a capability ontology, it can be materialized and added to the uplifted AutomationML information model in the form of RDF assertions, i.e., ABox axioms. Hence, the initial component's information model is complemented with additional knowledge derived from well-established domain ontologies. The materialization is important as it allows all the inferred knowledge about a component to be indexed by semantic data bases, so-called *triple stores* and retrieved through RDF query languages such as SPARQL.[3] Figure 6 illustrates how the uplifted AutomationML description of the Sick S30B component can be complemented with the inferred information.

This complemented description now contains classification information, i.e., it asserts that the concrete instance of a Sick S30B (represented via its UUID; see line 1) is a safety laser scanner (line 2) and so on. The metadata model also contains information about a component's capabilities ('{SafeMonitoringOf2DFields}', '{MonitoringOf2DFields}' and '{Monitoring}' (line 3), the purposes for which it

[3]SPARQL 1.1 Query Language. Technical report, W3C (2013).

```
1 <urn:uuid:f81d4fae-7dec-11d0-a765-00a0c91e6bf6>
2   rdf:type :2DSaftyLaserSensor , :LaserSensor , :Hardware Component ;
3   :hasCapability :{SafeMonitoringOf2DFields} , :{MonitoringOf2DFields} , :{Monitoring} ;
4   :hasPurpose :{HazardousAreaProtection} , :{AccessProtection} , :{PersonnelSafety} ;
5   :hasOperationEvironment :iIndoor, iOutdoor .
```

Fig. 6 An uplifted RDF description of a Sick S30B Laser Scanner component (concrete instance) with additional information derived by inferring additional and implicit information contained in domain ontologies

can be used by default together with information about its operation environments (the derivation of this entailments is not depicted in the example). With that complemented information models, software agents are then able to infer whether two components are compatible and able to collaborate in more sophisticated ways. More information together with detailed examples can be found in (Zander and Awad 2015), (Zander and Hua 2016) as well as in (Zander et al. 2016).

5 Summary

In this chapter we introduced the latest innovations in modelling knowledge using knowledge graphs and explained how those knowledge graphs enable value creation by making unstructured content, like text documents accessible by machines and humans. We covered different steps of the knowledge creation lifecycle including (manual) knowledge engineering and (automatic) understanding of unstructured content. Last, but not least, we have shown how semantic technologies help to make hard- and software components in cyber-physical systems interoperable.

References

Acosta, M., Simperl, E., Flöck, F., & Vidal, M.-E. (2015). HARE: A hybrid SPARQL engine to enhance query answers via crowdsourcing. In *Proceedings of the 8th International Conference on Knowledge Capture*. New York: ACM.

Acosta, M., Simperl, E., Flöck, F., & Vidal, M.-E. (2017). Enhancing answer completeness of SPARQL queries via crowdsourcing. *Web Semantics: Science, Services and Agents on the World Wide Web*.

Acosta, M., Vidal, M.-E., Lampo, T., Castillo, J., & Ruckhaus, E. (2011). ANAPSID: An adaptive query processing engine for SPARQL endpoints. *The Semantic Web–ISWC* 2011, 18–34.

Arenas, M., Bertails, A., Prud, E., & Sequeda, J. (2012). A direct mapping of relational data to RDF. W3C Recommendation. See https://www.w3.org/TR/rdb-direct-mapping/

Auer, S., Bizer, C., Kobilarov, G., Lehmann, J., Cyganiak, R., & Ives, Z. (2007). Dbpedia: A nucleus for a web of open data. *The Semantic Web,* 722–735.

Auer, S., Scerri, S., Versteden, A., Pauwels, E., Charalambidis, A., & Konstantopoulos, S., et al. (2017). The BigDataEurope platform–Supporting the variety dimension of big data. In *International Conference on Web Engineering*. Berlin: Springer.

Baader, F. (2003). *The description logic handbook: Theory, implementation and applications*. Cambridge: Cambridge University Press.

Bao, J. (2012, December). OWL 2 Web Ontology Language document overview. W3C Recommendation. *World Wide Web Consortium, 201*(2).

Björkelund, A., Malec, J., Nilsson, K., & Nugues, P. (2011). Knowledge and skill representations for robotized production. *IFAC Proceedings Volumes, 44*(1), 8999–9004.

Charalambidis, A., Troumpoukis, A., & Konstantopoulos, S. (2015). SemaGrow: Optimizing federated SPARQL queries. In *Proceedings of the 11th International Conference on Semantic Systems*. New York: ACM.

Dong, X., Gabrilovich, E., Heitz, G., Horn, W., Lao, N., & Murphy, K., et al. (2014). Knowledge vault: A web-scale approach to probabilistic knowledge fusion. In *Proceedings of the 20th ACM SIGKDD International Conference on Knowledge Discovery and Data Mining*. New York: ACM.

Drath, R., Luder, A., Peschke, J., & Hundt, L. (2008). AutomationML-the glue for seamless automation engineering. In *IEEE International Conference on Emerging Technologies and Factory Automation, 2008. ETFA 2008*. New York: IEEE.

Färber, M., Bartscherer, F., Menne, C., & Rettinger, A. (2016). Linked data quality of DBpedia, Freebase, OpenCyc, Wikidata, and YAGO. *Semantic Web(Preprint),* 1–53.

Franklin, M. J., Kossmann, D., Kraska, T., Ramesh, S., & Xin, R. (2011). CrowdDB: Answering queries with crowdsourcing. In *Proceedings of the 2011 ACM SIGMOD International Conference on Management of Data*. New York: ACM.

Görlitz, O., & Staab, S. (2011). Splendid: Sparql endpoint federation exploiting void descriptions. In *Proceedings of the Second International Conference on Consuming Linked Data-Volume 782*, CEUR-WS.org.

Kovalenko, O., Wimmer, M., Sabou, M., Lüder, A., Ekaputra, F. J., & Biffl, S. (2015). Modeling automationml: Semantic web technologies vs. model-driven engineering. In *2015 IEEE 20th Conference on Emerging Technologies & Factory Automation (ETFA)*. New York: IEEE.

Krötzsch, M., Simancik, F., & Horrocks, I. (2014). Description logics. *IEEE Intelligent Systems, 29,* 12–19.

Krötzsch, M., Vrandecic, D., & Völkel, M. (2006). Semantic mediawiki. In *International Semantic Web Conference*. Berlin: Springer.

Lehmann, J., Isele, R., Jakob, M., Jentzsch, A., Kontokostas, D., & Mendes, P. N. (2015). DBpedia–A large-scale, multilingual knowledge base extracted from Wikipedia. *Semantic Web, 6*(2), 167–195.

Lenzerini, M. (2002). Data integration: A theoretical perspective. In *Proceedings of the Twenty-first ACM SIGMOD-SIGACT-SIGART Symposium on Principles of Database Systems*. New York: ACM.

Marcus, A., Wu, E., Karger, D. R., Madden, S., & Miller, R. C. (2011). Crowdsourced databases: Query processing with people, Cidr.

Motik, B., Grau, B. C., Horrocks, I., Wu, Z., Fokoue, A., & Lutz, C. (2009). OWL 2 web ontology language profiles. *W3C Recommendation, 27,* 61.

Park, H., & Widom, J. (2013). Query optimization over crowdsourced data. *Proceedings of the VLDB Endowment, 6*(10), 781–792.

Paul, C., Rettinger, A., Mogadala, A., Knoblock, C. A., & Szekely, P. (2016). Efficient graph-based document similarity. In *International Semantic Web Conference*. Berlin: Springer.

Rudolph, S. (2011). Foundations of description logics. In *Reasoning Web. Semantic Technologies for the Web of Data* (pp. 76–136). Berlin: Springer.

Schleipen, M., Pfrommer, J., Aleksandrov, K., Stogl, D., Escaida, S., Beyerer, J., & Hein, B. (2014). Automationml to describe skills of production plants based on the ppr concept. In *3rd AutomationML User Conference*.

Schlenoff, C., Prestes, E., Madhavan, R., Goncalves, P., Li, H., & Balakirsky, S., et al. (2012). An IEEE standard ontology for robotics and automation. In *2012 IEEE/RSJ International Conference on Intelligent Robots and Systems (IROS)*. New York: IEEE.
Schwarte, A., Haase, P., Hose, K., Schenkel, R., & Schmidt, M. (2011). Fedx: Optimization techniques for federated query processing on linked data. In *International Semantic Web Conference*. Berlin: Springer.
Singhal, A. (2012). Introducing the knowledge graph: Things, not strings. https://googleblog.blogspot.co.at/2012/05/introducing-knowledge-graph-things-not.html 2016.
Souripriya, D., Seema, S., & Richard, C. (2012). R2RML: RDB to RDF Mapping Language, W3C Recommendation.
Tenorth, M., & Beetz, M. (2013). KnowRob: A knowledge processing infrastructure for cognition-enabled robots. *The International Journal of Robotics Research, 32*(5), 566–590.
Vrandečić, D., & Krötzsch, M. (2014). Wikidata: A free collaborative knowledgebase. *Communications of the ACM, 57*(10), 78–85.
Welty, C., Barker, K., Aroyo, L., & Arora, S. (2012). Query driven hypothesis generation for answering queries over nlp graphs. In *International Semantic Web Conference.* Berlin: Springer.
Zander, S., & Awad, R. (2015). Expressing and reasoning on features of robot-centric workplaces using ontological semantics. In *2015 IEEE/RSJ International Conference on Intelligent Robots and Systems (IROS).* New York: IEEE.
Zander, S., & Hua, Y. (2016, Feburary). Utilizing ontological classification systems and reasoning for cyber-physical systems. In *Karlsruhe Service Summit Research Workshop*.
Zander, S., Merkle, N., & Frank, M. (2016). Enhancing the utilization of IoT devices using ontological semantics and reasoning. *Procedia Computer Science, 98,* 87–90.
Zhang, L., & Rettinger, A. (2014). X-LiSA: cross-lingual semantic annotation. *Proceedings of the VLDB Endowment, 7*(13), 1693–1696.
Zhang, L., Thalhammer, A., Rettinger, A., Färber, M., Mogadala, A., & Denaux, R. (2017). The xLiMe system: Cross-lingual and cross-modal semantic annotation, search and recommendation over live-TV, news and social media streams. *Web Semantics: Science, Services and Agents on the World Wide Web.*

Author Biographies

Achim Rettinger, PD Dr. is an Research Group Leader at AIFB. Before joining KIT, he did his Ph.D. studies in Computer Science at the Technical University of Munich and at the Siemens AG in Munich, Germany. His research interests and publications are in combining machine learning with semantic technologies. He published extensively on cross-lingual and cross-modal representation learning, text and structured embeddings, cross-lingual semantic annotation, cross-lingual and multimodal retrieval, knowledge representation and statistical relational learning.

Stefan Zander, Prof. Dr. is a Full Professor for Web technologies and User-centered Design at Darmstadt University of Applied Sciences. He is also Guest Researcher at the FZI Research Center for Information Technology at KIT, where he was a research group leader before. His research interests cover topics related to knowledge representation, semantic technologies and its application in different domains such as the Internet of Things or cyber-physical systems in general.

Maribel Acosta, Dr.-Ing is a Postdoctoral Researcher at AIFB. She conducted her Ph.D. studies at the Karlsruhe Institute of Technology (KIT), Germany. Her research interests include data management over semantic data, with a special focus on data integration, data curation, and query processing. Her publications include works on enhancing Semantic Web technologies with human-based computation via crowdsourcing, adaptive techniques for Linked and Big Data management, and querying federated data sources on the web.

York Sure-Vetter, Prof. Dr. is Full Professor for Web Science at the Karlsruhe Institute of Technology (KIT), Germany. He is the spokesperson of the Institute AIFB, Director at the FZI Research Center for Information Technology at KIT and Director at the interdisciplinary Karlsruhe Service Research Institute (KSRI). His research interests include web science, knowledge representation, machine learning, Industry4.0 and the Internet of Things.

MEDICINE 4.0—Interplay of Intelligent Systems and Medical Experts

Hans-Peter Schnurr, Dominik Aronsky and Dirk Wenke

Abstract Healthcare professionals often have to take decisions under time constraints within a highly complex patient situation. This risky and error-prone process is fuelled additionally by an information overload due to sensor data, guidelines and ongoing updates of clinical information. Healthcare professionals need all of their experience and a lot of good luck to manage their decisions in this complex context. Acting under serious time pressure means having not enough time to gather, analyse and combine existing information. Suboptimal or wrong decisions may occur. A solution to guide and support healthcare professionals are Clinical Decision Support (CDS) systems. Today, there are many isolated CDS systems in a clinical environment causing tremendous maintenance efforts. This is one of the main drivers to centralize the authoring, maintenance and use of clinical knowledge with the help of Clinical Knowledge Management (CKM). Digitization, Artificial Intelligence (AI) applications and CKM also involves new knowledge processes, job roles and organization principles. There are new ways how experts, knowledge engineers and information technology interacts. This article describes the components of a CKM and the interplay of related job roles, limitations and challenges, and the implications of AI, CDS and CKM systems for healthcare organisations and healthcare professionals.

1 Introduction

More and more hospitals face the challenge of implementing the medical quality required by law in daily clinical practice. At the same time, it is a growing challenge for physicians to have all current guidelines with the valid standards for the diagnosis and therapy of the various disease pictures in mind. Healthcare professionals often have to take decisions under time constraints within a highly complex patient situation. This risky and error-prone process is fuelled additionally by an infor-

H.-P. Schnurr (✉) · D. Aronsky · D. Wenke
Semedy AG, Zug, Switzerland
e-mail: schnurr@semedy.com

K. North et al. (eds.), *Knowledge Management in Digital Change*, Progress in IS,
https://doi.org/10.1007/978-3-319-73546-7_3

mation overload due to sensor data, guidelines and ongoing updates of clinical information (e.g. medication lists). In medicine, knowledge has grown rapidly in recent years. In 1950, the knowledge doubled after 50 years, then doubling time was only 3.5 years by 2010. In 2020 the knowledge could be forecast to have doubled after only 73 days (Zwack and Lott 2017). In addition, an increasing amount of data about a patient is available, which must be evaluated and assessed correctly. Healthcare providers face the challenge to manage and use the available data and knowledge in an efficient and effective way to improve their decision-making quality. Healthcare professionals need all of their experience and good event timing to manage their decisions in this complex context. Acting under serious time pressure means having not enough time to gather, analyse and combine existing information. Suboptimal or wrong decisions may occur. In healthcare, the potential for cost savings through better decision-making quality, and cost avoidance through fewer follow-up treatments, is huge.

Some examples to illustrate the potentials:

- Approximately 18% of Medicare patients in the US returned to hospital within 30 days after discharge from the hospital.[1]
- Approximately 20% of the patients discharged from the hospital have to visit the doctor again due to drug side effects. According to United Health, a health insurer, a large proportion of these cases could have been avoided if smart CDS solutions alert the problem before dismissal from hospital.[2, 3]
- The adjustment to medication of a patient with multiple diagnosis and multi-medication may take several hours after hospitalization. A prescription suggestion through a smart CDS solution could shorten this time to minutes.

2 Intelligent Systems in Healthcare

Four factors will change medicine in the near future: personalization, digitization, mechanization and telemedicine. These factors will contribute to the fact that the medicine will change more over the next 10 years than in the last 100 years (Wikipedia). The digitization includes Big Data and Artificial Intelligence (AI). The combination of both will enable reliable diagnoses through software programs in the near future. Programs for the interpretation of the results of imaging methods and the diagnosis of oncological or certain rare diseases are already very advanced. The combination of Big Data and AI will be an important tool for evaluating therapies.

---

[1]Center for Medicare and Medicaid Services. USA.

[2]United Health Group, US Insurance Company.

[3]International Institute for the Safety of Medicines, Basel, Switzerland.

One of the focus application areas that include a combination of innovative AI technologies are Clinical Decision Support Systems. These systems help physicians to arrange the right examinations and therapies, while at the same time taking on a kind of control, so that no important step will be missed. Clinical Decision Support refers to procedures for improving clinical decisions by providing evidence-based medical information at the time of the doctor-patient contact or at the time of the treatment decision. This can be general clinical knowledge, decision-making, patient-specific data, or a mixture of both. There is a consensus among users that a critical technical assessment of the patient's case can't be replaced by IT systems.

On the other hand, there is a high probability to reduce unnecessary or even harmful medical services in the clinical decision-making processes by means of the stronger integration of guidelines and scientific evidence. In addition, Clinical Decision Support Systems offer the possibility to transfer the steadily growing number of medical publications and research results into clinical practice.

Examples of electronic expert systems to support clinical decisions are:

- Context specific links to quality approved information such as evidence based guidelines, systematic reviews, and other reliable sources from the hospital information system. In United States, these features are known as "InfoButton" applications.
- Individualized recommendations as well as warning and reminder functions through automatic linking of documented patient data with recommended guidelines in the hospital information system.
- Automated updates and validation checks of order sets e.g. in case of changes of the medication list.

Due to the underdeveloped digitalization of the hospitals in Germany, Clinical Decision Support (CDS) Systems in Germany are relatively little established in international comparison. In many cases, patient records are still paper based, which makes electronic evaluation practically impossible (Wikipedia).

In the United States, the use of knowledge management and CDS systems has been promoted for a number of years in the context of the so-called "Meaningful Use" Initiative. This initiative is a part of the American Recovery and Reinvestment Act of 2009 (ARRA) and the Medicare and Medicaid Electronic Health Record (EHR) Incentive Programs that were established to encourage healthcare professionals and hospitals to adopt, implement and demonstrate meaningful use of certified EHR technology.

The existing broad implementation base of CDS systems in combination with the Meaningful Use initiative led to new challenges. Today, there are many isolated CDS systems in a clinical environment causing tremendous maintenance efforts. Many information sources like drug and vaccination databases provide frequent updates of their content. They have to be imported and updated within many different clinical applications. Links and relations between existing information sources have to be updated when changes occur. The complexity of the maintenance task gets even worse with every new implementation of a CDS system that also uses clinical knowledge based on rules, guidelines and other information sources. This huge challenge is one of the main drivers to centralize the authoring, maintenance and use of clinical knowledge with the help of Clinical Knowledge Management (CKM).

In this context, the clinical knowledge management system (CKMS) of the company 'semedy',[4] which supports the processes for collaborative creation, linking and efficient maintenance of clinical knowledge, is implemented in several US hospitals.

3 Clinical Knowledge Management Approach

Healthcare institutions build increasingly large amounts of clinical knowledge assets. They are responsible for the accuracy, transparency and updating of the content. The institutions are monitoring this content, as failing to do so can lead to inappropriate or sub-standard care. Inconsistent, incomplete, and outdated clinical knowledge assets represent unnecessary patient safety risks.

Knowledge assets are:

- leveraged from external vendor sources for direct consumption by software applications
- modified or adapted from external vendor sources
- locally created and maintained by central knowledge management/quality assurance/documentation departments
- locally created and maintained by individual departments/sites.

However, most organizations limit their Knowledge Management (KM) to a reactive and an ad-hoc approach and have neither a formal review and maintenance process, nor an appropriate system to manage knowledge embedded in Health Information Systems (HIS) and other systems.

- Most organizations do not systematically manage the clinical knowledge life cycle of knowledge: acquisition → incorporation → review → updating → retiring.
- Most organizations limit their knowledge management to individual systems or single sites, leading to uncoordinated and sometimes conflicting clinical information "knowledge islands".
- Even in organizations with a knowledge management team, knowledge assets are not located in a central place where everyone can access it. It is not available enterprise-wide for easy maintenance for all team members.
- Most organizations do not support a systematic and collaborative approach for acquiring such knowledge. Also, the knowledge assets are mostly not

[4]semedy AG, www.semedy.com.

consolidated and optimized at a group level (e.g. selection of the best of several redundant rules from various sources).

Proper knowledge management requires processes for regularly acquiring and integrating up-to-date knowledge including proactively reviewing, maintaining, and monitoring the embedded knowledge. Such monitoring enables providers to continuously improve the knowledge—and to lay the basis to assess the effectiveness of Clinical Decision Support (CDS) on clinical outcomes.

Knowledge is ultimately created by a collaborative process between knowledge engineers and clinicians who need to work with a Clinical Knowledge Management System (CKMS) to review and manage new and existing knowledge assets. Some of the main requirements are:

- need to confirm with practicing clinicians that the content and logic is appropriate for clinical care
- need for a formal review and vetting process so that only approved content moves forward to be published (i.e., content lifecycle management)
- need to transform knowledge assets (guidelines, rules, etc. efficiently into a machine-executable format, without requiring coding or programming skills
- need for an auditing function (e.g. to check later where the content came from, or who signed off on it) and versioning control.

Knowledge content should be tagged with metadata so that it is easily searchable. Meaningful use (MU) regulations also require that CDS in Electronic Health Records (EHR) be tagged with bibliographic citations and with information about the developer of the intervention, the funding source, the date of publication or revision, the references, and other, so that users, clinicians and knowledge engineers can understand the source and background of the CDS participate in improving it.

Knowledge users and engineers prefer a central place for all knowledge management processes rather than using many different systems to manage a specific CDS. Additionally, most content depends on other content. The knowledge engineers need to understand these dependencies and the impact on what may occur if one part changes; dependency management is complex and critical to avoid "broken links".

Linking knowledge assets to standardized and controlled terminologies should be supported while considering local vocabularies.

- Meaningful Use (MU) will increasingly dictate the use of structured data, e.g. the use of standard terminologies.
- The exchange of data between sites applying standard terminologies also facilitates interoperability that is semantically clear.

Finally, providers need to have the ability to retire content that is outdated or no longer applicable to clinical practice.

semedy's Clinical Knowledge Management System (CKMS) is a functional suite of modularized applications and services that improves the efficiency and reliability of clinical knowledge creation and management, thus leading to greatly enhanced collaboration, access and content utility for all users. The CKMS manages authoring, maintenance and use of all domains and types of clinical knowledge. It is based on a common platform that includes a flexible content repository, easy-to-use content editors, and a publishing portal. The expected benefits are:

- a centralized inventory of all clinical knowledge assets, thereby facilitating searching, retrieving, using, linking and maintaining knowledge
- improved efficiency and reliability of knowledge content creation and maintenance processes
- eliminate redundant data entry
- streamline communication between knowledge engineers and subject matter experts (SME)
- standardize and unify content authoring workflows
- proper management of content dependencies
- appropriate use of reference content sources
- implementation of automated content validation processes
- increase transparency and trust on knowledge content
- improve overall knowledge content accuracy, completeness, and maintainability
- reduce any potential risks to patients due to incorrect, inconsistent, and/ or outdated content.

4 Interplay Between Intelligent Systems and Healthcare Professionals

Medicine is a science of uncertainty and an art of probability. (William Osler)

Already today, we use medical applications of artificial intelligence in many different ways. When prescribing medications, algorithms cross-check the prescription with patient data for allergies and intolerances and send alerts in case of risks. Natural language processing helps to transform unstructured knowledge into structured and machine executable data. This can help to identify symptoms, diagnoses and medications automatically in doctor's letters. The most advanced development today is in the area of medical imaging, including radiology, pathology and dermatology. Algorithms already detect breast cancer, predict heart diseases, recognize osteoporosis, and identify the first signs of skin cancer with a security that is equal to human physicians. Already in the near future, AI-based decision support systems will have more knowledge than a human expert. This leads to the fact that diagnoses and therapy proposals are increasingly being created by computers, which are first checked and released by physicians and therapists. In the future AI systems will in a fully autonomous manner.

To get this done, healthcare experts have to feed the necessary clinical knowledge to the intelligent systems. This knowledge generation and maintenance is an enormous undertaking. One of the most challenging issues in healthcare relates to the transformation of raw clinical data into contextually relevant information. Developing or modifying new clinical decision support content within an organization is always a challenge. A variety of people are usually involved in content development, including physicians, nurses, pharmacists, informaticians, software developers and quality improvement professionals. In order to better understand the current state of the art in Clinical Knowledge Management (CKM), (Sittig et al. 2010) developed a survey of potential CKM tools and techniques. All of the organizations studied had one or more of the following types of people involved in the CKM process:

- Pharmacists with formal informatics training (e.g., Masters or Doctorate in Medical Informatics or Informatics fellowship) or extensive clinical informatics experience to develop and maintain pharmacy content
- Physicians with informatics experience to translate clinical guidelines and study protocols into CDS interventions
- Doctoral-level Medical Informaticians
- Registered Nurses (RNs) with informatics experience
- Dedicated Software Developers and Project Managers without a clinical background.

Bringing all of these participants together for in-person meetings is challenging. Allowing content developers and users to have an asynchronous discussion regarding the pros and cons of specific CDS interventions can be a valuable method of both developing new clinical content and gaining organizational consensus. Typically, each piece of clinical content (i.e., alert, order set, and patient education material) has an individual responsible for monitoring the underlying clinical knowledge and maintaining the CDS intervention. The study (Sittig et al. 2010) sums up that all organizations used some sort of distributed CKM maintenance system. For example, all organizations had a pharmacist informatician responsible for developing and maintaining the content related to medications (indications, interactions, side effects, monitoring, default dosing, preferred medications and formulations). In addition, they each had physician informaticians with broader clinical oversight roles. Often content specific to a particular clinical department (for example, procedure or problem-based order sets) is managed by a clinical champion or medical informatician in that department.

Based on this situation, semedy developed a knowledge lifecycle process (Fig. 1) in close interaction with their customer, a large US based hospital group. This knowledge lifecycle explains the interaction of specific tools with people that are involved in this process. Specific tools need to support the lifecycle process to ensure consistency among knowledge assets and guarantee process efficiency. semedy's Clinical Knowledge Management System (CKMS) interacts with clinical applications and Business Intelligence (BI) tools to support and

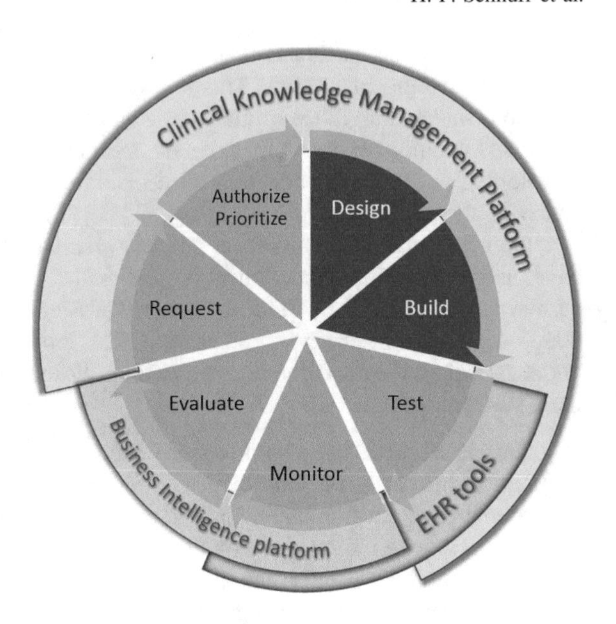

Fig. 1 The knowledge lifecycle process

streamline the complete knowledge asset lifecycle. Focusing on the design and build phase, CKMS supports also request, authorization/prioritization and test phases, while interacting (exporting/ importing knowledge assets) with Electronic Health Record (EHR) and BI systems to monitor and evaluate the utilization of published knowledge assets.

The knowledge life cycle steps in more detail:

- **Request**: The knowledge lifecycle process typically starts with a request. Any user may request a new or updated CDS intervention, such as adding a new guideline to the repository, or modifying the content of an existing knowledge asset. Another use case includes the request for creating a completely new domain or new type of assets. Request forms support the user to specify the intended recipients, use and rationale for making the request.
- **Authorize/Prioritize**: When receiving new requests, the knowledge management board (or other governing boards like a CDS committee) has to decide whether the requests are valid and the requested modifications to the knowledge repository should be made, or if the request should be disregarded. Board members discuss the requests, vote and decide following prioritization criteria, (e.g. if the request will objectively improve the safety, quality and efficiency of patient care delivery). All accepted requests have to be prioritized and are processed accordingly. Once a request is approved the knowledge lifecycle process transitions into the design phase.
- **Design**: During the design phase the scheme for the domain of interest is developed, including all the different types of assets, their defining properties, and the respective metadata. The phase is often characterized by a large number of model modifications. Defining the scheme is a collaborative and iterative

effort involving the knowledge engineers (KEs) and subject matter experts (SMEs) and often starts on a whiteboard.

- **Build**: After the domain model scheme has been finalized in the design phase, the model is populated with real-life knowledge assets during the build-phase. This means that the knowledge assets need to be represented in a way that conforms to the selected scheme. Existing content has to be transformed into the new scheme, typically using an "Extract, Transform, Load" (ETL) process. New content is directly created using the designed scheme. Knowledge engineers and application coordinators need to collaborate during the build phase to ensure alignment of the representation with the target applications.
- **Test**: Knowledge engineers, analysts and application coordinators write and execute test cases in the knowledge engineering or the target application test environment, so that new assets can be tested exactly within the same target application that clinicians or end users interact with.
- **Monitor**: After publishing new knowledge assets and exporting them to the target application their utilization is monitored in the target application, such as an EHR. Knowledge engineers and reporting team members can analyze utilization data using BI tools.
- **Evaluate**: The knowledge management board evaluates monitoring results to decide whether additional interventions or changes to knowledge assets are needed. Decisions may result in new request or authorization/prioritization tasks.

This interplay of diverse domains and roles in medicine, computer science and knowledge management can only work if both the tools and the appropriate staff are available. This is the biggest challenge for the future success of intelligent systems in medicine. In addition to the issues surrounding the general lack of use of existing tools to support the work of clinical knowledge management, there are also several organizational issues that must be addressed. For example, the lack of money to hire additional appropriately trained clinical informaticians to serve as knowledge engineers results in existing personnel being forced to work in unfamiliar territory without the necessary time or understanding to "do the job right". Additionally, the involvement of users who are clinical specialists on content development teams is often hard to maintain without sufficient monetary incentives. Further, the rapidly expanding regulatory reporting and compliance requirements along with increasing emphasis on quality measures are placing tremendous demands on development of human resources.

5 Implications for Healthcare Professionals

The new techniques of Artificial Intelligence bring about a "democratization" of medicine: the access to expert knowledge and the availability of cheap diagnostic devices is made available to all physicians and even to laymen—an enormous gain for the general medical practitioner and a possible loss for the medical specialist in ambulatory care. What cannot be treated on an outpatient basis, however, requires ever more complex diagnostic and treatment pathways, which can only be used in highly specialized centers.

5.1 Implications on Ambulant Care (Uhlig 2017)

- Sensors in wearables such as watches, rings, headbands, clothing or patches will provide data on heart rate, blood pressure, ECG, EEG, body temperature, oxygen saturation and physical activity. More detailed information can be obtained via insideables, such as glucose measurement in contact lenses or implants with sensors.
- There will be a large number of new diagnostic devices, which enable further examination with simple devices which can be connected to a smartphone, e.g. Otoscopes, stethoscopes or ultrasound equipment.
- With cameras available in smartphones at any time, photos, voice recordings and films are created that can be evaluated automatically or used in telemedicine.
- Enhanced pattern recognition methods and artificial intelligence will enable extensive prevention and diagnostics on the basis of these data.
- The analysis of all types of images (e.g., radiography, MRI, CT, sonography, eye background) will be reliably performed by programs.
- Patients may be assisted by physicians or nurses with AI support.
- Prevention will take place on several levels. Continuous data collection is suitable both as an early detection program and to avoid unnecessary hospitalization as well as to monitor healthy behavior (e.g., movement, posture, food intake)
- Home visits. Many diagnoses and prescriptions can be made online (also with bots) or telemedically without the patient visiting a doctor. The patient also has the choice of services such as Uber Health (similar to already established for taxi business) to select and order available doctors or care givers. Such services are entirely new part-time working models.
- Outpatient medical visits. Many doctor's visits will be omitted in this way. In addition to this, the practicing physicians are gaining an enormous competence gain through cost-effective, modern, powerful and ever-increasing devices (for example ultrasound, coherence tomography), powerful image analysis software, AI-supported diagnostic programs and telemedical support. This will lead to a revaluation of general practitioner activities and thus make specialist expertise widely affordable.
- The digital media is suitable for therapy support: telemedical monitoring, emergency warnings, reminders of medication intake. Many control visits are therefore superfluous.

5.2 *Implications on Hospital (in-Patient) Care (Uhlig 2017)*

- The medicine of the future in the hospital sector will be highly technicalized and personalized. Innovations such as stem cell therapies, intelligent prostheses from the 3D printer, artificial organs, Augmented Reality to support operations such as surgery, sensor implants in the brain or nanotheranostics will significantly improve the existing therapies.
- The individual diagnostics is followed by the individual therapy e.g. the appropriate administration and dosage of medication.
- The advanced technology not only causes high costs for acquisition and maintenance, but also requires ever more highly qualified personnel. The individualization of the medicine brings enormous amounts of data, so that the profession of the Medical Data Scientist will arise.
- Robots will assist the medical staff in their work. Examples include support for physical work such as patient repositioning or physiotherapy but also for automated disinfection of rooms or sterilization of surgical instruments.
- Artificial Intelligence will help to significantly improve patient pathways and treatment plans.
- With bioinformatics, the knowledge will dramatically increase and the clinician will be overwhelmed. IT solutions with Clinical Decision Support become indispensable.

5.3 *Implications on Knowledge Processes in Hospitals*

- The widespread adoption of artificial intelligence and clinical decision support in healthcare implies a centralization of the authoring, maintenance and use of clinical knowledge with the help of Clinical Knowledge Management. It also has effects on new knowledge processes, job roles and organization principles of healthcare professional.
- The search for relevant information becomes faster and more reliable as institutional (approved) knowledge is used (this also brings legal protection). The knowledge is also comprehensible as further information (rationality) is made available.
- Knowledge representation must be comprehensible and decision-making proposals must be substantiated and supported by literature, otherwise acceptance is much less. Good and current CDS rules are very well accepted by clinicians. Examples include complex and often changing recommendations, such as immunization regimes of children and adults, decisions for the correct investigation in complex imaging, antibiotic therapy for infections acquired in the hospital, as a non-exhaustive example list.

- New work roles are emerging in the context of relatively "new" operational knowledge management in the health sector. Knowledge managers, terminology experts, Medical Data Scientists and knowledge engineers are more and more found in the IT departments of large hospitals, usually in the CMIO (Chief Medical Information Officer) team.
- It is expected that these knowledge engineers will "translate" the expert knowledge to make it machine executable. The experience with the early use of expert systems shows that this knowledge acquisition is becoming the bottleneck of intelligent systems development. A system for the collaborative support of acquisition and maintenance of knowledge, as the Clinical Knowledge Management System (CKMS), helps to eliminate this knowledge acquisition bottleneck.

6 Conclusion

Today healthcare professionals often have to take decisions under time constraints within a highly complex patient situation. This risky and error-prone process is fueled additionally by an information overload due to sensor data, guidelines and ongoing updates of clinical information. Suboptimal or wrong decisions may occur. A solution to guide and support healthcare professionals are Clinical Decision Support (CDS) systems. Today, there are many isolated CDS systems in a clinical environment causing tremendous maintenance efforts. This is one of the main drivers to centralize the authoring, maintenance and use of clinical knowledge with the help of Clinical Knowledge Management (CKM) also involving new knowledge processes, job roles and organization principles. Especially, multidisciplinary teams responsible for creating and maintaining the clinical content will tightly collaborate.

The medical care of the future could have three levels (Uhlig 2017): (1) trivial diseases are treated at home via internet medicine or "Uber" doctors; (2) many other diseases are treated as outpatients at the general practitioner with the help of telemedicine and AI; (3) highly specialized services are provided in excellently and extensively equipped hospitals.

In times of demographic change and an imminent relative and absolute shortage of healthcare professionals, the future of care depends on whether we are able to develop techniques that lead to a noticeable relief of routine practitioners. Artificial intelligence can make a great contribution here. The technology does not have fatigue, and is available around the clock on Sundays and public holidays. Patients become more autonomous, doctors are relieved, diagnoses are faster and more accurately, and therapies are tailor-made. In combination with knowledge management methods enabling people and computers to generate, distribute and use the clinical knowledge the best way, a new level of healthcare IT will be created. This is Health 4.0.

References

Sittig, D. F., Wright, A., Simonaitis, L., Carpenter, J. D., Allen, G. O., & Doebbeling, B. N. (2010). The state of the art in clinical knowledge management: An inventory of tools and techniques. *International Journal of Medical Informatics, 79*(1), 44–57.

Uhlig S. e. a. (2017). Die Zukunft der digitalen Gesundheitsversorgung in Deutschland. DIV Report Spezial Digitale Gesundheit.

Wikipedia. Vivek Wadhwa. Retrieved June 15, 2017, from https://en.wikipedia.org/wiki/Vivek_Wadhwa.

Zwack, L., & Lott, K. (2017). Big Data und Clinical Decision Support—wissensbasierte Systeme zur Unterstützung klinischer Entscheidungen. DIV Report Spezial Digitale Gesundheit.

Author Biographies

Hans-Peter Schnurr, MS (Dipl. Wirt.-Ing.) is Co-founder and CEO of semedy AG. Formerly Mr. Schnurr was Co-Founder and Managing Director (CEO) of a software development company in the field of ontology and semantic software and worked as a Management Consultant. His expertise and interests include methods, implementation and evaluation of knowledge management, artificial intelligence applications, and semantic technologies.

Dominik Aronsky, MD, Ph.D., FACMI is the Chief Medical Information Officer and a member of the Board of Directors of semedy AG. His expertise and interests include the development, implementation, and evaluation of clinical information systems with a special emphasis on clinical decision support system, knowledge management, and the application of artificial intelligence to support real-time patient care.

Dirk Wenke, MS obtained an MS in Business-Mathematics from the University of Karlsruhe, Germany (2000). Dirk developed a variety of applications within the semantic web domain. Since 2012, Dirk is Head of Software Development at semedy AG, responsible for all aspects of the Clinical Knowledge Management System. Dirk is an author on several publications and conference proceedings. Dirk has given numerous instructional trainings on the foundations and applications of ontologies and knowledge management systems in the engineering and clinical domain.

Data Driven Knowledge Discovery for Continuous Process Improvement

Michael Kohlegger and Christian Ploder

Abstract Knowledge is recognized as an organizational resource for business value creation. The work with knowledge—knowledge work—is thus an important part of value-adding processes in organizations. The ability of knowledge workers to analyze complex phenomena, interpret them and develop meaningful actions is one central part of knowledge work. The advancements of digital aids and especially the ability to analyze big amounts of data is a new phenomenon that is increasingly seen in organizations. In this work, we assume that there needs to be an interplay between digital aids and knowledge workers to allow new, deep insights into phenomena and support business value creation. We develop a model that describes how this interplay could look like and critically discuss it using real-world cases. From that, we find that it is crucial (1) separating data-driven and expert-based analysis in knowledge discovery, (2) clearly describing the problem that should be solved by the analysis, (3) understand the particular domain that analysis is applied to, (4) complement data-driven with expert-based analysis and (5) understand the entanglement of analysis and action implementation.

1 Introduction

The digital has become an important part of how people, who are extensively engaged in the use and production of abstract knowledge as part of their professional work—commonly addressed as knowledge workers (Pyöriä 2005; Maier et al. 2009)—fulfill their professional tasks. Knowledge workers are strongly involved and supported by information systems (Maier 2007), which have changed with the rise of digitalization and enable new forms of knowledge discovery,

M. Kohlegger (✉)
University of Applied Sciences Kufstein, Kufstein, Austria
e-mail: michael.kohlegger@fh-kufstein.ac.at

C. Ploder
Management Center Innsbruck, Innsbruck, Austria
e-mail: christian.ploder@mci.edu

K. North et al. (eds.), *Knowledge Management in Digital Change*, Progress in IS,
https://doi.org/10.1007/978-3-319-73546-7_4

which is widely recognized as being a key-competence of knowledge workers (Schultze 2000, 2004; Pyöriä 2005). Recent advancements in the area of Business Intelligence (BI), which is often referred to as techniques, technologies, systems, practices, methodologies and applications that analyze critical business data to help an enterprise better understand its business in dedicated markets and make timely business decisions (Chen et al. 2012), support this shift. Although many topics that BI addresses are not new (e.g. Wu et al. 2008), the arbitrary availability of large amounts of data and computing power, today, have led to a renaissance of this topic. The new tools and practices that BI can offer today put pressure on knowledge workers as some of their skills are competing with the abilities of knowledge discovery (Fayyad et al. 1996) algorithms, which are BI tools (Chaudhuri et al. 2011). This data driven approach promises new opportunities for gaining knowledge with major shortcuts in the process (Davenport 2006). Although there are probably good arguments to counter this position, this might yield situations in which knowledge workers find themselves arguing for their existence.

However, the influence of BI on knowledge work (KW) is not only a limiting one. While some persons in the context of KW are obviously under competitive pressure, it also creates new niches for others. While some factual knowledge might get obsolete as it can be easily reconstructed from data, other forms of knowledge get increasingly important. Persons in the context of KW, who are using BI tools and practices, need to understand how these tools and practices work to put their results to good use and avoid misconception and misinterpretation. Organizations in general and organizational functions like knowledge management in particular need to understand these developments to properly support their KWs.

This work aims at understanding how data driven knowledge discovery in the context of KW can be described and what implications it has for organizations and individuals in the context of KW. We use the example of continuous process improvement to guide this analysis, as it is a widely implemented practice today with strong impact on organizational value creation (Rother 2010).

After the introduction (Sect. 1), we will define the theoretical foundations of this work (Sect. 2) and its study design (Sect. 3). We will then describe a theoretical model of data driven knowledge discovery in the context of KW (Sect. 4). Finally, we will introduce two real-world cases that we will use to evaluate the model (Sect. 5), discuss them (Sect. 6) and give the implications of this discussion (Sect. 7).

2 Theoretical Foundations

We use the term digitalization to refer to the transformation of analogous assets into electronic representations with the objective to allow storage, processing and sharing of these assets by electronic means, which strongly concerns the relationship of different representations of data. We understand data as simple facts about the world, information as a statement about a certain context with a clear objective and knowledge as facts represented in a mental structure that can be processed by consciousness, with a nonlinear, bidirectional relationship between them (Tuomi 1999). Besides this value chain model of knowledge, we also use the Autopoietic Model of knowledge (Parboteeah and Jackson 2011) to understand how knowledge can be pragmatically falsified through action.

In the age of digitalization, the discovery of knowledge from data is commonly understood as something critical. Therefore, we see models like the process of knowledge discovery in databases (Fayyad et al. 1996), that address this issue. The Fayyad et al. model comprises five steps—Selection, Preprocessing, Transformation, Data Mining and Interpretation/Evaluation—to describe how knowledge discovery from databases takes place without any particular application or domain focus.

Digitalization yet not only affects artifacts—e.g., data—but also affects the methods that are used to handle these artefacts. With the rise of digitalization, we, e.g. see changes in how KW is conducted as data about nearly any domain that KWers might be engaged with are getting increasingly available, while processing capacities get increasingly available and easy to use at low costs. Therefore, data driven knowledge discovery in the context of KW is rising (Pauleen and Wang 2017). Consequently, KWers will have to develop new processing skills for data.

3 Study Design

In this work, we are evaluating a model, using an approach that is informed by elements of design science (Hevner et al. 2004). We formulate a model from theoretical considerations and use two real-world cases to evaluate it.

The presented cases are constructed by taking a deep look into the field (Berg 1989) with the help of qualitative observation (Spöhring 1989; Bryman and Bell 2015) over an extended time period. In both cases, the authors have done this observation themselves, guided by the principles of qualitative observation—(1) done in the natural environment, (2) involving active participation to avoid subject-object-separation, (3) concentrating on bigger entities or systems, (4) being open for new observations and (5) combining behavioral and latent motivational structures (Bortz and Döring 2005). When writing-up the case for this work, individual observations were confronted with each other, aiming at developing two equal cases. We decidedly selected two cases from a larger case repository, to have the opportunity for providing a detailed case description. Based on the decision to use two cases, we selected the two most contrasting ones for presentation, to support a critical discussion. The documentation of the cases was done in a structured way, using a coding schema that was deduced from literature and inductively refined (Bryman and Bell 2015). The schema is presented in Sect. 5.

With respect to ontology—the study of being—we are taking the position of internal realism, recognizing reality as being an interpersonal construct allowing the description and analysis of phenomena but not as things in themselves (Archer 1988). Epistemologically—the study of knowledge—we follow a non-positivist approach, meaning that we recognize facts as being not definable by reference to an external reality, thus allowing only such knowledge that is made up of intertwined facts and values, (Archer 1988).

4 Model of Data Driven Knowledge Discovery

Based on the concepts and processes discussed in Sect. 2, we develop a theoretical model describing the phenomenon of data driven knowledge discovery in the context of KW. We have used a procedural layout for this model that can be divided into three main phases: (1) the bottom-up phase, (2) the top-down phase and (3) the feedback phase (informed by Fayyad et al. 1996).

Data driven knowledge discovery in the context of KW can have multiple starting points. Therefore, we assume that our model can have different triggers. The starting point for knowledge discovery will always be an observation that enables the discovery of knowledge (informed by Parboteeah and Jackson 2011). The observation in turn can be either triggered by a question (e.g., a person exploring how to improve a business process' cycle time out of curiosity) or by distinction (e.g. a person exploring ways to handle a malfunction in a business process in response to an observable symptom). We will describe the model in more detail in the following paragraphs and a graphical representation of the model is given in Fig. 1. We use the numbers in Fig. 1 to reference to specific parts of the model in the text.

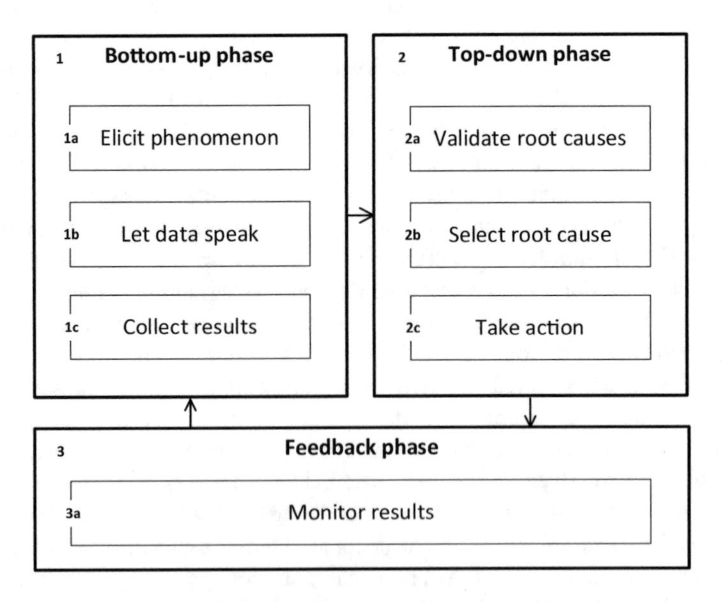

Fig. 1 Model of data driven knowledge discovery in the context of KW

In response to any new triggering event, the *bottom-up phase* (informed by Fayyad et al. 1996) is started (1), which can be divided into three subsequent steps. We assume that any trigger will be followed by an elicitation step (1a) where the phenomenon at hand is explored and documented (e.g., the overall quality of product ABC has decreased significantly over the last two weeks). As we are deliberately analyzing the phenomenon of data driven knowledge discovery, we further assume that the elicitation step is followed by a data analysis step (1b) that is used to explore potential root causes by letting the data speak (e.g., use log files from a process execution engine and sequential clustering to find pattern that indicate potential root causes for the quality decrease in product ABC). Finally, we assume that the bottom-up phase is ended by a documentation step (1c) where all potential root causes of the elicited phenomenon are collected and prepared for the top-down phase.

The bottom-up phase ultimately yields a collection of speculative root causes that have been produced by data analysis. These root causes are subsequently addressed in the *top-down phase* (informed by Fayyad et al. 1996; Tuomi 1999; Parboteeah and Jackson 2011) to interpret them in the light of practical application (2). In the first step of this phase (2a), we assume that all collected root causes of the phenomenon are closely analyzed with a praxis-oriented focus and validated in the light of reasonability and likelihood (e.g., process experts are confronted with the speculative root causes to judge them using their domain knowledge). After that, we assume that (2b) there is a selection step where one or a combination of several root causes is selected and (2c) used to set some mitigating action (e.g., one of the ingredients of product ABC is changed as this root cause was judged to be the one with the highest likelihood). Consequently, new knowledge about the phenomenon at hand is discovered and immediately validated with practical application.

The third part of the model is dedicated to secure the effectiveness and efficiency of all taken actions by establishing a *feedback phase* (3). This phase only contains one single step (3a), which concerns the monitoring of any action that has been taken, which can be a trigger for a new instance of the cycle (e.g., if a new questions arise from monitoring the results of the last cycle).

This new approach distinguishes from classical approaches that are purely based on reasoning using experience as it allows an open-minded approach to a phenomenon at hand, which is not limited by the amount of experience of a person or his/her ability to transfer experience to the phenomenon.

5 Empirical Cases

In this section, we present two real-world cases that we are using to discuss the introduced model. We will start by presenting the coding schema that was used to structure the case write-up. Codes 1 to 3 applies to the overall organization wherefore we will elaborate them in Sect. 5.1. Code 4 and 5 apply to the individual case and will therefore be elaborated in Sects. 5.2 and 5.3.

(1) *Overall facts* about the organization (e.g., industry, headcount and organizational structure) provide an insight of the company and its situation.
(2) *External and internal regulations* (e.g., norms and certification levels) are basic assumptions that are often triggered by quality management initiatives, which strongly determine how organizations work (Evans and Lindsay 2002).
(3) *Knowledge culture* (e.g., approach to retaining knowledge, valuing its importance and managing it) allow an insight into the working environment for KWers and how their work is valued in the organization (Davenport and Prusak 1998).
(4) *Level of formalization* (e.g., involved steps in phenomenon analysis) describes how well defined the currently executed process is and if there is room for experimenting. This category has been developed inductively.
(5) *Extent of involved parties* (e.g., extent of people needed to deal with the phenomenon currently) is used as a proxy for describing process efficiency in terms of resources. This category has been developed inductively.

In the following, we start to present the two organizations that our cases come from. After that, we describe an example instance of process improvement for each organization. Both example instances are currently done without the help of data-driven knowledge discovery. Thus, we will discuss how data-driven knowledge discovery could look like under the given constraints.

5.1 Two Real-World Organizations

We are going to split the description into three parts, giving an overview table for each part, followed by a summarizing narrative. We will highlight the introduced codes and their manifestations[1] as part of this presentation, also summarized in an overview table in Sect. 5.4.

As described in Table 1, *organization 1* is a global player with more than 3200 employees in nine production sites and more than thirty-five sales locations. Several mergers and acquisitions over the last years leads to different processes over all sites. This is the reason for only observing a particular process in the headquarter with around 1200 employees. The company is family owned but management driven. Organization 2 is the general secretary of an international non-government organization. They are geographically dispersed over all continents with approximately 600 employees in total. Organization 2 majorly provides services across

[1]The manifestations of *External/Internal Process Regime*, *Knowledge Culture* and *Level of Formalization* can be low (no or ad hoc setup of actions), medium (structured setup of action that is not consistently implemented across the organization) and high (structured setup of action that is consistent across the organization). The manifestation of the *Extent of involved parties* can be low (ad hoc organized, small group of people), medium (group of people that is organized with the help of communication standards only shared by the group) or high (large group of people, relying on formal communication standards that are implemented organization wide).

Table 1 Organizational facts

Dimension	Organization 1	Organization 2
Industry	Medical device manufacturing	Social responsibility
Description	Organization 1 is a private producer of dental products in the medical device sector with subsidiaries all over the globe. For this sample, we look at processes in it's headquarter	Organization 2 is the federation headquarter of a nongovernmental organization that works in the area of social responsibility with its focus in the area of child and family care
Structure	The organization is a global player with production sites in five countries all around the globe. Lots of mergers and acquisitions lead to diversity in business processes and understandings of how to manage production and fulfill the customer needs. A matrix organization is still implemented	The organization works on all continents with 130 member associations worldwide. The headquarter is structured along functional areas, which provide processes to all parts of the federation. The major mindset is process-oriented with no particular management approach in place
Headcount	Approximately 1200	Approximately 600

several functional areas to its member associations. That is why they have clear orientation towards business processes, however, with no particular process management approach in place. As a result, each department has its own understanding of how process should be managed.

Being part of the medical device industry, *Organization 1* works in a highly external regulated environment (see Table 2). Quality management processes are implemented based on strict external requirements. There is a low level of internal regulations due to (1) limited resources and (2) already very stable processes as response to external regulations. Key performance indicators are used to assess process performance. *Organization 2* works in an environment that is almost only determined by legal norms. They use CMMI as a guiding definition in the area of ICT support and are level three appraised. Internally, they use a self-defined process management system. There is, however, little to no compliance monitoring with regard to the internal framework although there is a large number of key performance indicators. This is majorly based upon the fact that performance measures are conceived by the individual functions.

As described in Table 3, organization 1 uses a knowledge management approach that is highly determined by codification (Hansen et al. 1999). They use standard operating procedures to codify a large amount of the process knowledge. The organization puts great emphasis on training employees. Personalization of knowledge is used in the context of process-internal activities only. *Organization 2* also recognizes knowledge as a central resource for value creation wherefore its management is institutionalized. Knowledge sharing is encouraged using internal guideline documents and supported with several internal IT systems—both using aspects of codification as well as personalization. Regarding knowledge baselines, the organization established several internal trainings (e.g., process management) that employees must attend to be allowed to work in certain positions.

Table 2 External and internal regulations as well as performance measurement in organizations

Dimension	Organization 1	Organization 2
External regime	*High*: Organization 1 with all the production sites is qualified by ISO 13485:2012 and FDA requirements. Additionally, some local needed certificates are necessary in some sites. A very high standard of external requirements is fulfilled	*Low*: Organization 2 does not comply with external standards or regulations. There is no reference model available for the domain in which organization 2 is working
Internal regime	*Low*: Organization 1 is bound to so many external requirements with a high number of products that there is no resource for additional internal requirements	*Medium*: Organization 2 has a process management system in place that defines internal process management guidelines but leaves it to the individual functions how they implement it. Consequently, internal rules are only moderately executed
Description of performance measures	Performance measures are in place for every site based on the external requirements. Financial measures are in place group wide	The organization uses around 100 performance measures to assess process quality in all areas

Table 3 Knowledge culture in organizations

Dimension	Organization 1	Organization 2
Knowledge culture	*Medium*: For medical device producers, it is standard to codify knowledge in standard operating procedures The knowledge is trained by experts and employees get the knowledge, which is particularly for their job roles. Based on the different sites and different processes it is a challenge to have short feedback loops in place. Knowledge is known as an important resource but knowledge sharing on top of external requirements is not pushed. Personalization is used for intra process activities	*High*: Knowledge is recognized as being a resource of central importance. The organizations formally support initiatives of knowledge sharing (there is e.g. a knowledge management function) and expresses this support in terms of policies. There is no knowledge baseline as all employees are specialists in their area. Knowledge exchange is often done among peers using provided tools. There is a strong attempt to codify knowledge across processes using various web-based platforms (e.g., MS SharePoint Server) and guideline documents. Personalization is especially used in intra-processes settings

5.2 *Case 1: Process Improvement in a Highly-Regulated Scenario*

Case 1 addresses the execution of change management in the context of continuous process improvement. *Currently*, as soon as a mal-function within an existing process is detected, a new change processes is started, involving a large number of different experts (*high* Extent of involved parties). This process will always be executed according to the standard operating procedure (*high* Level of formalization). A team of experts will evaluate possible root causes of the mal-function and evaluate its risk of occurrence. The initial analysis will be handed over to a change board. The members of the change board will discuss possible root causes and define mitigation actions. The process owner is responsible for action implementation and final evaluation. Not eliminating root causes leads to increases in the number of complaints. This will lead to higher service costs and in the worst case, even patients could be harmed.

Organization 1 is already documenting all change management instances using a workflow management system with defined process baselines. This information could be used to start data driven knowledge discovery.

In the *future*, the process could be enriched by collecting metadata about change processes and comparatively analyses change processes with respect to e.g. product types, involved components, involved materials and/or involved production equipment. This would lead to more precise picture for the experts and it would be possible to get an integrated view on the topic instead of dealing only with the reported phenomenon.

5.3 *Case 2: Process Improvement in a Bottom-Up Scenario*

Case 2 addresses the improvement of fundraising activities within organization 2 as a response to suddenly occurring changes in funding performance. *Currently*, as soon as there are significant changes in the structure and amounts of incoming donations, a group of process experts will use external reference (e.g. customer feedback or market review) to get an overview of the situation. The involved steps can vary from instance to instance, as there is no standard operating procedure in place (*low* Level of formalization). They will use descriptive statistics but mostly their own reasoning to analyze the evidence. Together with the process owner (*medium* Extent of involved parties), they will discuss possible root causes and decide which measures could be taken to improve fundraising performance. If no appropriate measures are taken, the income from individual donors will drop which might lead into serious budget issues.

Organization 2 has a complete record of donors as well as a complete record of donor transactions in their CRM and accounting system. These data are already integrated into a multidimensional database for online analytical processing

(OLAP) and could be easily used to start data driven knowledge discovery. In addition to that, the organization could use other data sources such as e.g. social media to complement their donor records with narrative feedback on their fundraising effectiveness.

In the *future*, organization 2 could apply data mining instruments (e.g., time series analysis combined with clustering algorithms or random forests) on their existing OLAP data structures to analyze the behavior of donors. They could use this information to find correlations between fundraising performance and donor fluctuation. They could even use external data (e.g., social media posts, market reviews etc.) to complement this analysis with additional data and use it to address donors individually to increase donor loyalty.

5.4 Case Comparison and Summary

Comparing the two cases, we can clearly see that both organizations handle the described phenomenon very differently. While organization 1 has to comply with very strict external requirements, organization 2 only has some internal guidelines, which allow handling root cause analysis in a much more unstructured way. With the introduction of data driven knowledge discovery in the described cases, there would probably be no change in their processes for root cause detection. Still, organization 1 would have to make sure that their external requirements are met, while organization 2 has no rules to comply with. We therefore assume that the current way of handling will also be carried-on in the described future scenarios.

Table 4 is intended to give an overview of the introduced organizations/cases by summarizing the earlier introduced codes. Organization 1 currently exhibits both, a high level of formalization in handling the described phenomenon as well as a high extent of involved parties in their handling approach. In contrast, organization 2 currently shows a somewhat low level of formalization. In practice, the involved process experts can decide, which analysis measures they will take as well as how and if they would like to document their results. Their behavior can be characterized by a high level of autonomy and self-responsibility, which is supported by the organizational environment that they are working in and the organization's high knowledge culture that is generally supporting a rich analysis with knowledge discovery that can be shared with others. Organization 1 in contrast needs to make

Table 4 Comparison of cases with respect to handling of phenomenon

Domain	Dimension	Organization 1	Organization 2
Organization	(1) External process regime	High	Low
	(2) Internal process regime	Low	Medium
	(3) Knowledge culture	Medium	High
Case	(4) Level of formalization	High	Low
	(5) Extent of involved parties	High	Medium

sure that procedures are executed in a standardized way. This is largely due to the high external regime that they have to comply with.

With respect to the extent of involved parties in the act of handling the phenomenon, organization 2 can be classified as being on a medium level. The described phenomenon is currently handled by a handful of process experts in close reconciliation with the process owner. We describe this as being medium as the group of people is organized with the help of communication standards only shared by the group. Compared to organization 1, there is clear difference in phenomenon handling. While organization 1 dedicates a large amount of resources to root cause analysis, organization 2 deals with phenomena in a more unstructured and ad hoc way. Again, this is very likely due to the very different external and/or internal regimes that they have to deal with.

6 Discussion

We use this section to discuss the model from Sect. 4 with the help of the presented cases from Sect. 5. We will start by showing how the model can be used to guide each of the introduced cases. Based on this, we are discussing the model more deeply, abstracting from the single cases.

Considering how *organization 1* could approach their case in the future once more, we can try to apply the introduced model to this case scenario. Phenomenon elicitation (1a) can start either when an external party reports an error or when an error is detected internally. In both cases, it will be important to find as much information about the malfunction as possible to precisely describe it before entering the data analysis phase (1b). The more precise the malfunction has been described, the easier it is to create hypotheses that can be tested using data analysis on process metadata. In contrast to today's approach, these data are not parsed manually by process experts but by algorithms instead. These algorithms are used to detect patterns, which stick out. These patterns will be described (i.e., what are the correlating features) and collected (1c). While a classical data driven approach might stop here, our model emphasizes the necessity to go into the top-down elicitation phase to cover the plausibility of results. This plausibility check will begin with carefully evaluating the collected results (2a) in the light of their practical plausibility and likelihood, which will probably ultimately yield one result being favored among others, thus being selected (2b) and addressed with action (2c). This will typically involve process experts with strong domain knowledge discussing the collected results from 1c and rating them in the light of application. Here, we expect to see meetings of experts discussing the results of step 1b which possibly leads to new questions. This is exactly what the initial situation at organization 1 looks like. Experts meet to discuss the phenomenon at hand and come up with root cause and action. There is, however, a clear distinction between the two approaches. While the current approach only allows explanations that are within the experts' scope, the new approach also allows explanations that are within the experts' blind spot area. Therefore, they would probably never suggest this explanation themselves. The feedback loop through result monitoring (3a) finally helps to safeguard the improvement. Here, experts can dedicatedly improve their domain knowledge, which can be used in the next knowledge discovery cycle.

Looking into how *organization 2* could approach their case in the future, we can see some similarities but also some differences with respect to organization 1. Phenomenon elicitation (1a) will start, when a change in donor behavior is detected and will be used to describe the phenomenon as clearly as possible, to generate hypotheses for analysis (1b). As there is already an optimized analysis database in place, some hypotheses can be tested right away. In some cases, analysis could, however, be complemented with external data. In some cases, e.g., a sediment analysis on social media feeds might help to detect possible root causes more appropriately than considering accounting or CRM records. In both cases, however, the results of the analysis will be described and collected (1c) for the top-down phase. The bottom-up phases of organization 1 and 2 are very similar as they both follow the process of knowledge discovery in databases (Fayyad et al. 1996) very closely, with some minor variations in the used data, data storages and analysis procedures. Because of the different external and internal regimes in place, the top-down phase and the feedback phase of both cases are somewhat different. Experts in organization 2 will also meet to validate (2a) the collected results and select the most probable root cause (2b) to take action (2c). Their evaluation, however, will be less formal. There will be no guidelines on who needs to be involved or how the evaluation has to be done. Equally, monitoring (3a) in the feedback phase is a mandatory and well-formalized part in the external regime of organization 1. Organization 2 will also learn from observing the established mitigation measure, yet their approach will be rather ad hoc and might even change from instance to instance.

As can be seen from the elaboration above, the model was designed to offer a level of granularity that is detailed enough to reflect all major steps in the described procedures, yet still is flexible enough to allow for reflecting the variations in the cases. Different than in the Fayyad et al. model, the focus does not lie on analysis in detail, but on the interplay between analysis, evaluation and process implementation. Therefore, our model is less detailed in the bottom-up phase, than their model was. The chosen level of granularity also helps to describe both introduced cases although they use completely different triggers. In case 1, root cause analysis is started by a manual event like a complaint, while case 2 would also allow for a periodic trigger.

Equally, the model allows depicting variations in the execution of the knowledge discovery process by limiting its guidance to the most important cornerstones of the process of data driven knowledge discovery. While case 1 does not show any potential for expanding the used data sources, case 2 allows for enriching the used data to external sources. In the same way, the model supports different levels of guidance in the knowledge discovery process. While the top-down phase in case 2 will be very ad hoc with almost no formal guidance, the same phase in case 1 will be very precisely executed according to standard operating procedures. Yet, the proposed model is still able to reflect both scenarios.

A third important feature of the model is that it connects the traditional value chain model of knowledge (from data to information to knowledge) with the Autopoietic model of knowledge by explicitly integrating the aspect of justification. As shown in both cases, knowledge discovery can be started from data analysis in the bottom-up phase. In both cases, the bottom-up phase will only yield possible explanations for the phenomenon at hand, which are used in the top-down phase for evaluation and selection. Here, the potential knowledge is justified by domain experts and directly put to use.

Moreover, justification is even expanded beyond the top-down phase by explicitly integrating a feedback phase that allows long-term learning. By monitoring the taken measures, persons in the context of KW can learn about both the bottom-up as well as the top-down phase. This means, they can refine their approach both on the data driven side of knowledge discovery and the justification of potential knowledge. This aspect of the model is central in terms of knowledge management and organizational development.

The discussion of case 1, however, shows that there is a clear necessity to allow backlashes in the model, which was not foreseen so far. If, e.g., process experts discover a defect of the analysis model in the evaluation (2a), there was no possibility to jump back to the analysis step (1b) yet. The same situation existed for other steps in the model. Therefore, we are going to revise the model to allow going back to the bottom-up phase from the validation step (2a). We propose to backlink the validation step with the data analysis step (1b). From there it is additional possible to go back to phenomenon elicitation (1a), if it can be assured that the analysis model was correct and yet the problem apprehension must be incorrect. In the first place, however, a backlash from validation should trigger reconsideration of analysis. Therefore, we suggest revising the initial model as shown in Fig. 2. To improve readability of the model, we have rearranged the boxes in the figure compared to Fig. 1.

We have updated the original model by explicitly allowing backlashes between the steps of the bottom-up and the top-down phase as well as a backlash between phases. Looking at the revised model, we can summarize several implications for the use of data driven knowledge discovery in KW. In contrast to the Fayyad et al. model, where jumps between every two steps are possible, we decidedly limited the number of possible backlashes in our model to keep the model's guiding nature high.

7 Implications

We are finally going to use this section to close this work by presenting its major findings and summarizing them with respect to *individuals*, who are working in the context of data driven knowledge discovery and *organizations*, which have continuous process improvement strategies in place (Table 5).

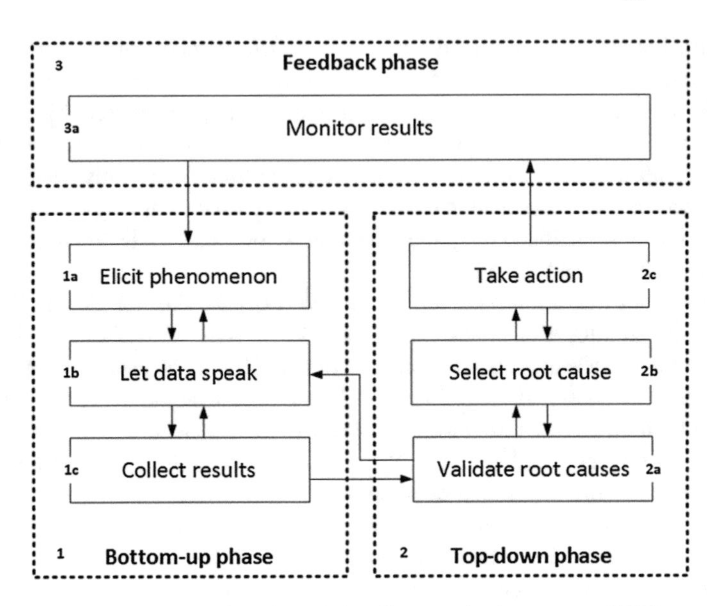

Fig. 2 Revised model of data driven knowledge discovery in the context of KW

Table 5 Implications of data driven knowledge discovery for individuals and organizations

Dimension	Suggestions for individuals	Suggestions for organizations
Separation of bottom-up and top-down	Make sure to have a good understanding of data analysis and its affordances as well as constraints	Make sure to combine domain experts and data analysis experts when doing root cause analysis. Support experts with a process model with clear responsibilities
Analysis needs a well described problem	Make sure to understand the goal of your analysis, independent of whether you are a data analyst or a domain expert	Make sure that analysis goals are well described and (if appropriate) linked to an operational/tactical objective or a strategic objective. Communicate analysis goals properly
Analysis needs to consider the particular domain	Make sure to have a knowledgeable understanding of the domain you are operating in	Make sure to have domain experts and data analysis experts working together, allowing for a transfer of domain knowledge
Top-down approach should be complemented with bottom-up approach	Make sure to challenge your basic assumptions about the phenomenon and consider other possible explanations	Support the critical reflection of commonly used assumptions by valuing a critical stance of KWers and providing them with the necessary resources
Analysis and action implementation are entangled	Make sure not to focus solely on the knowledge discovery process, but also consider action implementation and monitoring as crucial parts of what you are doing	Make sure to have measures in place to support continuous improvement of the knowledge discovery process

First, the findings of this work suggest that the separation of the bottom-up phase and the top-down phase increases efficiency in knowledge discovery. Domain experts who are commonly highly trained KWers are still involved in the process of knowledge discovery. They, however, can retreat to those steps in the discovery process that need strong domain knowledge, which saves resources. In addition to that, it ensures that each participant in the discovery process—analyst and domain expert—can unfold their full potential, as they do not have to grapple with tasks for which they have no training.

Secondly, the model advocates for a reasonable approach to data analysis that does not treat data analysis naively as an answer to any problem, but uses data analysis to find potential explanations to a well described problem, which is later evaluated in the light of practical applicability. Thoughtlessly speeding up the process of knowledge discovery by carelessly relying on analytics can lead to ill-founded decisions. In the light of these temptations, knowledge management will surely have to play its role in (i) advocating well-established and validated processes of knowledge discovery and (ii) harnessing the necessary knowledge in form of well educated KWers and decision makers. The model's feedback loop will help to do this.

Thirdly, the model can clearly help to understand, which knowledge is needed by persons who are doing data driven knowledge discovery. While earlier steps in the process clearly promote the need of strong meta-knowledge on how to work with data, later steps clearly promote domain knowledge that can be used to evaluate results. This information is important for knowledge management and organization development to create meaningful strategies.

Fourthly, the discussion of the cases suggests that a combination of bottom-up and top-down approaches is more robust against failure than a sole top-down approach as, e.g., currently used in organization 1. Domain experts can miss possible explanations of a phenomenon as these explanations are in their blind-spot area. A combination of bottom-up and top-down approaches, however, can help to broaden the spectrum of domain experts and assure that no possible explanation is missed.

Finally, the model clearly promotes the need for evaluating the bottom-up and the top-down phase by means of a feedback loop. This is an essential step in knowledge discovery since any knowledge building activity needs a form of justification. In our model, justification is done by means of applying the conceived action and monitoring its results. It is only this last step that allows for personal learning at the individual domain experts, and thus continuous improvement (e.g., Deming 2000).

Summing up we discussed the theoretical model for data driven knowledge discovery with the help of two real-world cases. After the discussion, we refined the model and carved out explicit implications of data driven knowledge discovery for individuals and organizations.

As the discussion is only based on two real-world cases, which are self-documented, we suggest a follow-up validation of the model using a larger amount of cases with complementary data to explore the phenomenon of data driven knowledge discovery in KW more closely.

References

Archer, S. (1988). 'Qualitative' Research and the epistemological problems of the management disciplines. In A. Pettigrew (Ed.), *Competitiveness and the management process* (pp. 265–302). Oxford: Basil Blackwell.

Berg, B. (1989). *Qualitative research methods for the social sciences*. Boston: Allyn and Bacon.

Bortz, J., & Döring, N. (2005). *Springer-Lehrbuch, Forschungsmethoden und Evaluation: Für Human- und Sozialwissenschaftler; mit 70 Tabellen* [in German] (3rd edn.), Heidelberg: Springer.

Bryman, A., & Bell, E. (2015). *Business research methods* [in English]. Oxford: Oxford University Press.

Chaudhuri, S., Dayal, U., & Narasayya, V. (2011). An overview of business intelligence technology. *Communications of the ACM, 54*(8), 88–98.

Chen, H., Chiang, R. H. L., & Storey, V. C. (2012). Business intelligence and analytics: from big data to big impact. *MIS Quarterly, 36*(4), 1165–1188.

Davenport, T. H. (2006). Competing on analytics. *Harvard Business Review, 84*(1), 98–107.

Davenport, T. H., & Prusak, L. (1998). *Working knowledge: How organizations manage what they know*. Brighton: Harvard Business Press.

Deming, W. E. (2000). *Out of the crisis* (2nd ed.). Cambridge: MIT press.

Evans, J. R., & Lindsay, W. M. (2002). *The management and control of quality*. South-Western Cincinnati, OH.

Fayyad, U. M., Piatetsky-Shapiro, G., & Smyth, P. (1996). Knowledge discovery and data mining: Towards a unifying framework. In *Proceedings 2nd International Conference on Knowledge Discovery and Data Mining Portland OR*, pp. 82–88.

Hansen, M. T., Nohria, N., & Tierney, T. (1999). What's your strategy for managing knowledge? *Harvard Business Review, 77*(2), 106–116.

Hevner, A. R., March, S. T., Park, J., & Ram, S. (2004). Design science in information systems research. *MIS Quarterly, 28*(1), 75–105.

Maier, R. (2007). *Knowledge management systems: Information and communication technologies for knowledge management* [in English]. Berlin: Springer.

Maier, R., Hädrich, T., & Peinl, R. (2009). *Enterprise knowledge infrastructures* [in English], 2nd edn, Berlin: Springer.

Parboteeah, P., & Jackson, T. W. (2011). Expert evaluation study of an autopoietic model of knowledge. *Journal of knowledge management, 15*(4), 688–699.

Pauleen, D. J., & Wang, W. Y. (2017). Does big data mean big knowledge? KM perspectives on big data and analytics. *Journal of Knowledge Management, 21*(1), 1–6.

Pyöriä, P. (2005). The concept of knowledge work revisited. *Journal of Knowledge Management, 9*(3), 116–127.

Rother, M. (2010). *Toyota Kata: Managing people for improvement, adaptiveness, and superior results*. New York: McGraw-Hill.

Schultze, U. (2000). A confessional account of an ethnography about knowledge work. *MIS Quarterly, 24*(1), 3–41.

Schultze, U. (2004). On Knowledge Work. In C. W. Holsapple (Ed.), *Handbook on knowledge management 1: Knowledge matters* (pp. 43–58). Berlin: Springer.

Spöhring, W. (1989). *Qualitative Sozialforschung*. Stuttgart: Teubner.

Tuomi, I. (1999). Data is more than knowledge: Implications of the reversed knowledge hierarchy for knowledge management and organizational memory. *Journal of Management Information Systems, 16*(3), 103–117.

Wu, X., et al. (2008). Top 10 algorithms in data mining. *Knowledge and Information Systems, 14* (1), 1–37.

Author Biographies

Michael Kohlegger, Dr., holds a Ph.D. in Information Systems and is a trained Business Analyst. After several years in praxis, doing requirements engineering, data engineering and data science, he has recently joined the University of Applied Sciences FH Kufstein Tirol and holds a Professorship in Web Business & Technology. His main research focus is on the (un-)intended consequences of digitalization, the support of ill-structured routines with information technology as well as the philosophical discussion of ontology, axiology and epistemology.

Christian Ploder, Dr., worked for over ten years as international Project Leader for ERP system implementations, process improvement and quality management. All companies were global player in the paper industry or medical device producers. He has several years of leadership experience, is qualified lead auditor for ISO 13485:2012 and holds a Ph.D. in information systems. He recently joined the Management Center Innsbruck (MCI) teaching ERP systems, process management combined with current challenges of the information systems. Quality driven business process improvement based on ERP systems combined with digitalization is his main research focus.

Digital Change—New Opportunities and Challenges for Tapping Experience and Lessons Learned for Organisational Value Creation

Edith Maier and Ulrich Reimer

Abstract Digital change and Industry 4.0 do not erase the need for human insight or experience. This has been shown by a recent survey conducted among managers in the German-speaking world who still consider experience a highly valuable asset. Digital change, however, has shifted the focus from products to customers and implies new roles for employees such as supervising machines and processes, and assessing data analysis results. At the same time, new digital trends and tools open up new opportunities for automatically capturing, exchanging and preserving lessons learned, and offer support that is both context-aware and situation-specific. Since they should not require any additional effort, digital trends and tools may also help remove a key obstacle to innovation, i.e. the failure to learn from mistakes.

1 Introduction

Digital change is driven by big data, a dramatic drop in communication costs and sensor prices as well as production strategies such as agile manufacturing and mass customisation. This results in a fundamental transformation of the economy which is often subsumed under the label "Industry 4.0". It holds out the promise of smart factories manufacturing products by largely autonomous systems that exchange data across the entire value chain (Ganschar et al. 2013).

Is there still a role for human insight and vision in an era of self-organising and self-adapting 'knowledgeable' manufacturing systems (Yan and Xue 2007), deep learning and data-driven trend spotting? Will big data override experience and intuition, i.e. the largely tacit knowledge harboured by experts, when it comes to taking decisions in the future? Will Industry 4.0 therefore spell the end of decisions based on experience and domain expertise and replace them with decisions based on data and text mining to discern trends, market developments or hidden patterns

E. Maier (✉) · U. Reimer
University of Applied Sciences, St. Gallen, Switzerland
e-mail: edith.maier@fhsg.ch

K. North et al. (eds.), *Knowledge Management in Digital Change*, Progress in IS,
https://doi.org/10.1007/978-3-319-73546-7_5

or correlations? And how can we make sure that lessons learned are shared across the collaborative networks that are emerging as a result of the virtualisation of process and supply chains?

Whilst increasing digitalisation will no doubt lead to the loss of certain jobs that can be taken over by machines, robots or algorithms, we still need experts who can ask the right questions, solve problems in the case of failures or deal with critical incidents. We need people who understand the problems and have the relevant experience and insight to solve them. They have to be able to analyse and interpret the results from mining data from various sources, such as sensors, and take decisions faced with incomplete information or when confronted with unforeseen events or crises.

As a result, there is an increasing need for knowledge management tools and techniques to assist people in these new roles. In this article we will investigate the implications of digital change for organisational knowledge creation with a particular focus on the tacit dimension of knowledge which is rooted in experience, insights, vision, commitment, ideals, values etc. (Nonaka and Takeuchi 1995).

This paper sets out by defining our understanding of the concept of digital change or transformation and Industry 4.0 as well as terms such as experience, tacit knowledge and lessons learned (Sect. 2). Then we investigate if experience-based or tacit knowledge still plays a role when faced with trends such as the Internet of Things (IoT), deep learning and artificial intelligence (Sect. 3). To answer this question, we can leverage the results of a recent survey carried out among managers in Germany, Austria and Switzerland (Sect. 4). The survey showed considerable discrepancy between the great importance assigned to experience-based knowledge and the lack of systematic support given to its exchange and preservation. Section 5 describes how this could be remedied by tapping the potential of new technological trends. We end with a brief look back to debunk the myth of unprecedented accelerated change and the importance of learning from mistakes for fostering innovation. Finally, we sum up the key messages for managers about how they can respond to the challenges posed by digital change and harness it for value creation (Sect. 6).

2 Definitions, Concepts and Models

There is no consensus with regard to terms such as digitalisation and digital change or digital transformation or Industry 4.0. A common definition is the one coined by Bounfour (2016), namely 'the change associated with the application of digital technology in all aspects of human society', which is also the one adopted by Wikipedia. But as shown by the numerous "edit" requests in the Wikipedia entry, there is a great deal of uncertainty about this definition. If you look it up on Google, you might be taken to the website of I-Scoop.eu and its online guide to digital business transformation. The consulting and publishing company has written extensively on topics such as digitalisation, transformation and organisational processes, IoT etc. They regard digital transformation as "the profound and accelerating transformation of business activities, processes, competencies and models to fully leverage the changes and opportunities of digital technologies and their impact across society in a strategic and prioritized way, with present and future shifts in mind."[1]

Experts, however, do agree that digital change implies a central shift from a focus on improving products and processes towards a focus on the needs and expectations of customers. In most discussions, we can also observe a move towards more experimental, collaborative and fluid approaches to doing business.[2] As we understand it, digital transformation goes beyond the use of digital technologies to support or improve processes and existing methods. It is a way to alter and even build new business models, using digital technologies.

The concept of Industry 4.0 was coined in 2000 by the German Research Centre for Artificial Intelligence DFKI and only after having attracted attention in the US under the label "Industrial Internet" has the concept seen widespread dissemination in the German-speaking world. It has become the central element of the high-tech strategy of the German government and boasts a dedicated research platform that brings together industry and academia.

The term *experience* or *experience-based knowledge* is closely related to terms such as good or best practice, lessons learned, tacit knowledge, knowledge-in-use etc. As early as 1958, Polanyi explored the distinction between tacit and explicit knowledge (Polanyi 1966) and thus laid the foundation for Nonaka and Takeuchi (1995) who made major contributions to knowledge management (KM) theory. They state that whereas explicit or codified knowledge is objective, easily communicated and transferred without requiring in-depth experience, tacit knowledge is subjective, context-specific, personal, and difficult to communicate. It consists of cognitive elements such as cultural beliefs and viewpoints as well as technical elements, i.e. existing know-how and skills.

Experience management (EM) can be defined as a special form of KM and an *Experience Management System* (EMS) as a socio-technical system established for managing, reusing and recording experience or lessons learned (Nick et al. 2007). Research in EM therefore deals with methods and technologies suitable for collecting them from various sources and for documenting, sharing, adapting and distributing experience. It also includes the organisational and social measures required to assure that these are integrated into business processes (see also Bergmann 2002).

In our industry-related projects, especially when it comes to succession planning, which is a major topic in small- and medium-sized companies, our clients and partners are mostly concerned about passing on lessons learned and good practices. *Lessons learned* can be defined as experience distilled from projects that should be actively taken into account in future projects so as to reduce or eliminate the

---

[1]See www.i-scoop.eu.

[2]See e.g. the roundtable discussion organised by the Economic Council (see www.wirtschaftsrat.de).

potential for mishaps or failures. But even large and renowned organisations such as BP and NASA have issues with lessons learned from projects as has been revealed in audit reports and reviews (Duffield and Whitty 2015). Actually, NASA today uses the BP Deepwater Horizon disaster as a case study to illustrate how communication deficiencies, disregard of data as well as of lessons learned from previous incidents may lead to such accidents (Duffield and Whitty 2015). Similarly, lessons are often ignored and the same mistakes repeated in large public sector projects despite extensive guidance available (e.g. Chesterman 2013).

When companies try to turn inherently tacit knowledge into explicit knowledge they often encounter pitfalls. Xerox is an example that is often quoted in the literature, e.g. (Hansen et al. 2005). They attempted to embed the know-how of its service and repair technicians into an expert system that was installed in the copiers and expected that technicians responding to a call could be guided by the system and complete repairs from a distance. That is not what happened. Rather the copier designers discovered that technicians learned from one another by sharing stories about how they had fixed the machines. The expert system could not replicate the nuance and detail that were exchanged in face-to-face conversations. This finding is in line with organisational knowledge creation theory, which considers knowledge conversion not only an individual but also a social process (Nonaka and Takeuchi 1995; Lam 2000).

Recent studies by Duffield and Whitty (2015) or O'Dell and Hubert (2011) confirm the widespread trend of failing to learn from past experiences despite the ready availability of lessons learned models, guides and tools to apply them. This is surprising since we are increasingly faced with incomplete knowledge in a world that is characterised by great uncertainties and imponderables as a result of disruptive innovations brought about by digitalisation. The experience we have accumulated over time may help us deal with these challenges, crises and conflicts. One would therefore expect companies and their managers to make the best use of the know-how of their employees as well as the lessons learned from previous projects and activities.

Lindner and Wald (2011) have pointed out that there is actually a gap in project management practice and suggested that there is a need for more research in understanding the role knowledge management plays in project management methodologies. In this respect, it may be worth mentioning the so-called "Syllk" model, which stands for Systemic Lessons Learned Knowledge model. According to its proponents (e.g. Duffield and Whitty 2015) it could assist in identifying the knowledge management barriers that need to be overcome for an effective transfer of lessons learned. Others such as Leal-Rodríguez et al. (2014) have demonstrated how the Syllk model can support knowledge sharing and integration between an organisation and its suppliers, customers and partners. As is the case with experience and its transfer, the human factor plays a major role in the studies on as well as applications of the Syllk model because it recognises that for organisations to learn, people and systems, processes and technology have to be working together closely (Virolainen 2014).

3 The Role of Experience in Times of Digital Change

As far as the impact of digital change on employment is concerned, studies cover the whole spectrum of scenarios from a widespread loss of employment, e.g. 50% of all jobs according to Frey and Osborne (2017) on the one hand, to studies that predict an increase of 390,000 jobs in Germany alone (Rüßmann et al. 2015) on the other. Pfeiffer and Suphan (2015) point out that the distinction between routine and non-routine work, which lies at the basis of most of the pessimistic forecasts, is methodologically faulty and does not do justice to the actual activities performed by industrial workers. Using the highly automated automotive industry as an example, the authors point out that the work of the employees is far from routine but that the efficient supervision and control of machines, robots or technical processes requires a high degree of technical know-how and flexibility.

Peinl (2017) puts forward a similar argument by citing electrical engineers as an example, who these days are not only responsible for wiring lights and switches, but have to install building automation systems for which they need substantial IT know-how as well as knowledge about data security. In the case of failures, malfunctions or stoppages they have to use their own judgment and either find a solution themselves, or get support from a maintenance specialist. At the same time, it takes a great deal of experience to anticipate potential problems and intervene to prevent failures (Pfeiffer and Suphan 2015).

Pfeiffer and Suphan (2015) argue that experience actually plays an eminent role in highly complex and automated digitalised work environments. They have developed an index to measure a person's ability of dealing with complexity and imponderables—the so-called "Arbeitsvermögen-Index"—and conclude that at least in Germany about three quarters of employees are "fit for digital change".

When transferring the insights from IT trends to the activities of knowledge workers in the age of Industry 4.0, (Peinl 2017) reaches similar conclusions. According to him they will increasingly be responsible for drawing the right conclusions from data analysis and teaching heuristics about when to trust the machine and when to better trust one's own experience. Judging the trustworthiness and authority of information is also seen as a key competence for modern knowledge workers by Thornley et al. (2016). Similarly, Pfeiffer and Suphan (2015) argue that in critical situations, intuition based on long-term experience may help as much as logical thinking.

In her discussion of work in the future, Holtgrewe (2014) forecasts an increasing need of non-technical skills such as being good at team working, communication and finding creative solutions as well as paying attention to customer demands and market developments. Self-controlling systems actually require employees to take on the role of coordinators and problem-solvers in case of unforeseen incidents or failures. Apart from social and entrepreneurial competencies, analytical skills and independent judgment, an openness towards other fields is mentioned as a desirable quality (see, for example, the Report of the Swiss Government, Schweizerischer Bundesrat 2017). This is corroborated by Holtgrewe (2014) who stresses the need for experts—and managers—to have deep knowledge in a specific domain combined with shallow knowledge about related areas such as regulatory requirements, product safety, psychology or IT trends.

Digital change has not only an impact on the future skillset and qualification requirements of employees and managers, but is associated with technological trends that offer new opportunities for managing experience-based knowledge. However, it seems that this potential has not yet been sufficiently recognised by companies and their managers, as testified to by the survey.

4 Experience Survey

To find out about managers' attitudes towards experience, we—the universities of applied science of Cologne (RHFH), St. Gallen and Burgenland—recently conducted a survey in Austria, Germany and Switzerland under the aegis of METIS, a research alliance dedicated to experience and its social and entrepreneurial implications (Maier et al. 2016). How do managers nowadays document, exchange, manage and maintain this valuable resource?

For the survey, a questionnaire was developed aimed at obtaining an overview of attitudes towards practices, instruments and methods with regard to the role of experience and its management and transfer in the corporate German-speaking world. Since the survey targeted senior and middle managers, the role of leadership in the management of experience was another important issue raised in the questionnaire. Overall, we received 829 filled-in questionnaires out of which 359 came from Germany, 147 from Switzerland and 51 from Austria.

The questionnaires were collected and analysed by the computing centre of the RHFH Cologne and interpreted by experts at the three universities of applied sciences. Univariate and bivariate statistical analysis were carried out to: (a) describe the attitudes of the total sample towards experience using a seven-part Likert scale; and, (b) to test for significant differences between subsamples, e.g. respondents from larger versus medium-size companies, using chi-squared and Mann-Whitney U tests which both allow the analysis of ordinal scaled non-normal data. At a significance level of less than or equal to 0.05 the null hypothesis, i.e. that the sub-samples (e.g. middle vs. senior managers) show the same distribution for a concrete variable, was rejected.

For comparing the three country subsamples we performed a Kruskal-Wallis H test in SPSS. It turned out that respondents from the three countries constitute three significantly different subsamples with regard to socio-economic attributes (e.g. age, gender, education, position), which makes any meaningful comparison of national differences difficult. Besides, random sampling was not possible because we do not know the total number of managers in Germany, Austria or Switzerland. Therefore, we had to make do with a convenience sample and cannot make any representative statements about the total management population.

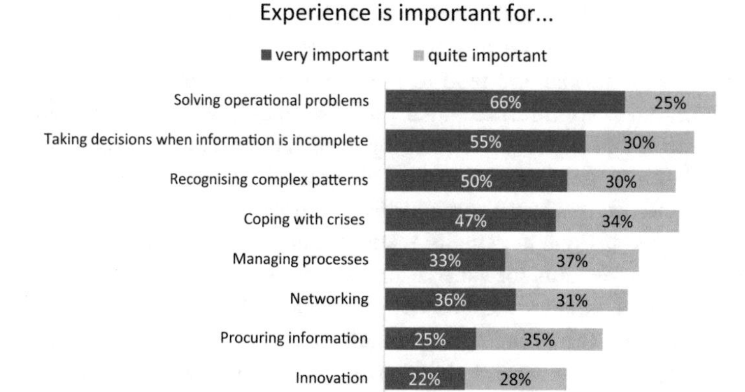

Fig. 1 Areas/activities where experience is (very) important

The results have shown that the majority consider experience an important organisational asset, especially for the areas and challenges listed in Fig. 1. They further show that company size and position rather than age or gender play a role when it comes to preferences, attitudes or practices for capturing, exchanging and using employees' informal knowledge and know-how. The survey also shows great discrepancies between methods considered useful versus those in regular use (see Fig. 2). In addition, it shows that many respondents have considerable reservations with regard to knowledge management techniques such as world cafés, lessons learned workshops or storytelling, networking approaches such as communities of

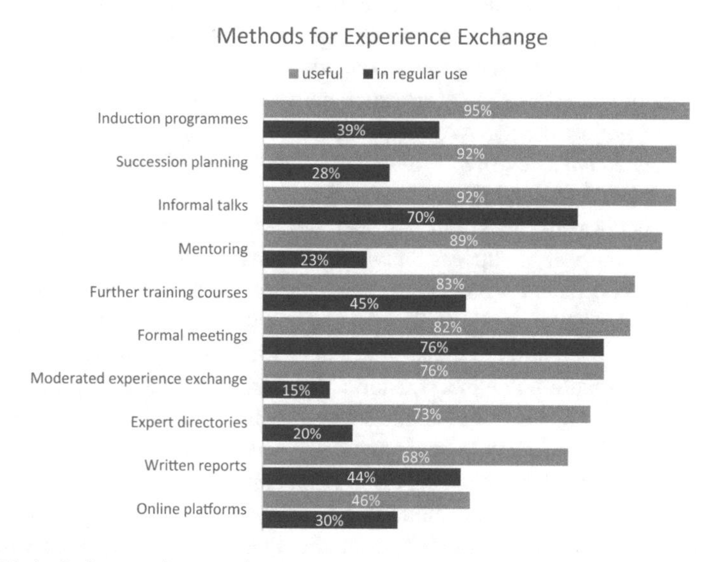

Fig. 2 Methods for experience exchange

practice as well as social media platforms or intranets. They see them as ineffective and/or do not use them on a regular basis. Even younger managers are sceptical with regard to such tools and tend to prefer traditional management and communication tools such as informal talks and meetings. What is interesting is that women on the whole appear to be more open with regard to the possibilities offered by online platforms or social networks (Maier et al. 2016).

We can conclude that whilst organisational know-how and experience is held in high esteem, little is done to actually manage and cultivate it and companies rarely offer incentives or rewards for such tasks. When asked for the reasons in follow-up interviews, lack of time and resources were cited most frequently (Schellhammer 2016). We interpret these findings as a call for action because it is clear that tacit knowledge, especially in the form of *lessons learned, is attributed great importance* but that there is a lack of know-how and support about how best to exploit this valuable resource.

In the following section we will discuss new approaches and tools that can assist employees in extracting, exchanging, disseminating and preserving lessons learned, especially with regard to the areas considered most important in the survey described above.

5 New Opportunities Offered by Digital Change

More than half of respondents in the survey considered 'solving operational problems', taking decisions when information is incomplete' and 'recognising complex patterns' as the most important areas for experience management (see Fig. 1). Both the survey and the preliminary results from the follow-up study, however, show that people will not engage in knowledge conversion, i.e. the interaction between tacit and explicit knowledge (Nonaka and Von Krogh 2009), if it implies additional effort. Therefore, such activities have to be integrated into the workflow and project management approaches so that *experience—in the form of best practices, lessons learned,* etc.*—can be automatically provided within the context and when it is needed.*

This idea is not really new, but has been voiced before, for example in various studies about how best to support knowledge-intensive work, resulting in approaches to process-oriented knowledge management and just-in-time knowledge delivery (e.g. Abecker et al. 2000; Reimer et al. 2001). However, these approaches have received little attention in the last decade or so. This is probably due to the fact that their implementation requires considerable effort. First, business processes have to be modelled, then the initial models have to be maintained and models that describe which kind of knowledge is typically needed within the various process steps have to be created. With digital change new opportunities are emerging which allow to automate parts of the modelling tasks. *Process mining* can be used to automatically derive process models from the event and activity logs which are increasingly created by today's business information systems, workflow systems and in production environments (van der Aalst 2016). With process mining it becomes even possible to *anticipate the next activities* of an employee.

Still needed for just-in-time delivery of relevant knowledge, are models that describe which kind of knowledge is needed within each knowledge-intensive process activity. Only then can a support system proactively *make relevant knowledge available in a way that takes into account the context and the specific situation*. Unfortunately, it is much more difficult to automate the creation of such models. One approach is to utilize process mining to identify which *organizational roles or actors* are associated with each knowledge-intensive activity. Whenever an employee requires support for solving a particular problem, the system can access relevant knowledge and point out which colleagues might be able to help. Another approach makes use of algorithms for *information extraction from texts* (Aggarwal and Zhai 2012) to find clues in text documents for which kind of knowledge-intensive activity they might be relevant. Text documents may be derived from discussion threads on social media platforms, pulled from the intranet of an organisation or be (automatically generated) transcripts of meetings of troubleshooting teams (cf. knowledge capturing below).

Apart from the above-mentioned possibilities for supporting *knowledge reuse*, the extraction *of experience-based knowledge* in terms of lessons learned and best practices, can also be partly automated. For example, process mining can be used to *derive good practices* from event and activity logs by identifying which sequences of activities have led to the best outcomes. Similarly, event and activity patterns that typically lead to problems can be identified and be captured as valuable insights to be provided in similar situations so as to prevent such problems from occurring again in the future.

Also, it is generally recognised that social media as well as the dramatic advance and widespread use of mobile devices, social software and online social networking are having a positive impact on knowledge management (O'Dell and Hubert 2011). Software for social exchange such as instant messaging, blogging and micro-blogging, social networking and collaboration are very suitable for the exchange of—especially ad hoc—experience.

Software engineers, in particular, consult blogs when they encounter a tricky problem. Many technology firms also offer Q&A sections where users can find answers to problems. Similarly, people frequently turn to online communities or fora when seeking advice for health problems. These platforms contain a plethora of valuable insights, which can be extracted using *information extraction* and *text mining* approaches (Aggarwal and Zhai 2012). Subsequently they can be stored in an experience base and made available, e.g. by using case-based reasoning (Bergmann 2002), when needed. Furthermore, new natural interfaces which include speech and gesture recognition can be used to automatically capture relevant knowledge (Hannola et al. 2016).

To sum up, knowledge conversion activities, especially those related to lessons learned and best practices, should be integrated closely with project and workflow management. To overcome the current reservations with regard to potentially effective methods for knowledge capture and reuse, we therefore suggest looking further into how to integrate experience and its management into project and process management practice as an automatic part that does not require any additional effort. By employing approaches for automatically extracting knowledge and making it automatically available when needed, we will be able to *close the loop* in the conversion of tacit or experience-based knowledge into organisational knowledge.

To ensure acceptance, such methods will have to provide added value e.g. in terms of facilitating troubleshooting in case of failures or preventing problems in the first place. Only then will companies be able—and willing—to tap the full potential of tacit knowledge for value creation.

6 Key Summary

Whilst we do not want to belittle the challenges associated with digital change, we would nevertheless point out that we need to be wary of over-dramatizing its effect. "For 200 years, people have held the belief that they are living in times of accelerated change", the historian Rödder is quoted in the June issue of *Technology Review*. Technological progress has actually slowed down according to the well-known US economist Robert Gordon (2017). Similarly, Wolfgang Wahlster, the director of DFKI, argues that disruption only comes as a surprise to those who don't care about scientific trends.[3]

Still, the new technological trends associated with digital change such as data analysis, process mining, text mining etc. offer new opportunities of reviving earlier ideas about externalising tacit knowledge and they provide new tools such as natural interfaces that facilitate the recording of experience. Also, the IoT can actually enable systems to better understand what users are doing at a particular moment and what they might need. As a result, they can offer context-aware support by connecting the physical environment and digital world with each other.

Thus, these new trends and tools may help remove one of the biggest barriers to the exchange and dissemination of lessons learned, i.e. the additional effort involved in this endeavour. Therefore, they will also help *remove a key obstacle to innovation*, namely the absence of any systematic review of lessons a company might learn from mistakes or failed projects. In a recent issue of the *Harvard Business Review* (Birkinshaw and Haas 2016), the authors suggest to rigorously extract value from failure to come up with innovative solutions. Their approach involves a three-step process:

[3]See https://www.heise.de/tr/artikel/Die-Maer-vom-rasenden-Fortschritt-3716643.html.

1. Learn from every failure, for example, the insights one has gained about customers or markets or in terms of one's personal growth as well as the liabilities (e.g. costs in time and money, reputation).
2. Share the lessons across the organisation, e.g. by means of regular reviews for sharing lessons including informal approaches such as capturing critical lessons with stories.
3. Review one's pattern of failure from a bird's eye view, e.g. is our organisation learning from unsuccessful endeavours?

Apart from the last point, IT could well contribute to facilitating steps one and two by drawing on the wealth of data generated continuously in today's businesses and apply data analysis approaches such as process mining and text mining to make sense of this data. IT tools can also help reuse experience-based knowledge by automatically providing it in a context- and situation-specific way. Apart from fostering a learning organization and a culture tolerant of failures, managers will furthermore need an open mind towards new technologies emerging with digital change so as to be able to harness them for value creation.

References

Abecker, A., Bernardi, A., Hinkelmann, K., Ku, O., & Sintek, M. (2000). Context-aware, proactive delivery of task-specific information: The knowmore project. *Information Systems Frontiers, 2*(3–4), 253–276.

Aggarwal, C. C., & Zhai, C. (2012). *Mining text data*. Berlin: Springer Science & Business Media.

Bergmann, R. (2002). *Experience management: Foundations, development methodology, and internet-based applications*. Berlin: Springer.

Birkinshaw, J., & Haas, M. (2016). Increase your return on failure. *Harvard Business Review, 94* (5), 88–93.

Bounfour, A. (2016). *Digital futures, digital transformation*. Berlin: Springer.

Chesterman, A. (2013). "Models of what processes?" Translation and interpreting studies. *The Journal of the American Translation and Interpreting Studies Association, 8*(2), 155–168.

Duffield, S., & Whitty, S. J. (2015). Developing a systemic lessons learned knowledge model for organisational learning through projects. *International Journal of Project Management, 33*(2), 311–324.

Frey, C. B., & Osborne, M. A. (2017). The future of employment: How susceptible are jobs to computerisation? *Technological Forecasting and Social Change, 114,* 254–280.

Ganschar, O., Gerlach, S., Hämmerle, M., Krause, T., & Schlund, S. (2013). *Produktionsarbeit der Zukunft-Industrie 4.0*. Stuttgart: Fraunhofer Verlag.

Gordon, R. J. (2017). *The rise and fall of American growth: The US standard of living since the civil war*. Princeton: Princeton University Press.

Hannola L., Heinrich P., Richter A., & Stocker A. (2016). Sociotechnical challenges in knowledge-intensive production environments. ISPIM Conference Proceedings, the International Society for Professional Innovation Management (ISPIM).

Hansen, M. T., Nohria, N., & Tierney, T. (2005). What's your strategy for managing knowledge. *Knowledge Management: Critical Perspectives on Business and Management, 77*(2), 1–10.

Holtgrewe, U. (2014). New new technologies: The future and the present of work in information and communication technology. *New Technology, Work and Employment, 29*(1), 9–24.

Lam, A. (2000). Tacit knowledge, organizational learning and societal institutions: An integrated framework. *Organization Studies, 21*(3), 487–513.

Leal-Rodríguez, A. L., Roldán, J. L., Ariza-Montes, J. A., & Leal-Millán, A. (2014). From potential absorptive capacity to innovation outcomes in project teams: The conditional mediating role of the realized absorptive capacity in a relational learning context. *International Journal of Project Management, 32*(6), 894–907.

Lindner, F., & Wald, A. (2011). Success factors of knowledge management in temporary organizations. *International Journal of Project Management, 29*(7), 877–888.

Maier, E., Bruns, W., Eschenbach, S., & Reimer, U. (2016). Experience-the neglected success factor in enterprises? LWDA.

Nick, M., Althoff, K.-D., & Bergmann, R. (2007). Experience management. *Kunstliche Intelligenz, 21*(2), 50–52.

Nonaka, I., & Takeuchi, H. (1995). *The knowledge-creating company: How Japanese companies create the dynamics of innovation*. Oxford: Oxford University Press.

Nonaka, I., & Von Krogh, G. (2009). Perspective—Tacit knowledge and knowledge conversion: Controversy and advancement in organizational knowledge creation theory. *Organization Science, 20*(3), 635–652.

O'Dell, C., & Hubert, C. (2011). *The new edge in knowledge: How knowledge management is changing the way we do business*. New York: Wiley.

Peinl, R. (2017). Knowledge management 4.0—Lessons learned from IT trends knowledge management. Tagungsband der 9. Konferenz Professionelles Wissensmanagement (Professional Knowledge Management), Karlsruhe, Germany.

Pfeiffer, S., & Suphan, A. (2015). Der AV-Index. Lebendiges Arbeitsvermögen und Erfahrung als Ressourcen auf dem Weg zu Industrie 4.0, Working Paper 2015#1 (draft v1. 0 vom 13 April 2015) Internet: http://www.sabinepfeiffer.de/files/downloads/2015-Pfeiffer-Suphan-draft.pdf (zuletzt abgerufen am 28 November 2015).

Polanyi, M. (1966). The logic of tacit inference. *Philosophy, 41*(155), 1–18.

Reimer, U., Novotny, B., & Staudt, M. (2001). *Micro-modeling of business processes for just-in-time knowledge delivery. Industrial knowledge management* (pp. 283–297). Berlin: Springer.

Rüßmann, M., Lorenz, M., Gerbert, P., Waldner, M., Justus, J., Engel, P., & Harnisch, M. (2015). Industry 4.0: The future of productivity and growth in manufacturing industries. Boston Consulting Group 9.

Schellhammer, F. (2016). Stellenwert und Bewertung von Erfahrung in Unternehmen. Master Thesis.

Thornley, C., Carcary, M., Connolly, N., O'Duffy, M., & Pierce, J. (2016). Developing a maturity model for knowledge management (KM) in the digital age. European Conference on Knowledge Management, Academic Conferences International Limited.

van der Aalst, W. M. (2016). *Process mining: Data science in action*. Berlin: Springer.

Virolainen, T. (2014). Learning from projects: A qualitative metasummary.

Yan, H., & Xue, C. (2007). Decision-making in self-reconfiguration of a knowledgeable manufacturing system. *International Journal of Production Research, 45*(12), 2735–2758.

Author Biographies

Edith Maier is a Professor and Senior Researcher at the Institute for Information and Process Management of the University of Applied Sciences in St. Gallen, Switzerland. Before she joined academia, she worked as a consultant for telecom companies, public authorities and NGOs in information policy and knowledge management. She has a background in applied linguistics, social anthropology and information science. More recently, her research focus has been on mobile health, behavioural change and the role of tacit knowledge.

Ulrich Reimer, Dr., is a Professor and Senior Researcher at the Institute for Information and Process Management of the University of Applied Sciences in St. Gallen, Switzerland. He has long-standing experience in knowledge management both from an applied as well as from a research-oriented perspective. He has been responsible for many projects including application of ontologies to knowledge capture and reuse.

Socializing with Robots

Anja Richert

Abstract The term *Industry 4.0* symbolizes new forms of technology and artificial intelligence, which will soon be embedded within production technologies. Smart robots are the game changers within smart factories, and they will work with humans in indispensable teams within the value chain. With this fourth industrial revolution, classical production lines are going through comprehensive modernization, which is commonly oriented to in-the-box manufacturing. Humans and machines will work side by side in so-called "hybrid teams." Thus, the success of these future production concepts will strongly depend on the successful implementation of direct cooperation between humans and robots. Hybrid teams will, more than ever, support demographic and diverse team structures. The difficulties behind physical limitations of workers are already being compensated through human-robot-cooperation, for example, through robots assisting with heavy lifting or physical duties. As a step further, robots should be able to identify and adapt to individual strengths and weaknesses and take over the role of a workmate, helping to construct knowledge in social, teamwork-oriented processes. What is necessary to change the role of a robot from a tool to a workmate? Can appearance and behaviour of the robot influence the team building processes? This chapter seeks to blend human demands of communication and cooperation in teams with empirical results of an experiment in a virtual factory of the future. The empirical study researches if the appearance of the robot and its behaviour influences the reception of the robot as a partner and the human cooperation behaviour, for instance, in terms of a shared understanding.

A. Richert (✉)
RWTH Aachen University, Aachen, Germany
e-mail: anja.richert@ima-zlw-ifu.rwth-aachen.de

K. North et al. (eds.), *Knowledge Management in Digital Change*, Progress in IS,
https://doi.org/10.1007/978-3-319-73546-7_6

1 Introduction: Hybrid Teams as a Consequence of Industry 4.0

The capacity of robots in working environments will change rapidly within the next years, from purely automated machines and programs to companions that make suggestions, provide advice or assist in physical tasks. These Human-Robot-Teams must communicate efficiently, should be flexible and broadly applicable (Schwartz et al. 2016).

At the same time, the industrial production methods and requirements are changing from top-down management and optimized systems to bottom-up, self-organized, modular production teams, demanding flexible, in-the-box production concepts rather than production lines. One unique opportunity, which is simultaneously a challenge of Industry 4.0, is the realization and shop floor implementation of new, flexible collaboration forms between humans, robots and virtual agents as hybrid teams.

A hybrid team, in the course of this chapter, is defined as a multi-agent system, consisting of at least two entities, of which one is human and the others are machines (e.g., robots) or virtual agents. By realizing a flexible collaboration of these entities, the human should become a creative designer with an active, cooperative role in the working world of the future (Schwartz et al. 2016).

Like in all teams, the idea of hybrid teams is to benefit from the different characteristics of the individual team members, and at the same time, make use of the fact that team members can substitute for each other temporarily in completing tasks when resources are running low. For purely human teams, this is completely natural behaviour; if a team member drops out, the team tries to compensate for this. However, industrial robots are still highly specialized in their tasks so that a new level of flexibility and universality is required to make a robot capable of temporarily substituting for a human or robotic team member (Schwartz et al. 2016).

If we talk about industry 4.0, we also speak of a high amount of individualized products manufactured in the companies of the future. It is about small-scale production of individually customized products. In most cases, the production line, and maybe the entire process, must be changed into a more flexible, modular production system, the so-called “in-the-box” production. The next logical step, therefore, is that mobile robots are used. They are not separated from humans, but work together in overlapping areas of work, such as manufacturing or intra-logistics. Mobile robotic systems are teamed with the worker on the industrial shop floor and can handle orders flexibly because of hybrid teamwork.

2 Human-Robot-Teaming: Implementation Demands

2.1 Human Factors in Human-Robot-Teams

To realize the teamwork of human and robots, "Human Factors" have to be considered. The term "Human Factors" is a collective term for psychological, cognitive and social factors in socio-technical systems and man-machine systems (Badke-Schaub et al. 2008). A common negative example for disregarded human factors in the teaming of humans and virtual agents is Microsoft's virtual agent Karl Klammer (Straube and Schwartz 2016), where most employers rejected the assistance system. From 1997 on, Karl Klammer (Clippy) had been in the service of Bill Gates. Six years later Microsoft finally gave into the pressure of its own customers and fired him. It is considered one of the worst inventions in the history of the company Microsoft. In MS-Office products, Clippy emerged unasked and inconvenienced the users with unwanted advice. Despite this, Microsoft did not finally bury the figure until 2007.

Since then, a lot of effort has been put into human-computer-interaction in general. A current example of a promising hybrid team system is the PART4you system of Audi, which is about direct Human-Robot-Collaboration (MRK) (Straube and Schwartz 2016). PART4you is implemented at the Audi factory in Ingolstadt. Adaptive to the workload of the employee, it is the first human-robot cooperation in the Volkswagen Group, which is used in final assembly. The PART4you Robot works hand in hand with the Audi employees and is equipped with a camera, as well as an integrated suction cup. Thereby, it can pick up components directly from the load carriers and pass them on to the employee, without protective separation, at the right time and in an ergonomically optimal position. The PART4you system chooses the correct component for the worker and holds it ready. Thus, long gripping paths or complicated bending are no longer necessary. "The robot becomes a production assistant that adapts to human tact - and not vice versa," says Johann Hegel, Head of Technology Development, Assembly at Audi.[1]

But as the description of the PART4you system shows, the robot in this case still holds a team-assisting function, which is clearly defined in the team setting. To change the role from an assisting "tool" to a workmate, it became apparent that the setting should give room for flexible role concepts within a hybrid team. One way to support the development of flexible roles is to design the teaming situation as a problem-oriented project task. Thereby the roles have to be defined in the beginning of the problem-solving process and the hybrid team members can develop a shared understanding of the tasks, which must be completed.

Shared understanding is an important factor for team functioning in several perspectives. For example, effective groups of humans often display shared conceptions of their expectations and rules (Bettenhausen 1991). When team members perceive shared understandings with other members, the positive effect and

[1]Pressemitteilung Audi AG: http://www.presseportal.de/pm/6730/2948444.

propensity to trust generated by such a discovery fuels performance improvement and bolsters group efficacy (Klimoski and Mohammed 1994). In an innovative simulation study, Carroll and Harrison (1998) found that length of service is positively related to a team's culture. Team members' personal characteristics shape their expectations of appropriate interaction rules, group efficacy beliefs, and group identity (Earley and Mosakowski 2000).

While robotic team members and virtual agents communicate with each other digitally in a real-time machine-to-machine communication manner, ways to include the human in the communication loop have to be defined. In the past, communication between humans and robots or virtual agents was mostly keyboard- or touchscreen-controlled. In recent years, the implementation of natural language processing in such systems has gained importance but has been, until now, only purely implemented in industrial robots. Non-verbal behaviour such as collision mitigation and Teach-in Procedures are at least a common starting ground for building up a relationship of shared understanding and trust. Further indicators for beginning team-building processes and a shared understanding are—according to human teams—the development and use of social rules and social signals, which is the focus of the next paragraph.

2.2 *Theoretical Approaches from Human Teams: Stages of Teamwork*

Hybrid interaction in close proximity can be shaped in different ways; a common differentiation is coexistence, cooperation, and collaboration (see Fig. 1).

Coexistence is the least form of human-machine interaction (Klimoski and Mohammed 1994). Onnasch et al. (2016) describe coexistence as an episodic meeting of humans and robots. The interaction between the two is very limited in time and space. The main motivation of the interaction is to avoid mutual obstacles and collisions. Cooperation involves "cooperative work, through the division of labour among the participants, as an activity in which every person is responsible for part of the problem solving" (Breazeal 2004). Collaboration refers to direct

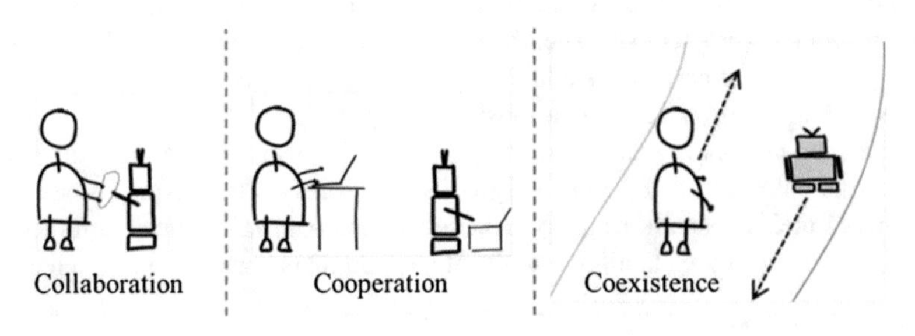

Fig. 1 Forms of human-robot interaction (Onnasch et al. 2016)

contact and coordination, regarding interaction (Klimoski and Mohammed 1994). Synergies are created and used.

The closer the collaboration is, the better the interaction should work. Communication and interaction should be intuitive for humans. The most intuitive form is human communication and interaction. Robots that have adapted human parameters are referred to as "social robots." Billard and Dautenhahn (1997) and Roschelle and Teasley (1995) have created this term. These are (autonomous) machines which interact on the basis of social rules and communicate with people. They can be humanoid or anthropomorphic and mobile (Ferrari and Eyssel 2016).

Regarding the human team building process, more detailed models have been introduced into staff and organizational development topics. A well-known model (beside the work of Tuckman, 1965) is the *stages of teamwork* according to Drexler et al. (1988). Allan Drexler et al. developed a comprehensive model of team performance that shows the predictable stages involved in both creating and sustaining teams. The model defines team development in seven stages, four to create the team and three to describe increasing levels of sustained performance:

1. **Orientation**: When teams are forming, the team members wonder why they are there, what their potential fit is and whether others will accept them. People need some kind of answer to continue.
2. **Trust Building**: Next, people want to know who they will work with—their expectations, agendas and competencies. Sharing builds trust and a free exchange among team members.
3. **Goal clarification**: The more concrete work of the team begins with clarity about team goals, basic assumptions and vision. Terms and definitions come to the forefront. What are the priorities?
4. **Commitment**: At some point discussions need to end and decisions must be made about how resources, time, staff—all bottom line constraints—will be managed.
5. **Implementation**: Teams turn the corner when they begin to sequence work and settle on who does what, when and where in action. Timing and scheduling dominate this stage.
6. **High Performance**: When methods are mastered, a team can begin to change its goals and flexibly respond to the environment.
7. **Renewal**: Teams are dynamic. People get tired; members change. People wonder, "Why should I continue?" It is now time to harvest learning and prepare for a new cycle of action (Drexler et al. 1988).

Within such human-oriented team building models, the interplay and the interaction of team members, as well as roles and tasks need to be defined, communication standards and social rules need to be followed by all and there needs to be a shared understanding of the aim of collaboration. The overall question of the empirical study, therefore, was: Can we observe any of the indicators, which display the beginning of a teambuilding process? The next section divides the questions into sub-questions and explains the empirical setting in detail.

3 Empirical Insights into Socializing with Robots—An Experiment

3.1 Research Questions and Sample

In the course of this chapter, three research questions to the empirical data can be formulated:

1. Are there indicators for beginning team-building processes and a shared understanding between a human and a robot like the development of social rules and social signals?
2. Can we find indicators for team development stages according to Drexler et al. (1988)?
3. Which level of hybrid collaboration can be achieved within the virtual setting?

3.2 The Experiment

Technical Setup: To get insights into hybrid team collaboration processes, a controlled experiment within the Virtual Theater was designed. The Virtual Theater (by MSE Weibull) is an immersive simulator that combines the natural user interfaces of the Oculus Rift Development Kit 2 (DK2) head-mounted display and an omnidirectional conveyor belt that allowed almost natural movement. Through a tracking system, the user's position and orientation in virtual space can be determined. The Virtual Theater was combined with a wireless presenter (Logitech Wireless Presenter R400), which served as an input device.

Furthermore, the experiment was completed with an online pre- and post-survey using SoSci Survey. While the pre-survey was designed to acquire insights into the participants' personal characteristics, the post-survey was designed to gain information of the individual, subjective assessment of the hybrid teamwork.

3.3 Task

The participants of the study were asked to fulfil a task that could only be achieved in a successful way via a cooperation with the mobile, voice-controlled robot Charles within the setting of a virtual hall (see Fig. 2). Further, the participants were provided with an exploration time where they could explore the virtual environment, including the production hall and the different machines (e.g. laths, bending machine) by pressing the corresponding buttons. The robot, which was located in the production hall, offered help by operating the machines through a textual interface that was read out aloud. The participants were informed that they could

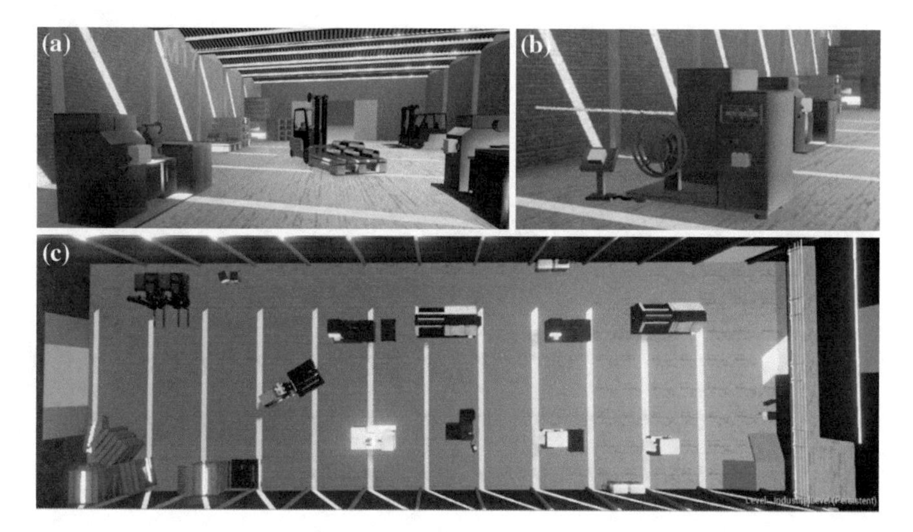

Fig. 2 Impressions of **a** the production hall, **b** the chain hoist and **c** the floorplan

navigate the robot by giving him instructions like saying "Go to machine…" (in German: *Gehe zu…*) or "Press button…" (*Benutze…*).

While the robot was only able to operate machines that had a red button available, the humans/participants were only able to operate machines with a green button, which they had to identify by trial and error. The machines were named with numbers that appeared above them and could thus be visually perceived. The participants received the actual task instructions when they operated a specific machine out of all the machines available; the task was to operate an electric chain hoist by either pressing the green or the red button. In order to secure the power supply, human or robot had to stand on a platform. Within five minutes they had to pull as much of the rope as possible, starting from the moment the instruction was activated. The average time within the Virtual Theatre was 15 min.

Robot Characteristics: The robot's characteristics were manipulated in a 2 × 2 design. The robot occurred either as an industrial or humanoid robot (see Fig. 2) and operated either reliably or imperfectly in response to the human operator. The participants were randomly allocated. The behaviour of the robot in both conditions was standardized by the specifications of an activity (see Fig. 3).

According to Onnasch, Maier, and Jürgensohn (Earley and Mosakowski 2000), the human-robot interaction, the robot and the team used in this experiment can be classified according to a Canvas diagram as described in Fig. 4, the classification of the reliability of the robot was added (Mueller et al. 2017 submitted).

The Sample: 153 people participated in the empirical study. The data from 31 participants had to be excluded from analyses because of technical problems or due to simulation sickness. The remaining 112 participants (91 male, 21 female) had a

Yes
Received command?
no
Stay and wait
Command: "Gehe zu"?
Command: "Benutze"?
Any other command
Yes
Yes
Go to the desired position
yes
Device is near by?
Not near by
Press the device's button
Stay and wait
Stay and wait
Scenario not done

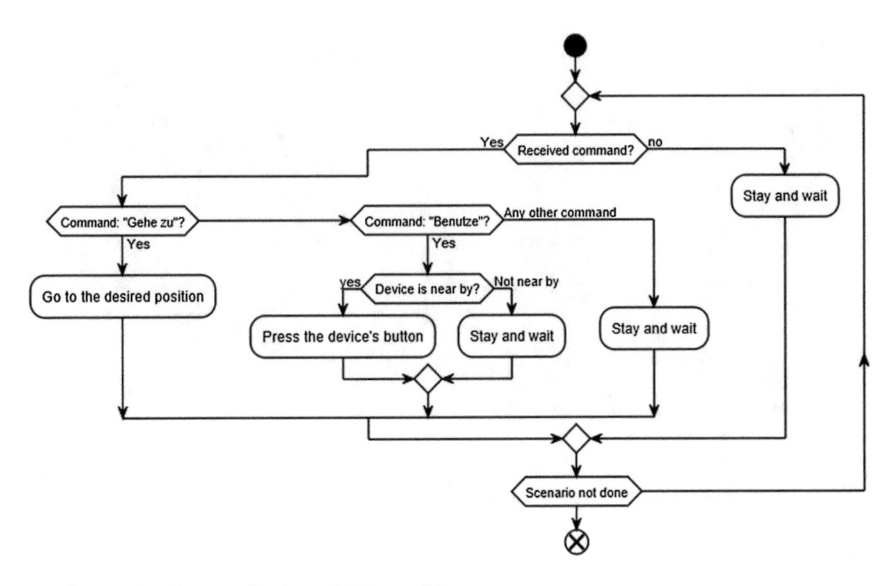

Fig. 3 Activity diagram for the reliable condition

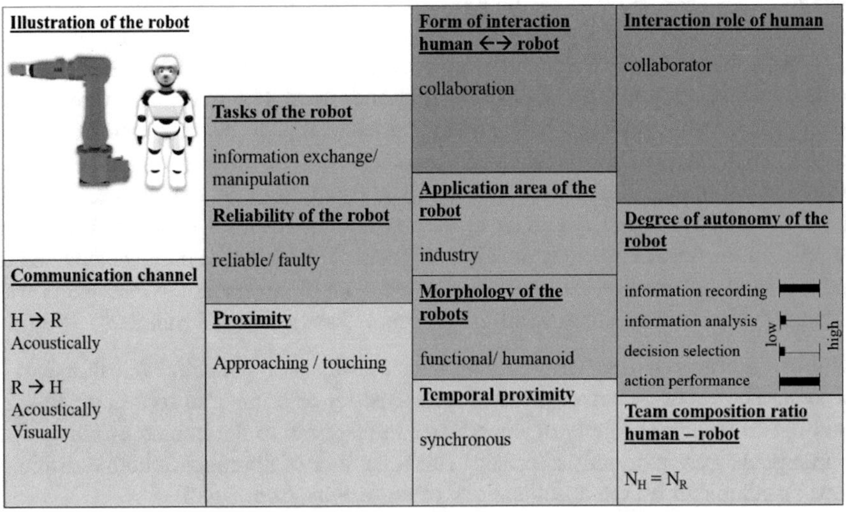

Illustration of the robot
Form of interaction human ←→ robot
collaboration
Interaction role of human
collaborator
Tasks of the robot
information exchange/ manipulation
Reliability of the robot
reliable/ faulty
Application area of the robot
industry
Degree of autonomy of the robot
information recording
information analysis
decision selection
action performance
low
high
Communication channel
H → R
Acoustically
R → H
Acoustically
Visually
Proximity
Approaching / touching
Morphology of the robots
functional/ humanoid
Temporal proximity
synchronous
Team composition ratio human – robot
$N_H = N_R$

Fig. 4 Interaction classification (dark grey), robot classification (medium grey), and team classification (light grey) of the hybrid interaction used in the experiment (Mueller et al. 2017, submitted)

range of age from 18 to 54 years ($\bar{x} = 24.53$, $SD = 5.84$). 95.55% of the participants had an academic background (49.10% engineering science, 16.96% information technology, 13.39% social science). 77 participants were students and 35 participants were full-time employees. They had been recruited via social media and posters at the University in Aachen, Germany. 34.8% of the participants had prior experience with an Oculus Rift or similar 3D-glasses. 91.1% reported being familiar with general game control. Almost half of the participants had modest visual impairment (47.3%), but this should not affect the reported results. The participants were assigned randomly to the experimental conditions (cf. 3.4) and the groups were nearly of the same size (N = 24 to N = 32).

3.4 Preliminary Results

Initial results of the quantitative analysis show that in general, most of the participants were able to solve the task and pulled the rope together with the humanoid or the industrial robot. Quantitative analysis showed that the industrial robot was the condition that brought a higher team performance and the humanoid condition brought more stability in solving the task. In terms of problem solving behaviour, hybrid teams with the humanoid robot were more stable, and people did not give up. In the hybrid team with the robot arm, people sometimes gave up if they got stuck in the task (Müller et al. 2017).

The increasing level of distress as well as the decreasing level of engagement show that the experiment was perceived as stressful and that no flow was created. As hybrid collaboration is new to the participants, working together might be perceived as mentally and physically exhausting. The time effect might also be a result of the simulation itself, as many participants reported simulation sickness. Besides the time effect, it must be noted that the level of distress both before and after was reported very low and the level of engagement was reported to be quite high (Mueller et al. 2017, submitted).

A qualitative analysis of a limited sample of (so far) eight video and interaction transcripts has shown that, although all eight teams have successfully mastered the task of pulling the rope together, the degree of performance varies significantly across the eight hybrid teams. In particular, it was found that two teams have performed excellently (pulling the rope five to seven times), that two teams performed weakly (pulling the rope two times each) and that four teams performed average (pulling the rope three to four times). Interestingly, the analyses of the data show that there are some characteristics of team work behaviour that are typical for each of the three groups, i.e., in terms of team work behaviour, the "excellent" teams share common ways of cooperative behaviour, the "weak" teams all behave similarly in some way and the same applies for the averagely performing teams. The behavioural characteristics for all three groups are outlined in the following, with a first glance at the two teams that performed excellently.

The two high-performing teams: VP_49 and VP_127

The first thing that the two high performing teams share is that the robots in both teams are industrial. The human team members in both teams, in the beginning of the "game," struggle to orient themselves and find out what the task is. The human team member in the most successful group, group VP_127 (7 meters of rope pulling), does not even approach the robot to ask for help until he is directed to do so by an external instructor. However, once the participants realize what their task is and how they can work together to solve this task, i.e. once the roles in the team are clear, the participants get straight into the task and work closely with the robot to master it successfully.

Another aspect that the two best teams share is that the participants continuously reflect on their own behaviour and communication style but also on the robot's behaviour in order to check if the robot is "understanding their requests and is doing what it is supposed to do". When the robot is not acting upon his commands, VP_49, for instance, is trying to find out if the robot's false behaviour is due to a faulty command directed to him and thus asks the instructor: "This was the right command, wasn't it?" However, this is the only question the participant directs to the instructor. Most of the time, he is trying to figure out the reasons for teamwork problems himself. Once, when the teamwork was not functioning, the participant, for instance, has a look around and diagnoses that the robot has not acted according to his previous command: "He is not at M13, is he? It says something else there." What is striking is that VP_49 is remarkably patient with the robot and, once he realizes that problems are not due to mistakes in his own manner of communication, keeps repeating the commands to the robot until he performs accordingly: "Go to M13, go to M13, go to M13… press button, go to M13, press button…"

Similar to VP_49, VP_127 also scarcely seeks the instructors help, and just as VP_49, he does so in order to make sure that he is using a proper style of communication: "Do I have to say ′go′ to machine…?" After he has understood how to communicate with the robot, the participant does not seek the instructors help any more, but engages into a close interaction with the robot. The participant strikingly often calls the robot by his name "Charles" and even uses the nickname "Charlie." Interestingly, the participant even extends the number of phrases available by adding further, colloquial, phrases such as: "Are you going?" or "Charles, press the button." Once the participant even praises the robot for his work, calling it "super." It appears that VP_127 has not only developed an effective way of communicating with the robot, but has even adapted a human-like communication style, i.e. a communication style that you would rather expect in human-human teams. Even when minor problems occur during the team-formation and performance stages, the participants are able to solve the task very successfully.

The two low-performing teams: VP_100 and VP_156

In contrast to the two high-performing teams, the two low-performing teams consisted of a human participant and a humanoid robot (instead of an industrial robot) each. Interestingly, the participants in both of these teams do not actively watch or follow the robot to check if he is performing correctly, but they keep a physical distance and turn to the external instructor in order to find out what Charles is doing. At first VP_100, for instance, turns often to the instructor, asking: "Is he gone?" and VP_154 asks the instructor: "What is Charles doing at the moment," "Is he still […]," or "Did he press […]?" The fact that the participants turn to the instructor so often makes it appear as though the participants are not only trying to keep a physical distance to the robot, but also an emotional one, as if they are not willing to accept the robot as an active team partner. On the contrary, it seems as though they are taking the instructor into the role of a team-partner, as someone who will help them make the robot—*a machine*—work properly. Therefore, it can only be concluded that the human and the robot are only coexisting in this setting and that no a hybrid team formation is not taking place. The underlying reasons need to be investigated further through correlating the video transcript data with the pre- and post-surveys, which will be done once the video transcription and analysis stage of the project is completed.

The average-performing teams: VP_109, VP_111, VP_116, VP_184

First, all average-performing groups of the limited sample consisted of industrial robots only. What is striking with the average-performing groups is that they, just as the lowest performing groups, noticeably often turn to the instructor for help. VP_116 for instance asks the instructor: "Is the other one [the robot] ready?", "So, could Charles press that?", "Where has Charles gone?", "Is he there?" and "Where does Charles need to go?". Similarly, VP_184 asks the instructor: "Did Charles actuate machine 13?" and "What does he actuate?".

What is also remarkable is that the participants in the average-performing teams tend to get annoyed by Charles when he does not follow their commands. VP_111, for instance, sighs "Oh, Charles…" and even calls him dumb: "Oh man, Charles is dumb…" when he does not perform the task he is supposed to. The fact that VP_111 belittles Charles: "… *(through laughter)* Cute! He is cute…" does not help to minimize the participant's annoyance. VP_116 even gets slightly aggressive when the robot does not perform rightly: "He does not find the machine, M5!" Similar to VP_111, the participant's feelings about the robot do not change through his human-like interaction with him. That the participant is using a human-like communication pattern with the robot for instance becomes obvious by the fact that he is mostly calling the robot by his name and also by the fact that he is using colloquial ways of speaking that would normally only occur in verbal interactions: "Then, press the button again". VP_111 is also using social signs in his communication when he, for instance, raises his voice against the robot angrily and says "M5", thereby meaning "go to M5". VP_184 goes even further when he angrily, and with a raised voice, tells the non-behaving robot: "Charles, I am not your friend" and uses the social signal of shouting at him "Charles? Actuate the machine!" VP_184 even expresses that he feels like Charles has let him down when he says: "Charles is not helping me".

In the following section of the chapter, the analysis results of the limited sample and their contribution to the research questions are discussed.

4 Discussion and Outlook

The qualitative analysis of the limited sample of eight video and interaction transcripts suggest that team-building processes do take place in hybrid teams and that humans and robots, just as humans and humans, go through several stages of team development. In addition, the analysis has shown that within the virtual setting, all three levels of hybrid collaboration can be achieved: collaboration, coexistence and cooperation. Which level is achieved, in particular, substantially depends on human factors such as the willingness to cooperate and communicate with the robot. In addition, the ability to act self- and task-reflective, i.e., to analyse the teamwork, diagnose and combat possible problems, and have patience was also found to be an important human factor. It became clear that a lack of patience towards the robots often leads to frustration, which in turn can hinder a high level of performance.

It may be said that all members in the teams go through some stages of team-development, orientation, trust building and goal clarification, until they are finally able to perform successfully. On their way through the team development stages they sometimes establish social signals (e.g. like nicknames, winking, grumbling about the robotic workmate). Those signals, as the preliminary results show, can be an indicator for the acceptance of the robot as a workmate. If the participants keep distant through extensively including the operator within the action, a team building process with insufficient handling of the stages (like e.g. a lack in trust building) was observable and the task performance tended to be low. Due to the limited sample size analysed so far, it is not yet clear, however, if this result holds general validity or can be traced back to certain character traits of the participants. To answer this question ultimately, further transcripts need to be taken into account and be analysed. Furthermore, the findings of this study need to be correlated with the participants' questionnaire-results to gain deeper insights into this matter. All in all, it needs to be pointed out that even though social signs were used by participants, it could not be found that the use of these signs in general leads to a higher level of performance.

It can be concluded that the participants' high-performing teams need some time to figure out the social rules and signals and make sure they know how to communicate with the robot. It takes them some time to become an effectively functioning team; in other words, they have to successfully master at least the three team-development stages according to Drexler et al. (1988) of orientation, trust building and goal clarification. Once the teams have passed through these stages, however, they perform excellently, even reaching the high performance stage (Drexler et al. 1988). Even when minor problems occur during the team formation and performance stages, the participants are able to successfully solve these through slight assistance from the instructors, patience, self-reflection, a reflection on the robot's behaviour, and effective communication with the robot. Taking the close interaction between human and robot in both teams into account, it can be concluded that a collaboration is taking place in these teams, rather than a cooperation or coexistence.

The participants in the "average performing" teams are not collaborating with each other, but are cooperating. Even if the cooperation is not entirely successful at all times, the robot and the human are still working together and interacting intensely with each other. However, the humans and the robots in the teams are still far from reaching Drexler et al.'s (1988) High Level of performance, but are rather stuck in the stages of team formation in which they have not entirely built trust, developed common goals and made clear decisions that are followed by both team members. Compared to the two most successful groups, it appears that the four average groups are lacking an effective style of communication and, closely related, patience. The creation and establishment of social signals and rules is at a rather limited level. This, in turn, appears to lead to frustration, which then hinders the teams' ability to diagnose and combat problems, and through that, reach a level of high performance. It also appears that the participants in the average-performing teams are not as self- and task-reflexive as the participants in the "best" teams, i.e., instead of questioning their own behaviour and analysing why the teamwork is not working, they often turn to the instructor for help or get frustrated, sometimes even resigning.

The results of the qualitative analysis are preliminary and show first findings of the huge data corpus of the empirical study. Therefore, all given insights are only hints that have to be validated and deeper investigated once the transcription is completed. Moreover, the study has some further limitations that should be considered while interpreting the results. First, due to technical issues and simulation sickness of some participants, comparatively few test subjects were included in each condition of the experiment; hence, the statistical power was limited. Second, the sample consisted mainly of students and it is possible that construction workers, who will be the actual end-users, will perceive hybrid collaboration in a different way. Third, habituation effects might occur, so that the described patterns might not remain over time. Fourth, a virtual environment setting was used to guarantee a safe interaction with a robot and to manipulate the robot's characteristics easily. However, it must be explored whether the findings are transferable to real production environments.

Overall, the empirical data strongly suggest, that the human-robot collaboration within hybrid teams could be fruitful when we overcome the tool-aspect of human robot collaboration. Furthermore the data show that shared understanding and social interaction behavior are necessary as well as tasks that foster a constant communication process. Within current, running projects like the BMBF-Project "Work in the industry of the future (ARIZ)",[2] such sample work places will be built, and the empirical studies in these physical workplaces will allow deeper insights into the teaming processes of humans and robots.

Acknowledgements The author would like to thank the RWTH Start-Up-Grant for the kind support within the project SowiRo (Socializing with Robots).

[2]http://ariz-ac.de/de/.

References

Badke-Schaub, P., Hofinger, G., & Lauche, K. (2008). Human factors. In *Human Factors* (pp. 3–18). Berlin: Springer.

Bettenhausen, K. L. (1991). Five years of groups research: What we have learned and what needs to be addressed. *Journal of Management, 17*(2), 345–381.

Billard, A., & Dautenhahn, K. (1997). *Grounding communication in situated, social robots*. Proceedings Towards Intelligent Mobile Robots Conference, Report No. UMCS-97-9-1, Department of Computer Science, Manchester University.

Breazeal, C. L. (2004). *Designing sociable robots*. Cambridge: MIT press.

Carroll, G. R., & Harrison, J. R. (1998). Organizational demography and culture: Insights from a formal model and simulation. *Administrative Science Quarterly*, 637–667.

Drexler, A., Sisbbet, D., & Forrester, R. (1988). Team building, blueprints for productivity and satisfaction. *Applied Behavioral Science and Pfeiffer & Company.*

Earley, C. P., & Mosakowski, E. (2000). Creating hybrid team cultures: An empirical test of transnational team functioning. *Academy of Management Journal, 43*(1), 26–49.

Ferrari, F., & Eyssel, F. (2016). *Toward a Hybrid Society*. International Conference on Social Robotics, Berlin: Springer.

Klimoski, R., & Mohammed, S. (1994). Team mental model: Construct or metaphor? *Journal of Management, 20*(2), 403–437.

Mueller, S., Stiehm, S., Jeschke, S., & Richert, A. (2017 submitted). *Subjective Stress in Hybrid Collaboration*. Proceedings of the Ninth international Conference on Social Robotics 2017, Lecture Notes in Artificial Intelligence, Berlin: Springer.

Müller, S. L., Schröder, S., Jeschke, S. & Richert, A. (2017). *Design of a Robotic Workmate*. International Conference on Digital Human Modeling and Applications in Health, Safety, Ergonomics and Risk Management. Berlin: Springer.

Onnasch, L., Maier, X., & Jürgensohn T. (2016). *Mensch-Roboter-Interaktion-Eine Taxonomie für alle Anwendungsfälle*, Bundesanstalt für Arbeitsschutz und Arbeitsmedizin.

Roschelle, J., & Teasley, S. D. (1995). *The construction of shared knowledge in collaborative problem solving*. Computer-supported collaborative learning.

Schwartz, T., Zinnikus, I., Krieger, H.-U., Bürckert, C., Folz, J., Kiefer, B., et al. (2016). *Hybrid teams: Flexible collaboration between humans, robots and virtual agents*. German Conference on Multiagent System Technologies, Berlin: Springer.

Straube, S., & Schwartz, T. (2016). Hybride Teams in der digitalen Vernetzung der Zukunft. Mensch-Roboter-Kollaboration. *Industrie 4.0 Management, 32,* 41–45.

Author Biography

Anja Richert, Prof. Dr., is Junior Professor at the Faculty of Mechanical Engineering and Managing Director of the Center for Learning and Knowledge Management (ZLW) at the Cybernetics Lab IMA/ZLW & IfU, RWTH Aachen University. Her main areas of research are work and society 4.0 with an emphasis on innovation and knowledge management.

Part II
Collaboration and Networking

IT Support for Knowledge Processes in Digital Social Collaboration

René Peinl

Abstract IT support for collaboration has gone through quite some change since the beginning of the century and is providing more and more support for knowledge workers. Although single systems are getting easier to use, they are often not replacing former systems but accompany them, which makes the overall system landscape harder to oversee for knowledge workers. Future information systems should therefore combine the existing building blocks under a consistent user interface and assist the user in storing information at the right place. Seamlessly switching between formats, so that the user doesn't have to decide upfront whether a blog entry, a wiki page or a text processor document is better suited for the information. This section discusses the development of digital collaboration solutions and shows how social software has changed them to better support knowledge processes.

1 Introduction

In the next wave of knowledge management (KM) initiatives, many of the established KM instruments (Peinl 2011a) and methods are still necessary and valid. Organizations should, for example, still decide whether codification or personalization strategy is more important for them, but foster both (Maier and Remus 2003). However, the focus of codification should be on creating digitally executable artefacts that are both helpful for the human reader, but also transport meaning for the machine (Peinl 2017). Examples are process models that can be directly executed in a workflow engine or at least only need a few technical enhancements to do so. Furthermore, personalization strategy should be implemented in a way to not only further communication between employees, but also to provide context for machines to better decide which information will be helpful in the current situation.

R. Peinl (✉)
Hochschule Hof, Hof, Germany
e-mail: rene.peinl@iisys.de

K. North et al. (eds.), *Knowledge Management in Digital Change*, Progress in IS,
https://doi.org/10.1007/978-3-319-73546-7_7

This chapter provides an overview of the development of collaboration software, especially business process management, intelligent systems and social media in the context of knowledge management and then exemplifies the proposed new focus of KM with the outcome of the project "Social Collaboration Hub" (SCHub) (Peinl and Ochsenkühn 2015).

2 From Information to Knowledge

When speaking about knowledge, we first have to briefly clarify what we denote with the term. An important distinction, that both splits the KM community as well as leads to different IT support is between knowledge as possession and knowledge as social practice (Newell 2015). The first perspective treats people as bearer of knowledge, which is based on experiences but separable from that experience. The second prefers to speak of knowing as something people do that is context-dependent, emerging and socially situated. The author believes that both perspectives add important aspects when thinking about knowledge and are not mutually exclusive. Newell also states that contexts, processes and purposes need to be considered whichever approach to knowledge one is adopting.

According to the knowledge ladder (North 2016), you need to add context, experience and expectations to get from information to knowledge. This is true for humans to derive knowledge from information, but it is also key for supporting knowledge processes based on information in computers. Therefore, the computer needs to collect context information so that it can conclude in which activity or process the knowledge worker is currently working.

> Context is any information that can be used to characterize the situation of an entity. An entity is a person, place, or object that is considered relevant to the interaction between a user and an application, including the user and applications themselves. (Abowd et al. 1999).

Simple context information like the application that is currently used are not very helpful for that, because most systems used are not specific to a certain process (e.g. email, text processor, wiki, ERP system). The question is therefore where to derive context information from. Possible sources are business process management systems for structured activities as well as social networks and contents in general for semi- or unstructured activities. However, the activities collected in an activity stream of a social network that seem similar to tasks in a process at first glance, do not provide the desired context information, since they are granular and generic data, e.g. "person 1 created new blog post about xyz". Unlike tasks in a process, they are missing the link to the encompassing larger activity. We need content analysis to extract hints about the context from the contents of social media. For documents in document management system, the case is a bit different if there are associated document templates and/or content types that can be used as context information.

> The term content type refers to the semantic meaning of a document like meeting minutes, project proposal or product data sheet in contrast to the document type which refers to technical aspects of a document and states the format like pdf, pptx or odt.

Finally, knowledge processes are discussed nearly as diversely as knowledge itself.

> Remus and Schub define knowledge processes as "service processes that support the exchange of knowledge between business units and business processes. Examples are processes that support the collection, organization, storing and distribution of knowledge as an outcome of business processes or processes that manage the allocation of skills and expertise to business processes or projects." (Remus and Schub 2003).

Other authors call them differently and collect different number of representatives. Probst et al. (2000) call them KM building blocks and collect eight of them: goal setting, identification, acquisition, development, distribution, usage, protection and evaluation (Probst et al. 2000). Holsapple and Joshi call them knowledge manipulation activities and distinguish five of them: acquisition, selection, assimilation, generation and emission (Holsapple and Joshi 2004). The famous SECI model with its knowledge conversion processes collects four processes: socialization, externalization, combination and internalization (Nonaka et al. 2006).

The following section shows the development of IT systems supporting knowledge processes.

3 Collaboration Systems for KM in the 21st Century

In line with the two perspectives of knowledge as possession or practice of knowing, Hansen et al. proposed that organizations should follow a codification or personalization strategy (Hansen et al. 2005). IT support for knowledge processes can be divided into repository systems (codification) and network systems (personalization) accordingly (Newell 2015). As a result of that, many KM initiatives at the beginning of the 21st century built on **document management systems** (DMS) to foster codification on the one hand and/or **groupware** to foster personalization on the other hand (Alavi and Leidner 2001). Since then, a large number of other systems and technologies arose (see Fig. 1) to accompany those two system categories in modern organizations (Sultan 2013). DMS have evolved into Enterprise Content Management Systems (ECM) that include both documents and web content. They were further extended into so called **knowledge portals** (Firestone 2000). These added personalization (in the sense of adapting contents to personal needs, not the personalization KM strategy) and access to a number of information sources (Priebe and Pernul 2003), and therefore fostered knowledge sharing (Van Baalen et al. 2005). One of the information sources usually was a more or less sophisticated **skill or competence management system**. Representatives of this category were reaching from simple yellow pages with the job title being the only hint for the skills of the employee, to full blown e-learning

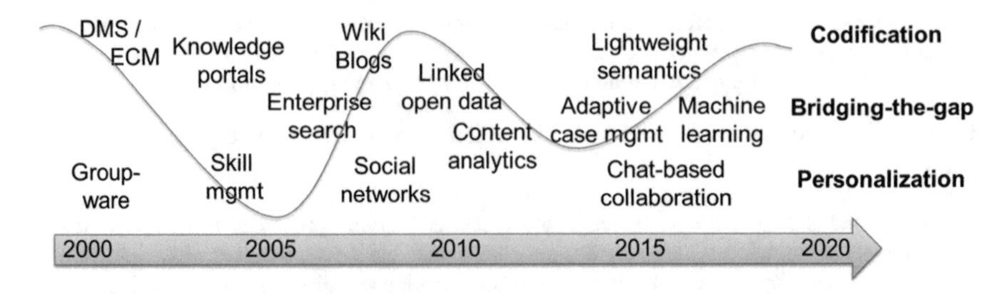

Fig. 1 Rough timeline of KM supporting technologies and system categories

or human resource management solutions that support not only identification, but also assessment, acquisition and tracking of competence usage (Berio and Harzallah 2005; Draganidis and Mentzas 2006). Starting as a central task of the human resource department, it was later on seen as a natural part of social networking and therefore should be decentralized in a form of collaborative skill management (Elia et al. 2008; Braun et al. 2010; Varshney et al. 2013). Public social networking sites like Linked.in made this approach popular by allowing so called endorsements where colleagues could state that a certain person has a skill. This could be either by confirming a skill the person has entered herself, or by suggesting completely new skills. Therefore, skill management systems partially compensated for the lacking advances in groupware systems and therefore technical support for the personalization strategy.

Disillusioning: Despite these technical advances between 2000 and 2005, the KM community came to the insight that human and organizational factors are more important than technology during the disillusioning phase of KM around 2005. The curve in the background of Fig. 1 illustrates the public perception of IT-driven KM initiatives. This was due to several reports of failed technology-driven KM initiatives (Malhotra 2004; Heisig 2009; Akhavan et al. 2012).

Social software: However, social networks on the one hand, as well as wikis and blogs on the other hand renewed the interest in technology-based KM initiatives around 2008 (Stocker and Tochtermann 2009; Yang 2009). They were perceived as an evolution of KM systems from static archives of information to being a connector of humans to information and of humans to each other (Yew Wong 2005). After the success of Facebook and Wikipedia, many companies had exaggerated expectations of what social software could achieve in their intranet. Just to provide a simple example, one of the author's clients was very disappointed that there was no content after the consultant had installed the wiki software. He had somehow expected that his organizations "corporate knowledge" would be provided together with the software out of the box. Nevertheless, there were also successful KM initiatives built on social software, mostly those that accepted that social software is more about changing organizations culture to an open, fault tolerant sharing culture. In contrast to that, many initiatives were failing because companies were trying to impose their hierarchical closed structures to wikis and social networks which resulted in chimaeras like wikis with fine-grained security control which totally thwarted the basic wiki principles.

Enterprise search: Already before social media and user-generated content made the amount of available content explode, enterprise search engines got into the focus of knowledge workers. Although connecting to a variety of different information systems, understanding and unifying their permission model as well as extracting text contents together with meta data from dozens of file formats is challenging enough, it turned out that the real challenge is perceived relevance of search results in companies' intranets to be on par with that of Google on the internet (Grudin 2006) (although the task seems easier due to a limited amount of data). However, the internet is much better interlinked than most intranets are and the pure mass of users on Google allows them to do a better result ranking. In addition to that, Google is investing a lot of effort in both providing shortcut answers for very popular questions, e.g. results of games during a championship and using structured semantic data to actually provide answers instead of a result list in their knowledge graph. For enterprise search projects on the other hand, it is still usual to introduce the search engine once and leave it without further optimizations for several years. However, enterprise search is still providing added value to knowledge workers, e.g. by incorporating structured information. This started with employee profiles that allowed quick finding of experts, especially with the help of faceted search. Later on it expanded to include product, customer and other structured data, so that the search engine became the single point of access to information, that portals claimed to be before (Peinl 2011b).

Linked open data: Google is able to provide answers instead of result lists because both Semantic Web as well as language processing technologies matured and the combination of those with social media (Schaffert 2006; Levy 2009) laid the foundation for Linked Open Data (LOD) or the Web of data (Bizer et al. 2009; Westerski and Iglesias 2011). The challenge with human annotation of contents (Uren et al. 2006) was partly solved by new technology that is able to sufficiently well extract named entities from (English) texts and use these as meta data (Maynard et al. 2005; Hassell et al. 2006).

The formal models used are called ontologies and can enhance information systems supporting knowledge workers in many situations. Collaborative skill management systems for example suffer from the problem that the tags that people use as skills are not clearly described, have no relationship to other tags and are not machine understandable (Weber et al. 2009). When tags are linked to concepts in an ontology, these problems are overcome. However, the process of agreeing on an ontology in an organization is tedious and slow and therefore seldom beneficial due to the quick changes in the environment. **Lightweight semantics** are proposed as a compromise between formal ontologies and unstructured tags (Kammergruber et al. 2010). A thesaurus is a controlled vocabulary of terms that can be used as keywords. It includes a small number of predefined relations such as synonym, broader/narrower term and related. Converting the unstructured tags into a thesaurus is called gardening (Braun et al. 2012). It can be done by the community itself, e.g. embedded in the "search for experts" process, or by experts that review tags and add structure. Terms can also be linked to a wiki page where it can be further described, discussed and linked to interested people (Peinl 2015). This makes it easier to maintain the terms by adding explanation and also serves as a rectification ground.

This trend is also reflected in recent solutions for enterprise search which are now termed **content analytics** (Zhu et al. 2014) or cognitive search and claim to "deliver the new generation of search and knowledge discovery" (Curran and Gualtieri 2016). In order to cope with the sheer amount of user-generated content and gain insights from social communication, content analytics does not only enhance search, but also provides automatic summaries, charts on topics that are discussed and other advanced forms of data analysis. The LOD cloud can be seen as a huge distributed database containing structured information about all kinds of objects with public interest, from media like books, music and movies, over people like actors, politicians and athletes to locations like countries, cities and points of interest. Semantic technologies also influenced the advances in social software, namely with semantic wikis like the semantic media wiki or the efforts to integrate tag clouds to become folksonomies or even full blown ontologies (Braun et al. 2012). However, their adoption was not as widespread as expected, which is mainly due to technical aspects like performance issues and perceived complexity for developers. Nevertheless, they are a logical step in the development of information systems and recent developments like JSON-LD and schema.org that can be summarized as lightweight semantics are expected to provide for the necessary wide use.

Adaptive case management: In parallel to this development, process-oriented KM systems (Remus and Schub 2003; Woitsch and Karagiannis 2005) were suggested to bridge the gap between codification and personalization. Despite a strong research interest however, they were never widely adopted in corporations (and therefore not shown in Fig. 1). One reason for the lacking adoption might be that the strict task sequences that are focused by business process management (BPM) and workflows do not fit the typical tasks of knowledge workers which face a lot of variations. As a result of that, BPM systems evolved into adaptive case management (ACM) in order to better support knowledge-intensive, weakly structured business processes (Herrmann and Kurz 2011; Traganos and Grefen 2015). ACM is characterized by goal- and data-orientation, transparency, runtime flexibility, continuous improvement and integration of information systems (Hauder et al. 2014). It looks at tasks from a case perspective, which is familiar to doctors or lawyers. Within a case, tasks can be arranged in sequence, but will most of the time have no strict order. A case model can therefore be seen as a form of "checklist on steroids".

Existing BPM systems like Appian and Pegasystems, ECM systems like EMC Documentum and IBM FileNet, as well as enterprise social software like Jive and Yammer are trying to evolve into ACM solutions (Motahari-Nezhad and Swenson 2013). It is expected that future ACM solutions will excel in support for knowledge workers by providing better personalization, context identification and intelligent assistants. Similarly, Osuszek and Stanek (2016) argue that integration of decision support systems (DSS) and social networking in ACM systems is a major development of the last years. (Ariouat et al. 2016) state that support for emergent collaborative and flexible processes is an important requirement for future collaboration solutions and examine research prototypes from the areas of BPM, ACM, Computer Supported Collaborative Work (CSCW) and KM (Fig. 2).

They analysed 14 research prototypes and found that currently only a few of the systems incorporate functionality from several areas. Bonita and Caramba combine CSCW and BPM functionality. Promote and ProCollab combine KM with BPM. Cognoscenti was the only approach that includes both ACM and KM functionality. Not included in the study but also relevant here is "Knowlege Intensive Service Support" (KISS) and its prototypical implementation KISSmir (Brander et al. 2011). The study's authors themselves try to integrate social networking with ACM, but there was no system that amalgamates all four areas. In the second part of the chapter, SCHub is introduced, which tries to close this gap. It also stands out by using standards like CMMN (Case Management Model and Notation) of the Object Management Group (Kurz et al. 2015) instead of proprietary models, which provides manufacturer-spanning compatibility and plays well together with BPMN (Business Process Model and Notation) (Hinkelmann and Pierfranceschi 2014). CMMN covers all requirements for ACM besides visualization of ad hoc modification, visualization of current goals directly inside the model and visualization of a case and its progress during runtime (ibid). These aspects were included in SCHub in a proprietary way as part of the case UI.

Besides the increased flexibility compared to structured workflows, ACM also stresses the link between contents and people and therefore provides a means to bridge the gap between codification and personalization (Motahari-Nezhad et al. 2012; Osuszek and Stanek 2016). An additional aspect relevant to ACM is therefore finding the "right" colleagues for a given case. (Heil et al. 2014) try to recommend suitable colleagues based on the social network of the user and the provided skills

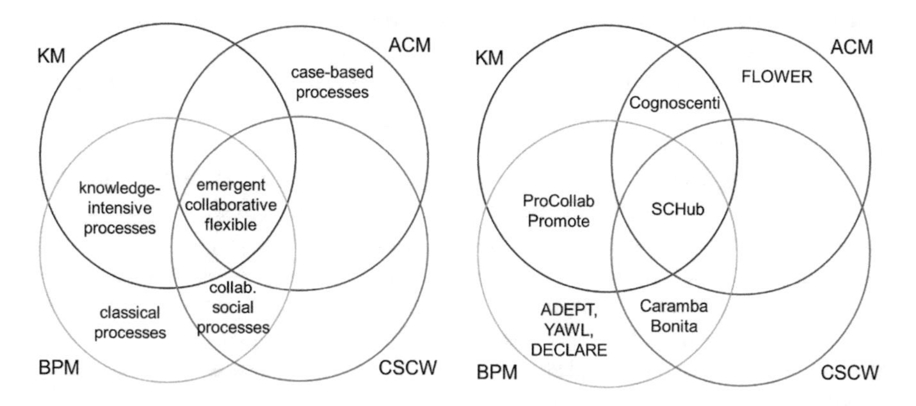

Fig. 2 Research areas for process-oriented digital collaboration (based on Ariouat et al. 2016)

in the social network. To achieve that, every case needs a goal and derived requirements that can be matched with the skills. A similar approach is presented in (Huber et al. 2015). Their prototype additionally makes recommendations for next steps and estimates the required remaining time until completion of the case. These recommendations also include hints on how strongly actions support the case goals and whether pursuing them will shorten the case duration or helps avoiding time-outs. The approach presented by Rangiha et al. (2015) is based on social BPM and recommends roles and tasks for ad hoc processes, based on tags and ratings of task results by the community. That means in return, that all tasks have to be tagged, hierarchized and rated by community members which can be a problem. In SCHub, case co-workers are recommended based on skills, skill-difference, role and organizational position (see below).

Case instances can also provide an umbrella for **chat-based workspaces** that is becoming more and more common in organizations. It can be seen as a mixture of Twitter, WhatsApp and email. Slack is probably its most prominent representative. Big players like Microsoft with its Teams solution as well as Cisco with Spark show the relevance of this category. Smaller players like Asana demonstrate how chat can be integrated with task and project management. The current interest in these kinds of solutions is a bit surprising since Google introduced its very similar solution Wave in 2009 without any success and closed it already in 2010. Chat-based collaboration also seems underrepresented in scientific articles. However, it seems a promising candidate to foster knowledge sharing in teams and organizations. Especially the integration to other tools like file sharing, real-time collaborative text processors, video chat and screen sharing makes a chat platform interesting for knowledge workers. The integration of chatbots that can currently be seen in instant messengers like Facebook messenger, Telegram and WeChat will surely spill over to chat-based workspaces in the near future, since it is part of the larger trend towards conversational user interfaces (McTear et al. 2016). Depending on the work situation, even consumer gadgets like Amazon Echo and Google Home with speech input and output may be used to get a more natural access to information in the future.

Both content analytics and chatbots will benefit from **machine learning** in the future by e.g. retrieving images based on recognized contents or better text understanding in context. ACM can also be enhanced with machine learning in order to detect patterns in work that can be suggested to less experienced workers later on (see practical example below).

Summarizing those developments, a modern collaboration solution for the digital enterprise uses social software, (lightweight) semantic technologies, natural language processing, machine learning and conversational user interfaces to flexibly support business and knowledge processes as well as projects in an ACM manner. Despite these modern trends, document management is still an important foundation and not obsolete. Instead of building single monolithic information systems with enormous complexity, the integration of smaller, easy to understand building blocks should be pursued.

4 Practical Example and Implications

Within the SCHub project (Social Collaboration Hub), a digital collaboration solution based on case management has been developed that incorporates several of the highlighted developments. The goal was a seamless IT support for knowledge processes for teams in larger organizations. SCHub is completely built on open source software and breaks the larger monolithic systems used as a foundation (Enterprise portal, ECM system, groupware) into smaller building blocks (document store, user profile, workflow engine, search engine, …) that can be reused in other contexts so that a service-oriented architecture according to the microservice or more precisely self-contained system approach arises (Peinl 2015). Before, both the portal and the ECM system had their own workflow engine and the groupware had none at all. After our modifications, all systems are using the same workflow engine, so that users don't have to learn different tools and workflows can be system spanning. The same principle applies to user profile, document store, search engine. Before, each system had its own search engine which only finds contents from this system. After our modifications, the search engine can find contents from all systems with a single search, while still retaining the functionality to search in a single system. The user interface was framed with a global navigation to switch between systems. The coloring and partly layout of the systems were adapted to look as similar as possible without larger changes to the underlying system so that it looks to the user as if it was a single intranet solution.

SCHub incorporates all the building blocks shown in Fig. 1 except Linked Open Data. It uses the knowledge maturing model as a foundation (Maier and Schmidt 2015) and directly addresses content maturing, semantic maturing and process maturing.

Content maturing is supported for innovation management. Ideas are expressed as blog posts, which can be shared, rated and commented. Promising ideas can be discussed in more detail using a forum and elaborated in a wiki page. Transfer of data from one system to the next one is done automatically. Finally, the matured idea can be semi-automatically transferred to a project proposal document and submitted for review in a workflow.

Semantic maturing is supported via collaborative competence management (see above). People profiles consist of centrally managed data, which is read-only and stems from the corporate directory. Other data can be edited by the user (e.g., project experience, skills). Fellow employees can be marked as colleagues so that an enterprise social network arises. Users can suggest skills for their colleagues (endorsing). The latter can accept the proposal or reject it. Other users can confirm the skills, once they are listed in the profile. Despite auto-completion, a large list of different terms for skills will arise in this way. In order to aid finding experts, e.g. for team staffing, an administrator can use these terms, map synonyms and bring them into a hierarchy, so that a taxonomy is formed (Lin et al. 2009). In order to foster a shared understanding of the meaning of a skill tag, each term is linked to a wiki page where it can be described (Elia et al. 2008). An embedded search query on the wiki page displays a list of users tagged with the respective skill, related documents as well as tweets about the topic. Finally, the tagged skills can also serve as a basis for periodical skill development discussions between employee and superior.

From the process maturing perspective, a case can be seen as a less mature form of a process. SCHub provides support for process maturing by semi-automatically creating CMMN (Case Management Model and Notation) case models on the one hand and social communication means that provide context for the machine on the other hand (Ochsenkühn and Peinl 2015). Activities across all systems are recorded in a central graph database compliant with the W3C activity streams standard. The competence profiles described above are stored there as well. Both serve as context information in complex collaborative document creation processes, e.g. of project proposals (see Fig. 3). The context is limited to the time frame from initial creation of the document in the system (t_0) until the document is published in the system in version 1.0 (t_1). All people interacting with the document are recorded together with their activities. Since they are likely working on more than one document or project, a content analysis regarding the similarity of electronic artefacts like e.g., blog posts, forum entries, wiki pages and emails to chapters of the final document leads to task candidates. A few creation case instances of documents with the same content type are necessary to allow the system to identify similarities in the creation activities and find task candidates. It then generates a CMMN case model out of the identified activities.

The resulting model can be collaboratively enhanced to capture formerly implicit knowledge about task sequences or dependencies using the Web-based modeller (see Fig. 4). The system further records the relation between the involved people based on organization hierarchy (imported from the corporate directory server/LDAP)

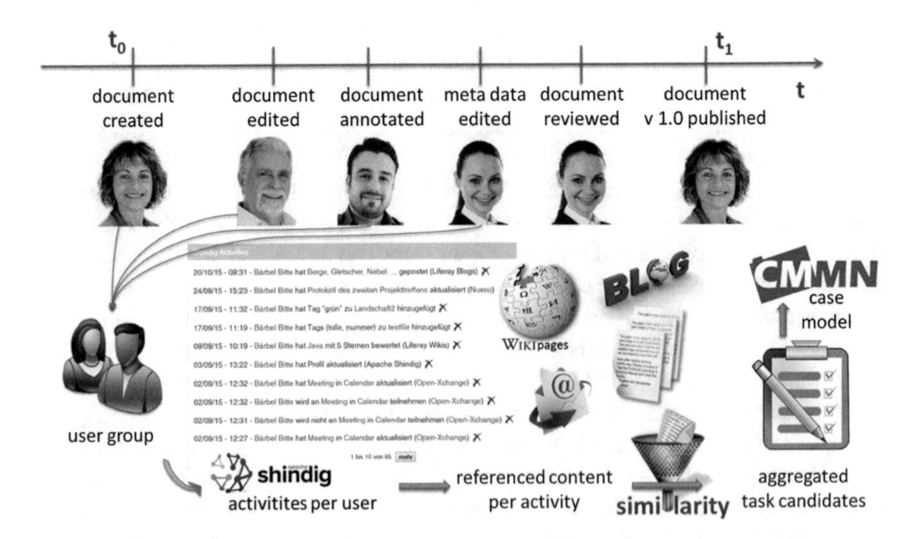

Fig. 3 Case model creation process explained (own illustration)

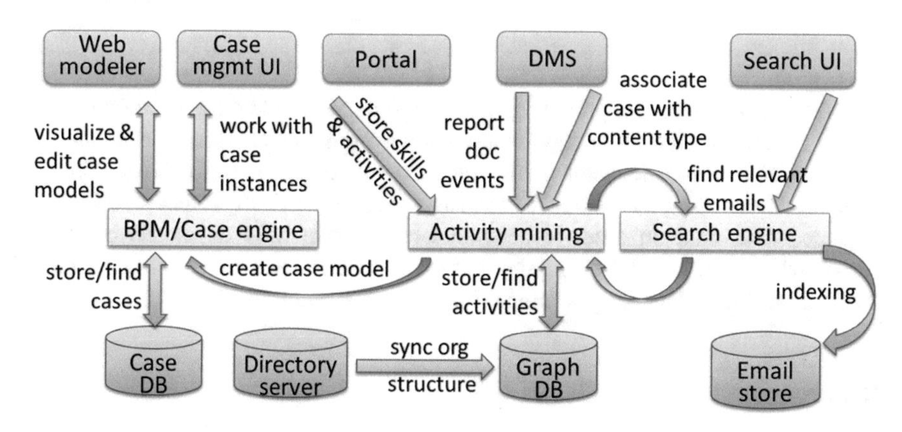

Fig. 4 Overview of system components and their interplay

and competence profiles (skills) and therefore links contents, processes and people. Since there are potentially many emails that are unrelated to the case, the search engine is used to retrieve only relevant emails using keywords from the document. Figure 4 summarizes the systems components. The user is interacting only with those in the upper part of Fig. 4. Underneath are a number of middleware components backed by different types of databases (lower part of the figure).

5 Conclusion and Outlook

The IT landscape has become significantly broader since the beginning of the century and it is not easy for employees to keep up with the speed of IT innovation. New systems may provide better support for operational tasks and especially knowledge processes, but they also require some effort to learn. SCHub supports several knowledge (maturing) processes like knowledge identification and creation using the idea management instrument (Peinl 2011a) and knowledge sharing and storing using meeting rooms with integrated task tracking. It therefore represents an enterprise knowledge infrastructure (Maier and Peinl 2017). However, the multitude of components included may make it difficult for occasional users. Although the described systems all have their strength and complement each other from a theoretical point of view, in practice people may find themselves lost when having to decide where and how to store information. Many employees are good in using text processors, spreadsheets or presentation software, but have no experience with document management in a sense of using versioning, check-in/out, metadata and publishing workflows. Therefore, the focus of future information and communication technology must be on providing natural interfaces and really helpful assistance. Speech interfaces like Amazon Echo or Siri might be one success factor. Virtual or Augmented Reality may be another, e.g. to make video conferences even more immersive and close to a face-to-face meeting which also means it can be more productive.

On the assistance side, providing machine-understandable information will help the IS in turn provide more support for knowledge workers. However, creating a case model is more demanding and requires higher skills from the employees compared to creating a checklist. That means that knowledge workers do also have to keep on learning in order to become what Gartner calls citizen developer, a person who had no formal IT training but is still able to hack a few lines of code or create a workflow including technical bindings to systems.

References

Abowd, G. D., Dey, A. K., Brown, P. J., Davies, N., Smith, M., & Steggles, P. (1999). Towards a better understanding of context and context-awareness. In *International Symposium on Handheld and Ubiquitous Computing* (pp. 304–307). Springer.

Akhavan, P., Jafari, M., & Fathian, M. (2012). Exploring the failure factors of implementing knowledge management system in the organizations. *Journal of Knowledge Management Practice, 6*.

Alavi, M., & Leidner, D. E. (2001). Knowledge management and knowledge management systems: Conceptual foundations and research issues. *MIS quarterly*, 107–136.

Ariouat, H., Andonoff, E., & Hanachi, C. (2016). Do process-based systems support emergent, collaborative and flexible processes? Comparative analysis of current systems. *Procedia Computer Science, 96,* 511–520.

Berio, G., & Harzallah, M. (2005). Knowledge management for competence management. *Journal of Universal Knowledge Management, 1,* 21–28.

Bizer, C., Lehmann, J., Kobilarov, G., Auer, S., Becker, C., Cyganiak, R., et al. (2009). DBpedia - A crystallization point for the web of data. *Web Semantics: science, services and agents on the world wide web, 7*(3), 154–165.

Brander, S., Hinkelmann, K., Martin, A., & Thönssen, B. (2011). Mining of Agile Business Processes. In *AAAI Spring Symposium: AI for Business Agility*.

Braun, S., Kunzmann, C., & Schmidt, A. (2010). People tagging and ontology maturing: Toward collaborative competence management. In *From CSCW to Web 2.0: European Developments in Collaborative Design* (pp. 133–154). Springer.

Braun, S., Kunzmann, C., & Schmidt, A. P. (2012). Semantic people tagging and ontology maturing: an enterprise social media approach to competence management. *International Journal of Knowledge and Learning, 8*(1–2), 86–111.

Curran, R., & Gualtieri, M. (2016). Forrester Brief: Cognitive search is ready to rev up your enterprise's IQ.

Draganidis, F., & Mentzas, G. (2006). Competency based management: A review of systems and approaches. *Information management and computer security, 14*(1), 51–64.

Elia, G., Margherita, A., Taurino, C., & Damiani, E. (2008). A Web 2.0 Platform Supporting Collaborative Development of Personal Skills. In *Database and Expert Systems Application, 2008. DEXA'08. 19th International Workshop on*, IEEE.

Firestone, J. M. (2000). Enterprise knowledge portals: What they are and what they do. *Knowledge and Innovation: Journal of the KMCI, 1*(1), 85–108.

Grudin, J. (2006). Enterprise knowledge management and emerging technologies. In *System Sciences, 2006. HICSS'06. Proceedings of the 39th Annual Hawaii International Conference on*, IEEE.

Hansen, M. T., Nohria, N., & Tierney, T. (2005). What's your strategy for managing knowledge. *Knowledge management: Critical perspectives on business and management, 77*(2), 1–10.

Hassell, J., Aleman-Meza, B., & Arpinar, I. B. (2006). Ontology-driven automatic entity disambiguation in unstructured text. In *International Semantic Web Conference*. Springer.

Hauder, M., Pigat, S., & Matthes, F. (2014). Research challenges in adaptive case management: A literature review. In *Enterprise Distributed Object Computing Conference Workshops and Demonstrations (EDOCW), 2014 IEEE 18th International*, IEEE.

Heil, S., Wild, S., & Gaedke, M. (2014). Collaborative adaptive case management with linked data. In *23rd International Conference on World Wide Web*, ACM.

Heisig, P. (2009). Harmonisation of knowledge management–comparing 160 KM frameworks around the globe. *Journal of knowledge management, 13*(4), 4–31.

Herrmann, C., & Kurz, M. (2011). Adaptive case management: supporting knowledge intensive processes with IT systems. *S-BPM ONE-Learning by Doing-Doing by Learning,* 80–97.

Hinkelmann, K., & Pierfranceschi, A. (2014). Combining process modelling and case modeling. In *8th International Conference on Methodologies, Technologies and Tools enabling e-Government MeTTeG14*.

Holsapple, C. W., & Joshi, K. D. (2004). A formal knowledge management ontology: Conduct, activities, resources, and influences. *Journal of the Association for Information Science and Technology, 55*(7), 593–612.

Huber, S., Fietta, M., & Hof, S. (2015). Next step recommendation and prediction based on process mining in adaptive case management. In *7th International Conference on Subject-Oriented Business Process Management*, ACM.

Kammergruber, W. C., Brocco, M., Groh, G., & Langen, M. (2010). Collaborative Lightweight Ontologies in Open Innovation-Networks. In *Competence Management for Open Innovation: Tools and IT Support to Unlock the Innovation Potential Beyond Company Boundaries, 30*, 93.

Kurz, M., Schmidt, W., Fleischmann, A., & Lederer, M. (2015). Leveraging CMMN for ACM: Examining the applicability of a new OMG standard for adaptive case management. In *7th International Conference on Subject-Oriented Business Process Management*, ACM.

Levy, M. (2009). WEB 2.0 implications on knowledge management. *Journal of knowledge management, 13*(1), 120–134.

Lin, H., Davis, J., & Zhou, Y. (2009). An integrated approach to extracting ontological structures from folksonomies. In *European Semantic Web Conference*, Springer.

Maier, R., & Peinl, R. (2017). Enterprise Knowledge Infrastructures for Organizational Resilience. In *12th International Forum on Knowledge Asset Dynamics (IFKAD'17)*, St. Petersburg, Russia.

Maier, R., & Remus, U. (2003). Implementing process-oriented knowledge management strategies. *Journal of knowledge management, 7*(4), 62–74.

Maier, R., & Schmidt, A. (2015). Explaining organizational knowledge creation with a knowledge maturing model. *Knowledge Management Research and Practice, 13*(4), 361–381.

Malhotra, Y. (2004). Why knowledge management systems fail: enablers and constraints of knowledge management in human enterprises. In *Handbook on Knowledge Management 1*, Springer: 577–599.

Maynard, D., Yankova, M., Kourakis, A., & Kokossis, A. (2005). Ontology-based information extraction for market monitoring and technology watch. In *ESWC Workshop" End User Aspects of the Semantic Web")*, Heraklion, Crete.

McTear, M., Callejas, Z., & Griol, D. (2016). The conversational interface - Talking to Smart Devices. Springer.

Motahari-Nezhad, H., Bartolini, C., Graupner, S., & Spence, S. (2012). Adaptive case management in the social enterprise. In *10th International Conference on Service-Oriented Computing*.

Motahari-Nezhad, H. R., & Swenson, K. D. (2013). Adaptive case management: Overview and research challenges. In *Business Informatics (CBI), 2013 IEEE 15th Conference on*, IEEE.

Newell, S. (2015). Managing knowledge and managing knowledge work: What we know and what the future holds. *Journal of Information Technology, 30*(1), 1–17.

Nonaka, I., Von Krogh, G., & Voelpel, S. (2006). Organizational knowledge creation theory: Evolutionary paths and future advances. *Organization studies, 27*(8), 1179–1208.

North, K. (2016). Die Wissenstreppe. In *Wissensorientierte Unternehmensführung* (pp. 33–65). Springer.

Ochsenkühn, C., & Peinl, R. (2015). Collaborative process maturing support by mining activity streams. In *15th International Conference on Knowledge Technologies and Data-driven Business*, ACM.

Osuszek, Ł., & Stanek, S. (2016). The Evolution of Adaptive Case Management from a DSS and Social Collaboration Perspective. In *Information Technology for Management* (pp. 3–16). Springer.

Peinl, R. (2011a). Knowledge management instruments. In *11th International Conference on Knowledge Management and Knowledge Technologies (iKnow11)*, Graz: ACM.

Peinl, R. (2011b). Unified information access. *Informatik-Spektrum, 34*(6), 594–597.

Peinl, R. (2015). Supporting Knowledge Management Instruments with Composable Micro-Services. In *Wissensgemeinschaften - ProWM 2015*, Dresden, Germany.

Peinl, R. (2017). Knowledge Management 4.0—Lessons Learned from IT Trends Knowledge Management. In *9. Konferenz Professionelles Wissensmanagement (Professional Knowledge Management)*, Karlsruhe, Germany.

Peinl, R., & Ochsenkühn, C. (2015). Social Media-integrated Collaboration Systems for Business Use. In *2nd European Conference on Social Media, ECSM 2015*, Porto, Portugal.

Priebe, T., & Pernul, G. (2003). Towards integrative enterprise knowledge portals. In *12th International Conference on Information and Knowledge Management*, ACM.

Probst, G., Romhardt, K., & Raub, S. (2000). *Managing knowledge: Building blocks for success*, J. Wiley.

Rangiha, M. E., Comuzzi, M., & Karakostas, B. (2015). Role and task recommendation and social tagging to enable social business process management. In *International Conference on Enterprise, Business-Process and Information Systems Modeling*, Springer.

Remus, U., & Schub, S. (2003). A blueprint for the implementation of process-oriented knowledge management. *Knowledge and Process Management, 10*(4), 237–253.

Schaffert, S. (2006). IkeWiki: A semantic wiki for collaborative knowledge management. Enabling Technologies: Infrastructure for Collaborative Enterprises, 2006. WETICE'06. In *15th IEEE International Workshops on*, IEEE.

Stocker, A., & Tochtermann, K. (2009). Enterprise wikis–types of use, benefits and obstacles: A multiple-case study. In *International Joint Conference on Knowledge Discovery, Knowledge Engineering, and Knowledge Management*, Springer.

Sultan, N. (2013). Knowledge management in the age of cloud computing and Web 2.0: Experiencing the power of disruptive innovations. *International Journal of Information Management, 33*(1), 160–165.

Traganos, K., & Grefen, P. (2015). Hybrid service compositions: when BPM meets dynamic case management. In *European Conference on Service-Oriented and Cloud Computing*, Springer.

Uren, V., Cimiano, P., Iria, J., Handschuh, S., Vargas-Vera, M., Motta, E., Ciravegna, F. (2006). Semantic annotation for knowledge management: Requirements and a survey of the state of the art. *Web Semantics: Science, Services and Agents on the World Wide Web*, *4* (1), 14–28.

Van Baalen, P., Bloemhof-Ruwaard, J., & Van Heck, E. (2005). Knowledge sharing in an emerging network of practice: The role of a knowledge portal. *European Management Journal, 23*(3), 300–314.

Varshney, K. R., Wang, J., Mojsilovic, A., Fang, D., & Bauer, J. H. (2013). Predicting and recommending skills in the social enterprise. In *AAAI ICWSM Workshop on Social Computing*. Workforce.

Weber, N., Schoefegger, K., Bimrose, J., Ley, T., Lindstaedt, S. N., Brown, A., et al. (2009). *Knowledge Maturing in the Semantic MediaWiki: A Design Study in Career Guidance*. EC-TEL: Springer.

Westerski, A., & Iglesias, C. A. (2011). Exploiting structured linked data in enterprise knowledge management systems: An idea management case study. In *Enterprise Distributed Object Computing Conference Workshops (EDOCW), 2011 15th IEEE International*, IEEE.

Woitsch, R., & Karagiannis, D. (2005). Process oriented knowledge management: A service based approach. *J. UCS, 11*(4), 565–588.
Yang, S.-H. (2009). Using blogs to enhance critical reflection and community of practice. *Journal of Educational Technology and Society,* ***12***(2).
Yew, Wong K. (2005). Critical success factors for implementing knowledge management in small and medium enterprises. *Industrial Management and Data Systems, 105*(3), 261–279.
Zhu, W.-D. J., Foyle, B., Gagné, D., Gupta, V., Magdalen, J., Mundi, A. S., Nasukawa, T., Paulis, M., Singer, J., & Triska, M. (2014). *IBM Watson Content Analytics: Discovering Actionable Insight from Your Content*, IBM Redbooks.

Author Biography

René Peinl Dr., leads the research group systems integration at the Institute of Information Systems and is a Full Professor at Hof University of Applied Sciences. He studied business information systems and got his Ph.D. in knowledge management with a dissertation about knowledge transfer. His research interests include web technologies, semantic web, cloud computing, digital transformation and the Internet of Things.

Digital Knowledge Mapping

Sebastian Kernbach and Sabrina Bresciani

Abstract In this chapter, we propose that visual knowledge mapping is a very effective way of sharing, integrating and creating knowledge and value for collaborative work in organizations. We present online visual collaboration tools that enable digital change in organizations and present the theoretical bases which explains the benefits of visualization for facilitating digital collaboration. We provide illustrative examples and further propose a classification of ten visual tools to show organizations what these tools are useful for, and which criteria are relevant to assess and select visual collaborative tools. We conclude with key learnings and a checklist for the integration of visual collaboration tools in organizations.

1 Introduction

The challenge for organizations to effectively manage information and knowledge, and enable knowledge-based value creation lies in the way organizations enable their employees to collaborate. In this article, we propose that visual knowledge mapping is a very effective way of sharing, integrating and creating knowledge and thus creating value in collaboration for organizations.

Collaboration is one of the key drivers for innovation, it is therefore critical for organizations to enable their employees to collaborate effectively and efficiently. Google, for example, considers collaboration as the key success factor for innovative organizations and aims to translate everything that is possible in physical meetings into virtual meetings through virtual collaboration. The evidence of the importance of virtual collaboration is further supported through new products, such as Google Jamboard, Microsoft Surface Hub and the Ricoh Cognitive Whiteboard in collaboration with IBM Watson technology. These products are large touch

S. Kernbach (✉) · S. Bresciani
University of St. Gallen, St. Gallen, Switzerland
e-mail: sebastian.kernbach@unisg.ch

K. North et al. (eds.), *Knowledge Management in Digital Change*, Progress in IS,
https://doi.org/10.1007/978-3-319-73546-7_8

screens with features to support online collaboration and to make meetings more productive. At the core of these products is the use of visualization for online collaboration which shows that visualization is and will be even more important in the future to support online collaboration.

However, those new devices (hardware) alone will not make collaboration more productive, it is the people and the new tools (software) that will make the most out of these devices. Also for the standard devices today such as laptops and projectors, it is important to understand the value of visualization for collaboration. We thus need to understand better how visual tools can support individuals, teams and organizations in collaborating visually and virtually.

2 The Value of Visualization

The digitalization of work enables collaboration in new powerful ways. We can hold meetings supported by digital collaborative systems that not only support remote collaboration—allowing people from different places to work together effectively—but also for co-located work, with all meeting participants in the same room (Eppler and Kernbach 2016). Digital knowledge sharing tools can provide guidance to the meeting by collecting input from participants, keeping a record of participants' contributions (that can easily be shared online or by email), assessing options with voting systems or mapping different opinions. In this context, visualization emerges as a powerful way to support knowledge work and through mapping thoughts, discussants place elements and relations in graphic space to convey concepts and their relations (Kernbach 2015). Doing so, they utilize visuospatial reasoning (Tversky 2005) to share their knowledge and to draw inferences. Visualizing concepts has the further advantage of reducing information overload by externalizing knowledge and thus offloading memory (Mengis and Eppler 2006).

The human brain processes visual information more efficiently compared to written information, as it is shown that when the same information is provided both in written and visual form (such as with a key-word and a corresponding icon), performances are enhanced (Kernbach, et al. 2015). According to Dual Coding Theory (Paivio 1991) this effect is due to the fact that our brain processes visual information and verbal/textual information in two different areas of the brain. When both channels are used together, performances are enhanced and people understand and remember a concept better compared to when a concept is given in only verbal or only visual format (Paivio 1991).

More specific to the organizational context, utilizing visual mapping to facilitate meetings—for example with visual templates, diagrams, sketches or Navicons (Eppler, et al. 2015)—can improve meeting productivity (Bresciani and Eppler 2009) and make knowledge sharing more precise (Bresciani and Comi 2017). A core benefit of visualization in collaborative knowledge work is that it provides Representational Guidance (Suthers 2001). According to this principle, important guidance for interaction is given by the ways in which a representation affords or constrains what can be represented. Suthers (2001) argues that "The visual presence of the knowledge unit in the shared representational context serves as a reminder of its existence and any work that may need to be done with it." (p. 7). Also, it is easier to refer to a knowledge unit that has a visual manifestation, so learners will find it easier to express their subsequent thoughts about this unit than about those that require complex verbal descriptions (Clark and Brennan 1991).

Despite the numerous and well known advantages of visualization, working visually is not yet mainstream in organizations, possibly due to the fact that managers often do not feel comfortable drawing in a business context or do not have the graphic skills to develop their own visuals. In addition, managers and knowledge workers often utilize software they are already familiar with, even for purposes for which the software is not intended. This common phenomenon is known as "reappropriation" and is discussed in Adaptive Structuration Theory (DeSanctis and Poole 1994). A typical example of this phenomenon is the widespread use of PowerPoint for meeting facilitation, for creating reports and for visualizing information and knowledge. Although Microsoft PowerPoint is a presentation software, knowledge workers often utilize it for a wide range of purposes because they are used to working with it and they don't want to take the effort to look for, and learn, a specific software for collaboration. However, the reappropriation of software not originally intended for collaborative work in organizations leads to inefficient practices (Kernbach, Bresciani and Eppler 2015).

Recent developments in technology now enable everyone with a laptop to utilize visualization for knowledge work, without having to learn specific skills in design. Knowledge mapping tools typically have pre-loaded templates, icons and provide built-in support for collaboration (i.e., tracing who contribute what, allowing annotation and comments on other people's contribution). These digital visual tools or apps promise disruptive change in the way we manage knowledge in organization by enabling novel ways of working visually.

In the next section we will provide some examples of such visual tools or applications that can be easily used on a laptop for enabling visually supported collaboration. In addition to briefly describing the tools, we will provide a list of criteria useful for comparing and selecting tools.

3 Digital Knowledge Sharing with Visual Tools

We review a number of illustrative and typical visual tools for organizational knowledge work that are currently on the market. The list is not exhaustive: the purpose is to give an overview of what visual apps can do, which types of collaboration they can support and according to which criteria can we describe and differentiate these apps. We have selected the following tools: (1) Conceptboard, (2) Groupmap, (3) Let's focus, (4) Lucidchart, (5) Mural.ly, (6) Pinterest, (7) PowerPoint, (8) Prezi, (9) Realtimeboard, and (10) Spacedeck.

We have selected the following criteria because we consider them as relevant for organizations to select, test and implement visual tools for their collaboration practices: (a) easy-to-learn/easy-to-use, (b) flexibility (one or multiple purposes), (c) cost, (d) collaborative support (multiple users can edit it, history, annotations, …), (e) visual guidance (pre-loaded templates), and (f) support for different formats. The criteria are relevant from a theoretical perspective as they provide a vocabulary for the description of visualization tools. They are useful from a practical point of view, as they can provide managers and knowledge workers with a key decision criteria per selecting the most appropriate tools for their meetings and collaborative work. In the next section, we provide a brief description of each visual app and an assessment based on these criteria.

4 Visual Tool Assessment

In this chapter, we assess ten visual tools to support visual collaboration. For each tool, alongside the previously identified criteria, screenshots and business examples illustrate the visual appeal and use of the visual tools as well as their strengths and weaknesses. The tools are presented in alphabetical order:

4.1 *Conceptboard (Conceptboard.Com)*

Conceptboard (Fig. 1) is a visual board to help teams bring different types of media on one empty canvas and collaboratively discuss the content by adding comments, giving tasks, moderating the content, having written, audio or video chats.

Fig. 1 Screenshots Conceptboard

a. Easy-to-learn/easy-to-use: Middle, the interface is clean and gives a good overview, the lack of templates makes it a bit tricky to start, the many options for collaborative support can be overwhelming at times.
b. Flexibility (one or multiple purposes): High, many purposes, in particular for co-creating and reviewing visuals.
c. Costs: 28 USD per month for three users, 590 USD per month for ten users, 1700 USD per month for 100 users.
d. Collaborative support: Very high, especially through comment function including tagging people, live pointers, see who is online, task management, alerts, chat function, live moderation function, audio chat, video chat, screen sharing.
e. Visual guidance (pre-loaded templates): Low, sticky notes, texts and files can be added by drag & drop but no templates are available.
f. Support for different formats: Very high—image, video, audio, any type of data, import from many sources such as Dropbox, Google Drive, One Drive, Box, ImageRelay, etc.

Strengths: Good video introduction and interactive tutorial, comment function incl. tagging other people, live pointers to see where everybody is at.

Weaknesses: The board can be overwhelming at times with many options to integrate content and the menu on the top and on the left side, lack of templates, very focused on images as input factor.

Business example (Fig. 2):

Building a children home in Bangladesh with the non-profit "Kids underneath the mangotree".

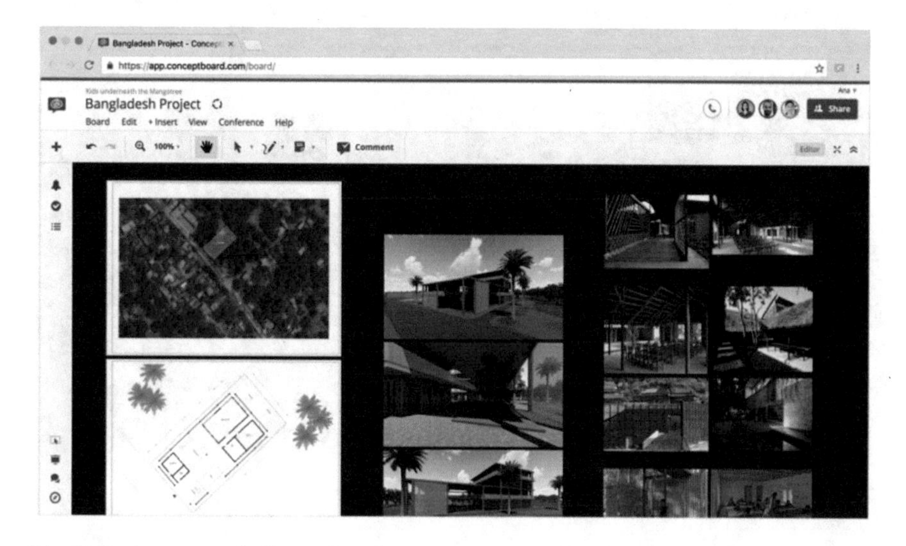

Fig. 2 Business example Conceptboard

4.2 *Groupmap (Groupmap.Com)*

Groupmap (Fig. 3) is an online collaboration platform which incorporates several templates and advanced online collaboration functionalities such as voting, grouping ideas, (partial) anonymous contributions, commenting, chatting and exporting the results for the creation of reports.

a. Easy-to-learn/easy-to-use: the app has several functionalities that needs to be learned.
b. Flexibility (one or multiple purposes): High.
c. Cost: 20$ month for 10 participants and basic design; 60$ month for 50 participants per map and advanced functionalities; from 100$ per month for organizations (unlimited participants).

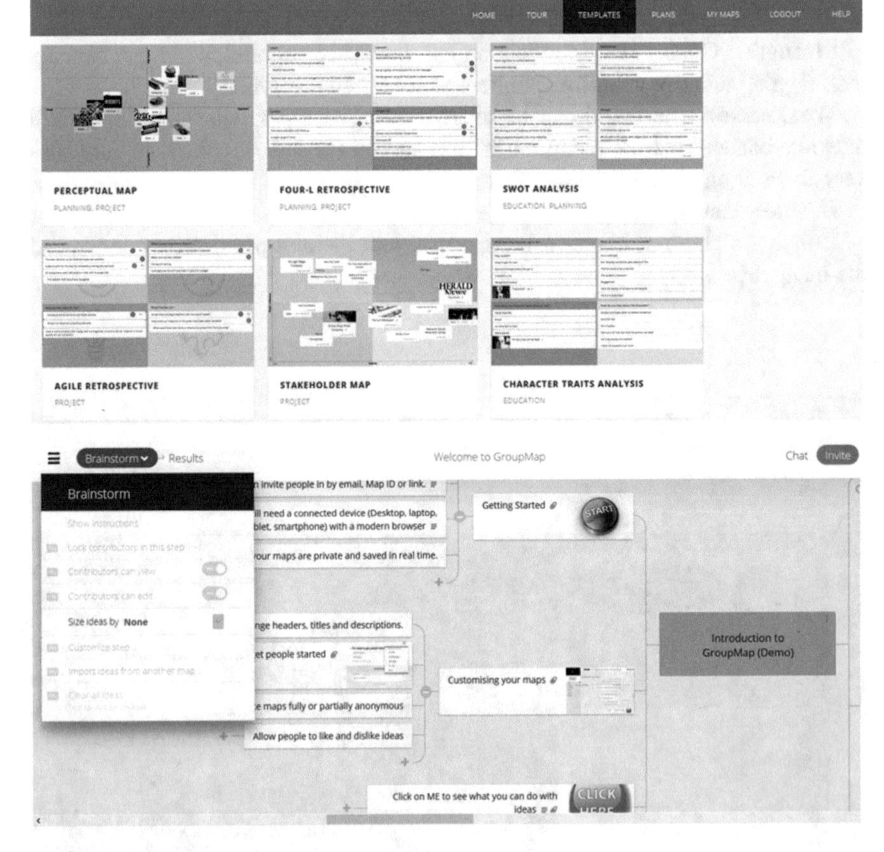

Fig. 3 Screenshots Groupmap

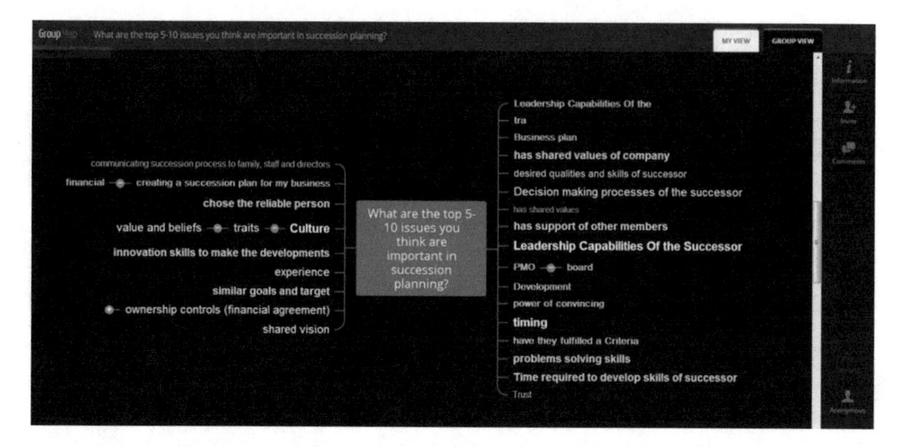

Fig. 4 Business example Groupmap

d. Visual guidance (pre-loaded templates): Several pre-loaded templates belonging to four categories: mind maps, charts, Canvases, lists. Each template is explained in details so that users understand how to use it.
e. Collaborative support: Advanced, users can comment, vote (like and dislike), dot voting, and decide actions for each entry; chat function; the facilitator can make parts of the map anonymous. It supports full and partial anonymity (i.e., only the facilitation can see who is writing what; participants can firstly see only their ideas and then all the group participants ideas).
f. Support for different formats: it supports only text entries eventually with attachments of documents (including Word): it's not possible to add free hand drawing or icons.

Strengths: Groupmap offers advance collaborative support and several templates; resulting maps can be exported as pdf, CSV or XLS.

Weaknesses: There is no free version.

Business example (Fig. 4):

Succession planning by John Broons—family business expert.

4.3 Let's Focus (Lets-Focus.Com)

Let's focus (Fig. 5) is a moderation and meeting facilitation tool that allows to make knowledge visible and discussable. It provides abstract and pictorial templates and allows to integrate text, icons, and arrows. This app helps individuals and teams in preparing, conducting, and documenting meetings as well as discussions, conferences, or work sessions.

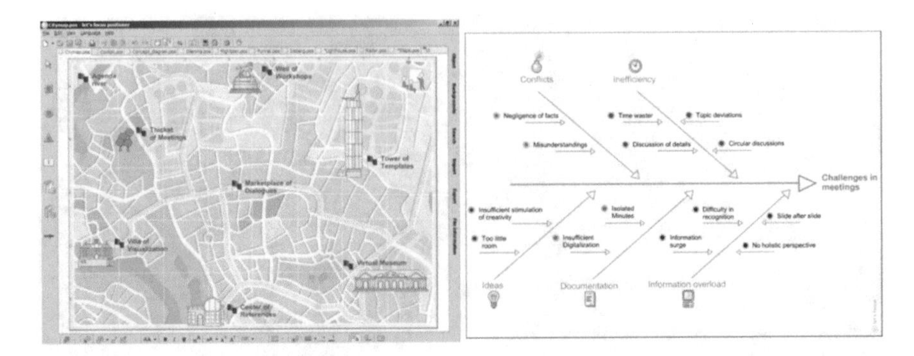

Fig. 5 Screenshots Let's focus

a. Easy-to-learn/easy-to-use: The functionalities are easy to find, the access to backgrounds and symbols is very intuitive.
b. Flexibility (one or multiple purposes): Medium, the app focuses on collaboration.
c. Costs: one license for 197.50 EUR, ten licenses for 1777.50 EUR.
d. Visual guidance (pre-loaded templates): Large number of abstract and metaphorical templates.
e. Collaborative support: Medium, in the cloud version remote collaboration is possible but the program is rather designed to help co-located teams to collaborate through a facilitator.
f. Support for different formats: Low, images can be integrated.

Strengths: Many templates, abstract and metaphorical templates, replay function to review the development of the visualization.

Weaknesses: Only few examples are given on how to use the templates, and the colorful style of the software and of the templates looks very playful for a business context.

Business example (Fig. 6):

Books on Visual Literacy, provided by visual-literacy.org

4.4 Lucidchart (Lucidchart.Com)

It is a business process mapping software (Fig. 7) which offers collaborative support. It has 8 million users (as of 2017). It is an advanced mapping software with several templates, which offers advanced integration with major software (i.e., Visio, MS Office, Google suite and Slack) but with basic collaboration functionalities.

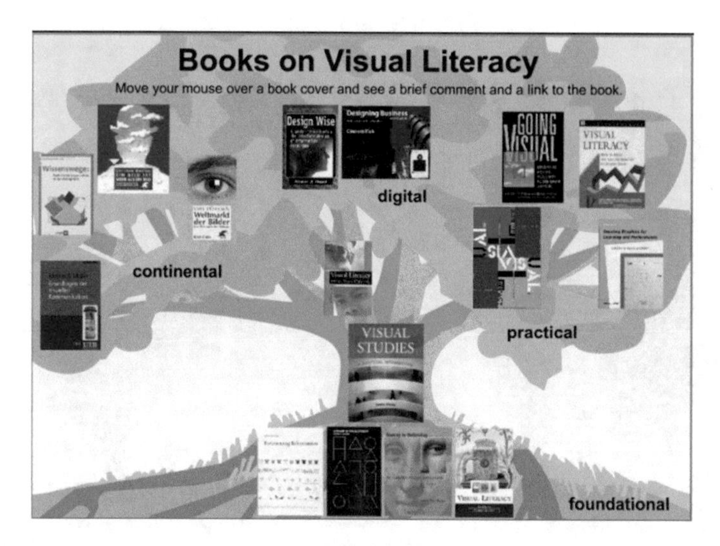

Fig. 6 Business example Let's focus

a. Easy-to-learn/easy-to-use: It is easy to use as the menu is self-explanatory and rich of options.
b. Flexibility (one or multiple purposes): the several templates provide support for a variety of tasks. The maps are easy to customize.
c. Cost: 4.95$/month/user (very basic) to 20$/month/3users.
d. Collaborative support: With the "Team" version there are team functions such as commenting and chat.
e. Visual guidance (pre-loaded templates): The software and apps offer a large number of templates belonging to several diversified categories (i.e., flowcharts, BCG matrix, argument map, mind map, Venn Diagrams, Building blueprints). Lucidchart also offers a number of shapes and color customization.
f. Support for different formats: The Team version integrates with Visio, Google suite, Slack, PDF, JPEG or PNG. There is also a mobile app to create diagrams on mobiles (Pro version).

Strengths: Several templates for different purposes; impressive integration with other software packets.

Weaknesses: Collaboration functionalities are basic.

Business example (Fig. 8):

Sales process switch.

Screenshots:

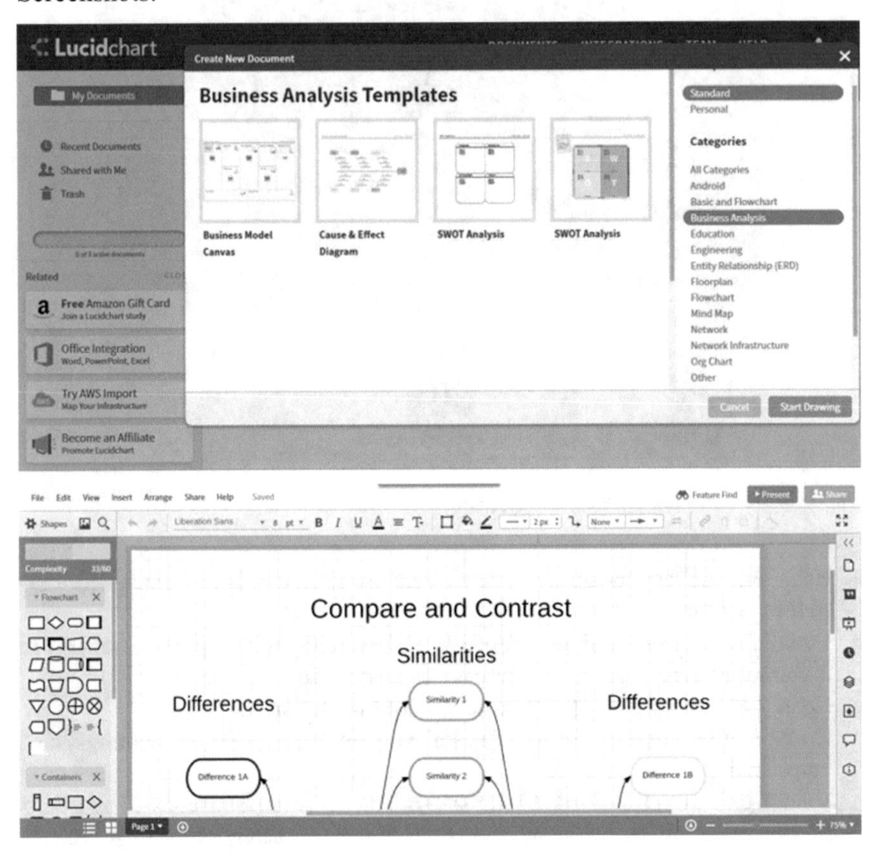

Fig. 7 Screenshots Lucidchart

4.5 *Mural.Ly (Mural.Co)*

Mural.ly (Fig. 9) is a large white canvas on which individuals and teams visualize their knowledge in many different ways, ranging from simple sticky notes, over images and icons to simple and sophisticated templates. Users can connect items with arrows and conduct voting sessions.

a. Easy-to-learn/easy-to-use: Very intuitive interface and all important functions easily accessible.
b. Flexibility (one or multiple purposes): High, you can use it for many purposes.
c. Costs: 12 USD per member per month.
d. Collaborative support: High, simultaneous collaboration, chat function, activity log, anonymous voting session.

Sales process switch

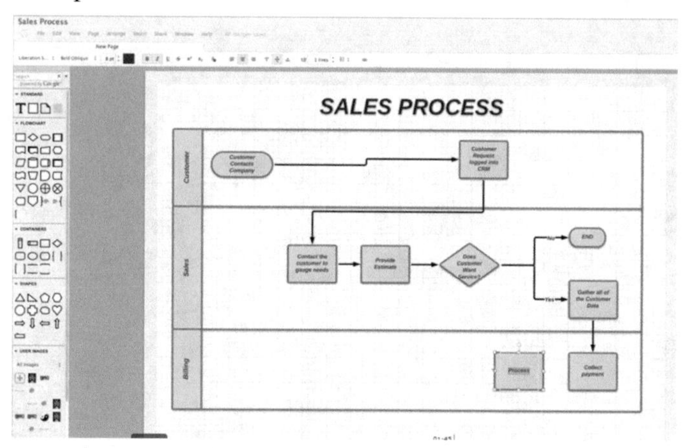

Fig. 8 Business example Lucidchart

e. Visual guidance (pre-loaded templates): Very high with a large number of templates categorized into layouts, design, agile, business, calendars.
f. Support for different formats: images, drag and drop any type of file, it appears as an icon and you can open it by double-clicking on it.

Strengths: Many ways to visualize knowledge, different types of sticky notes but also various templates, insert all types of media which appear as icons to open by clicking on them.

Weaknesses: The templates (called frameworks) are useful but there could be a user guidance to access the templates and guide use to choose the right template for the purpose at hand. The whiteboard is endless which can lead to an information overload when using it.

Business examples (Fig. 10):

Design thinking at IBM

MIT Student turned Global Entrepreneur.

4.6 Pinterest (Pinterest.Com)

Pinterest (Fig. 11) is the fastest growing social media platform and is possibly the most visual of all social media. Its main purpose is to allow people to share interesting visual content (images and videos) found on the web, while providing the direct link to the source. The platform utilizes the metaphor of the "Pinboard".

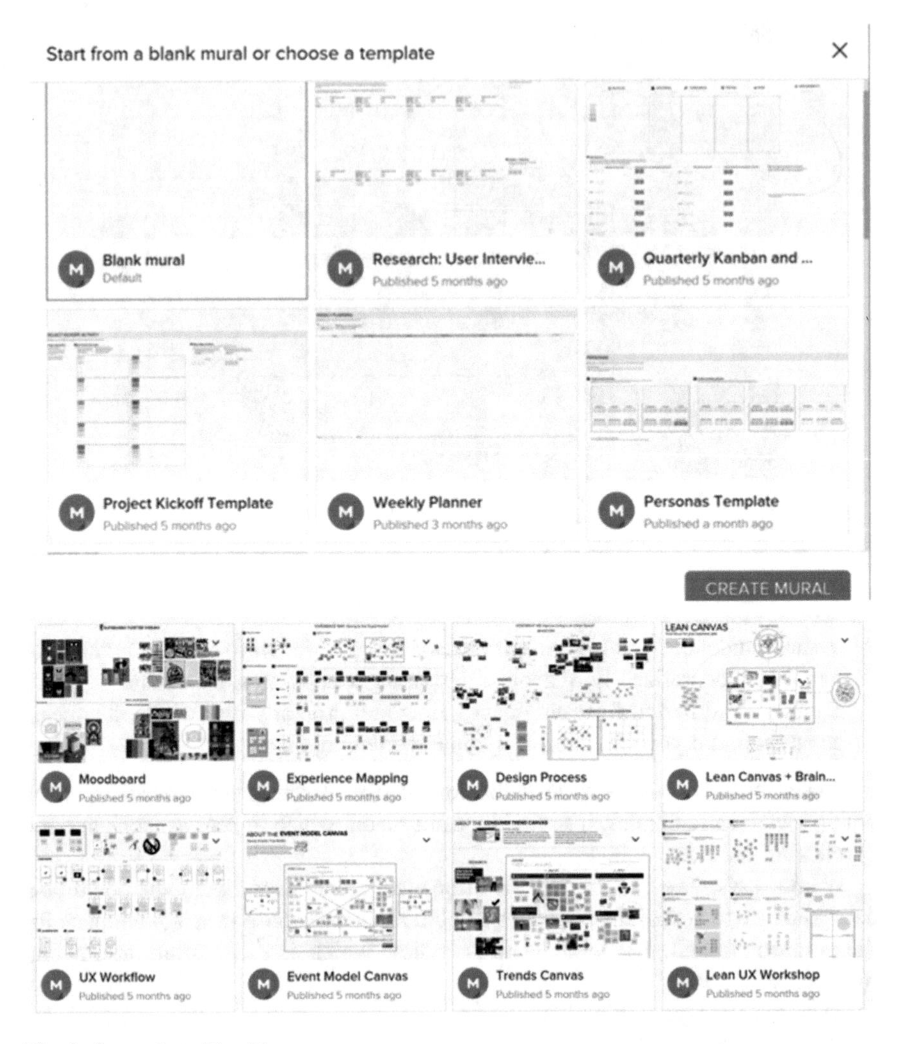

Fig. 9 Screenshots Mural.ly

Users can "pin" an image from the web or from another user's board on their board. For instance, a person might create a board on a specific topic, such as ideas for corporate events, and pin on that board images and videos from the web, from other Pinterest's pinboards and can even upload his or her images. Each image can have a description and tags, and other users can comment on it.

Organizations are re-appropriating (DeSanctis and Poole 1994) the platform by collaboratively creating pinboards with their colleagues to collect inspirations about specific topics, or to interact with clients asking their input or their opinion.

a. Easy-to-learn/easy-to-use: There are limited options that can be learned quickly. Users can create pinboards and pin a picture from the web, repin an image from

EXAMPLE: IBM

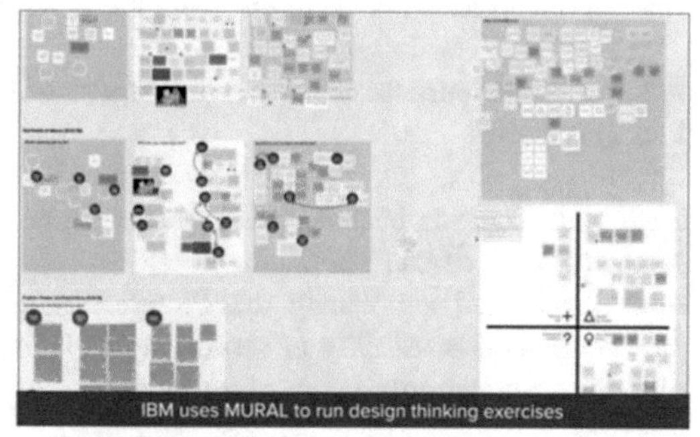

MIT Student turned Global Entrepreneur

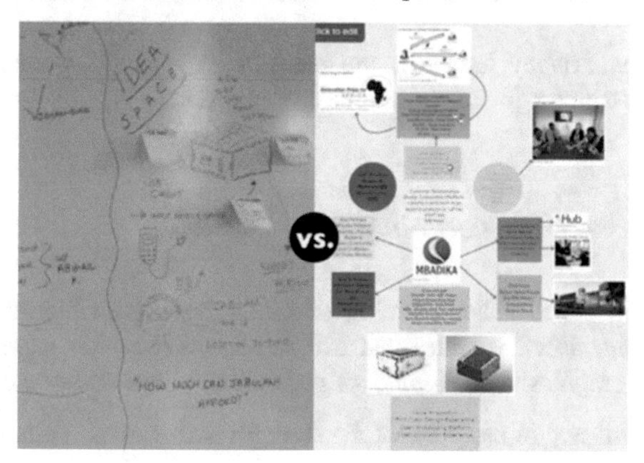

Fig. 10 Business examples Mural.ly

Fig. 11 Screenshots Pinterest

Pinterest images or upload their own. For each of their images they can add tags and a description, and they can comment on any image. Users can search images in Pinterest through the typical (textual) search engine or with the visual search tool.

b. Flexibility (one or multiple purposes): Low. The purpose of the app is to create pinboards on specific topics.
c. Cost: free
d. Collaborative support: There is basic support for collaboration; multiple users can edit the board (if invited) and comment.
e. Visual guidance (pre-loaded templates): the Pinterest algorithm arranges the images and the user has no possibility or reordering or laying out the content
f. Support for different formats: only images and videos

Strengths: The visual search. Pinterest developed a visual search engine to search picture similar to the source we provide or even a part of a picture. The "visual search tool" is currently far from perfect, but certainly a disruptive innovation.

Weaknesses: Privacy issues as the platform is a social network and is not conceived as a support for organizational meetings.

4.7 PowerPoint (Powerpoint.Com)

PowerPoint (Fig. 12) is a slide-based presentation tool that enables users to allocate their content into different slides that can be shown one after another. Slide can represent text, graphics and all kinds of media like video, audio and websites.

a. Easy-to-learn/easy-to-use: One of the strength is that users are familiar with the setup of PowerPoint.
b. Flexibility (one or multiple purposes): Low to moderate—its primarily function is presentations.
c. Costs: 110 USD for one computer.
d. Collaborative support: Low to moderate, asynchronous collaboration through comments and tracking of changes.
e. Visual guidance (pre-loaded templates): Low to moderate, SmartArt Graphics, diagrams and forms.
f. Support for different formats: Moderate to high, import of pictures, videos, audio files, web pages.

Strengths: Very widespread, the "standard" when it comes to presentation software, showing multiple types of data on the slides, online version, mobile app, easy-to-use.

Weaknesses: Content display limited to slides, no real-time online collaboration, and lack of integration of files from other sources such as Dropbox.

Screenshots:

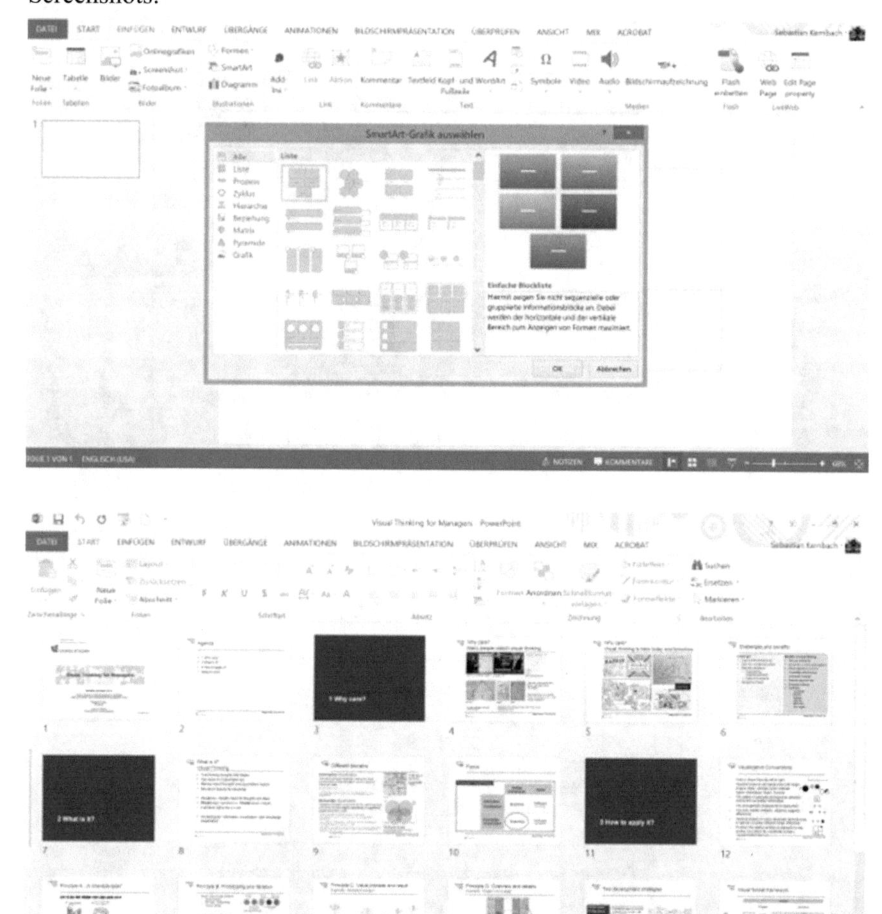

Fig. 12 Screenshots PowerPoint

4.8 Prezi (Prezi.Com)

Prezi (Fig. 13) is a zoomable presentation software. Its main aim is to enable users to create non-linar presentations, braking the convention of the slideshow. The software can be used offline or online, and each presentation can be created in collaboration with multiple Prezi users. Prezi presentations can be viewed without having to install a viewer.

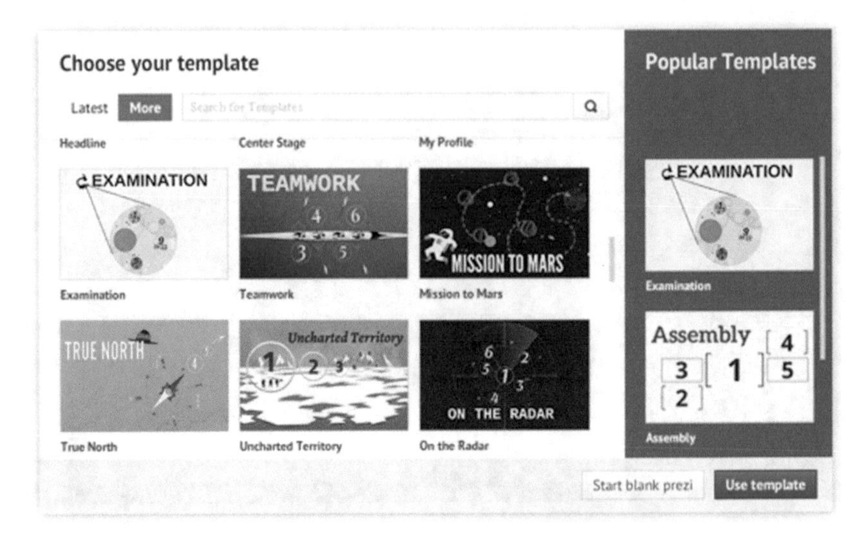

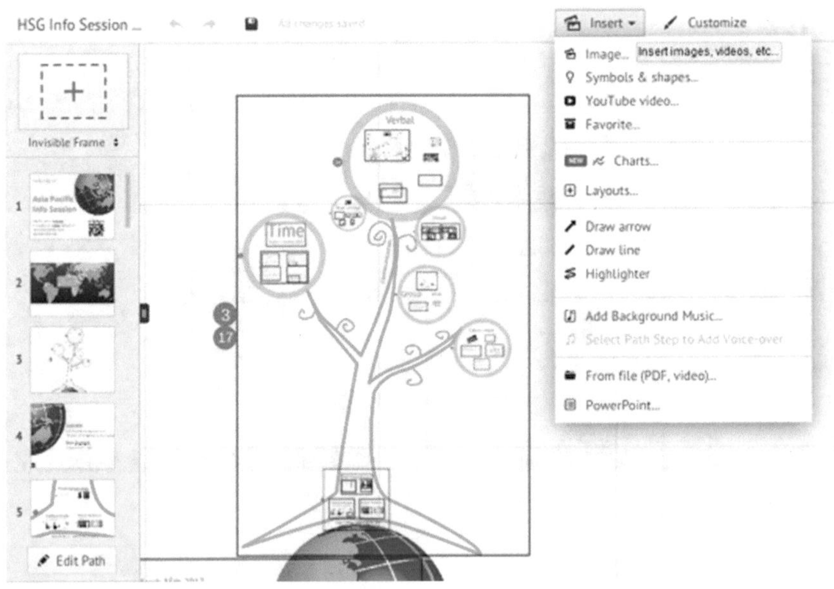

Fig. 13 Screenshots Prezi

a. Easy-to-learn/easy-to-use: Medium. Although it has a user-friendly interface, the logic of Prezi is "unusual" for the typical Office user and thus can pose challenges to novices.
b. Flexibility (one or multiple purposes): High. Prezi is flexible as it does not constrain the user to pre-defined visualization techniques. Personalized templates, icons and images can be imported, so that the user can create any type of visual.

c. Cost: Low (10–30 USD a month)
d. Collaborative support: Multiple user can edit the same file but they should have a Prezi account. There is no annotation or revisions system, nor a history of modifications.
e. Visual guidance (pre-loaded templates): Prezi provides a large number of different templates but they are not categorized by task or purpose, and there are no instructions on how they should be utilized.
f. Support for different formats: Images, videos, pdf, PPT, etc.

Strengths: Embedded image editing tool, large range of pre-loaded templates, icons and images.

Weaknesses: As the aim of the app is to create presentations, the templates are not specific for collaborative work in organizations. All contributing users need to have a paid account to be able to work together.

4.9 Realtimeboard (Realtimeboard.Com)

Realtimeboard (Fig. 14) is an online whiteboard for real-time collaboration which allows for ideation, project management and multiple other purposes. It allows you to drop information, comment on it and add files from various sources.

a. Easy-to-learn/easy-to-use: The use of templates makes it easy to not start from scratch, otherwise also very handy.
b. Flexible (one or multiple purposes): High, many purposes.
c. Costs: Free for up to three team members, 40 USD per month for five team members, company rate individually.
d. Collaborative support: High, comment function, chat, video chat, screen sharing, ability to present
e. Visual guidance (pre-loaded templates): Large number of templates, 72 at the time of review, organized in six categories, the category “popular” is interesting.
f. Support for different formats: Very high, direct import from many sources such as Dropbox, Google Drive, Capture Web Page, Goodle Image Search, Adobe Creative Cloud.

Strengths: Access through mobile device, good range of demo boards, creates graphs directly inside the board.

Weaknesses: The endlessness of the board can make it hard to keep the overview.

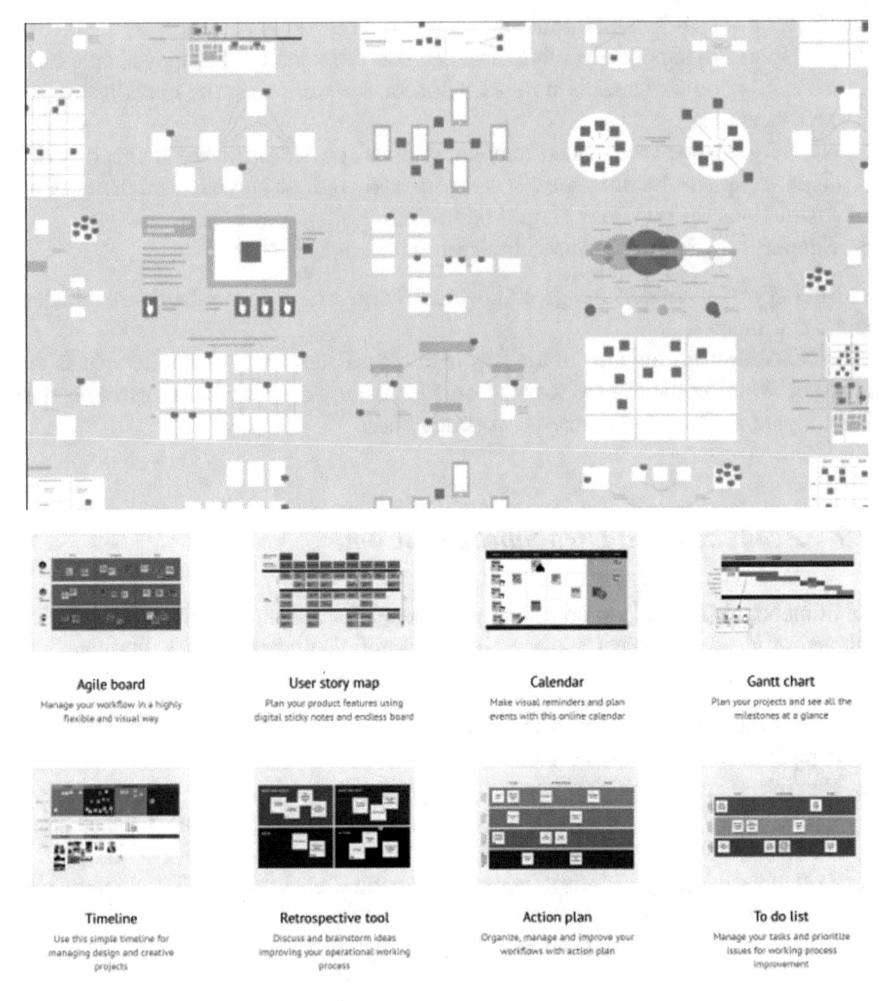

Fig. 14 Screenshots Realtimeboard

4.10 Spacedeck (Spacedeck.Com)

It is an online shared whiteboard (Fig. 15) that offers a web-based collaborative group-work platform. It was founded in Berlin in 2013. Users can type text, insert shapes, draw free-hand and import files (images, sound and video). It has a functionality for creating "zones" which allows to easily present the content of the whiteboard.

a. Easy-to-learn/easy-to-use: It is very easy to learn as the functionalities are rather basic.
b. Flexibility (one or multiple purposes): High, collaboration and presentations

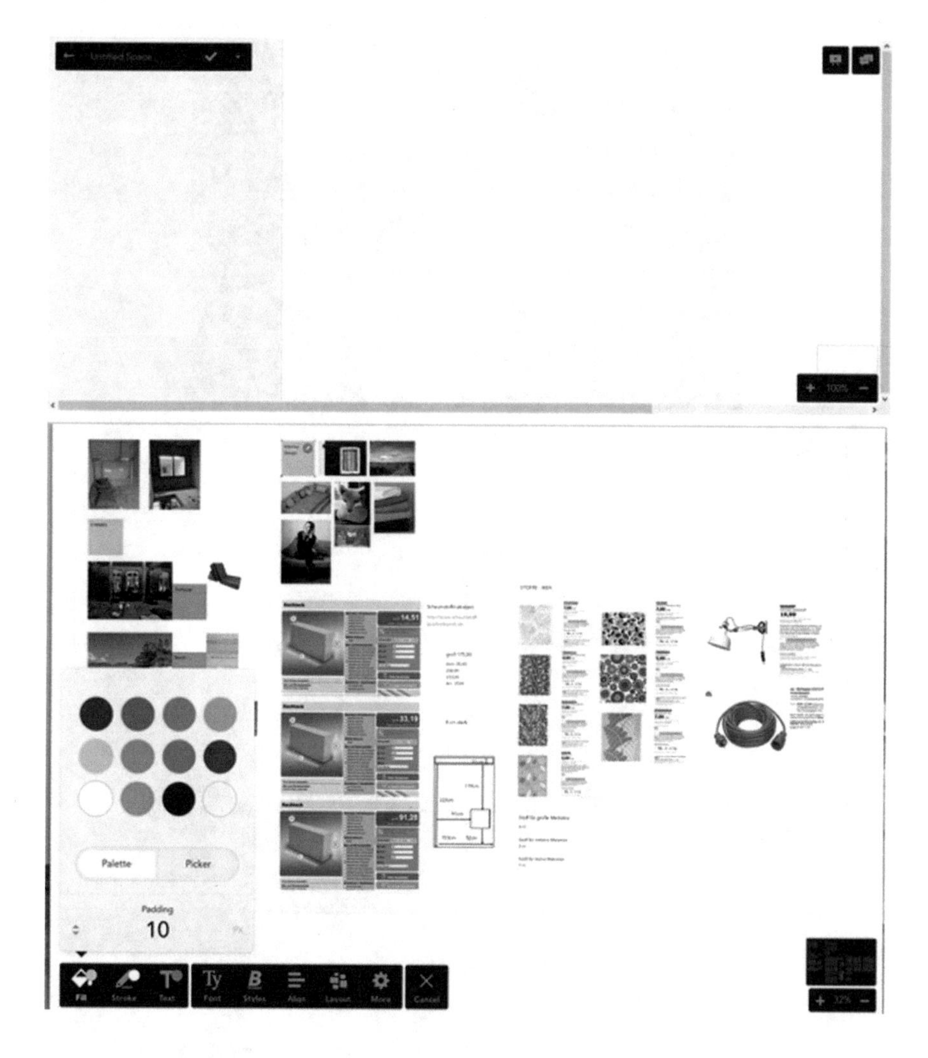

Fig. 15 Screenshots Spacedeck

c. Cost: Free version; 4.90 EUR per month for exporting files and removing watermark, additional space, importing custom background.
d. Collaborative support: simultaneous collaboration, chat function.
e. Visual guidance (pre-loaded templates): very low as there are no templates available. It allows freehand drawing and has a basic library of shapes.
f. Support for different formats: images, sounds and videos also via mobile.

Strengths: Presentation function with zones.

Weaknesses: There are no templates and it is not possible to insert or create tables or charts.

Business example (Fig. 16):

Brainstorming with Nike

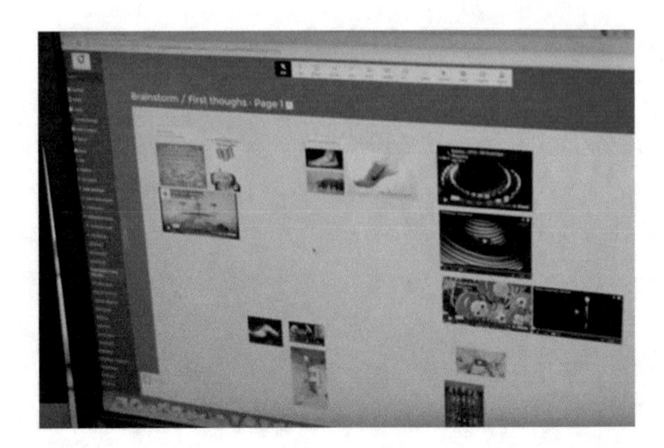

Fig. 16 Business example Spacedeck

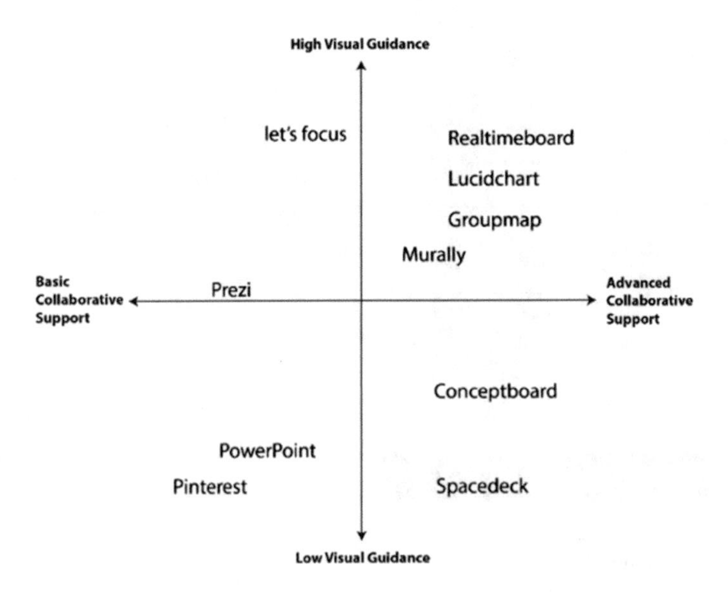

Fig. 17 Positioning of visual tools according to visual guidance and collaborative support

The assessment of those ten tools gives a first idea about how collaborating visually could look like, how organizations are using them and hopefully inspire organizations to think of more ways to collaborate visually.

In the next section, we provide an overview of the visual tools in terms of their visual guidance, which is the degree to which the tools give guidance to start collaborating visually straight away; e.g. through templates, and the collaborative support, which is the degree to which the tools provide basic or advanced support through chats, voting, screen sharing, among others (Fig. 17).

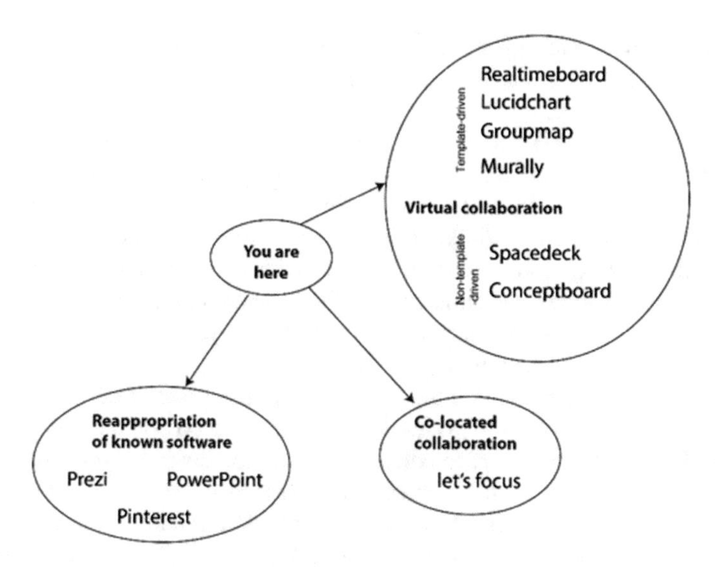

Fig. 18 Visual tool selection support

In addition, the following map (Fig. 18) helps to select a tool based on the type of collaboration, whether it is co-located or virtual and also the degree of visual richness.

After reviewing and positioning the ten visual tools, we conclude this chapter and provide practical implications including learnings for organizations for the use of visual tools for collaboration.

5 Conclusion and Outlook

Collaboration is a key driver for innovation in organizations. Since the amount of information will further increase, teams will be further distributed and the time to deliver results will decrease, the need for productive collaboration will increase. With this chapter, we want to show how visualization, in particular visual tools, can help to make collaboration more productive.

We hope to encourage the use of visual tools in organizations through the assessment of ten visual tools, by providing an overview for the selection of visual tools and by providing some theoretical background to explain why and how visualization can add value for collaboration.

However, often organizational constraints hinder the successful implementation of visual and other tools. Therefore, we provide organizations with a checklist of the ten things to support the successful integration and implementation.

Visual Collaboration Implementation Tool Checklist:

1. Using the right tool: In your next (virtual or co-located) group meeting, consider using a dedicated collaborative software rather than just speaking or using a software which was invented for a different purpose.
2. Using the tool right: Select your visual collaborative tools based on the collaborative support you need, is your meeting taking place with all participants in a room (i.e., co-located) at the same time or should people contribute from different location and at different time? Some software applications specifically support virtual synchronous collaboration with chat functions, while software for virtual a-synchronous collaboration have annotation. If you conduct mainly meetings with all participants in the same room, you might want to select a software specific for facilitating co-located work.
3. Thinking from the end—with the goal in mind: You should then consider which kind of topics you cover in your collaborative work: do you need to brainstorm and take decisions or rather to collect an inspiration board? Should the outcome of the collaboration be then shared and presented to others, or is the collaborative space used only for reaching a decision during the meeting? Some software have an embedded presentation mode, while others have a large number of templated to facilitate knowledge work (creativity, knowledge sharing and decision making) and voting.
4. Visual tool curator: Have someone in your organization with good knowledge about the problem and the visual tools to help solve them. This person can test different tools and can check requirements of the internal IT department to check how the implementation is possible within the IT system landscape.
5. Prototype the implementation: Have a small team who uses the new visual tool for 30 days and then debrief among the team and invite others to also benefit from the learnings about using the tool.
6. Informal exchange: Allow the prototype teams to share their best practices with larger audiences, have informal lunch introductions to the tool and experiences sharing.
7. Review issues and best practices on a regular basis: Make a monthly or bi-monthly meetup for sharing issues and best practices, make it fun, stand up meeting at 5 pm (with beers).
8. Consider pricing: Review and anticipate the costs involved with getting the license for these visual tools and make you have a pay plan that does not increase heavily if you add more users.

Do it as organizations such as the European Central Bank is doing it. Create a community of practice with employees from different departments, meet on a regular basis (e.g. once a month), and create a platform for more meaningful and memorable exchange in meetings and other collaboration forms. In this community, you can run pilots for online visualization collaboration, exchange about experiences and foster the use of visualization for collaborative practices.

The visual tools mentioned in this chapter have been chosen as they work easily, are low cost and can implemented in an easy way. The visual tools work on all devices such as mobile phones, tablets, computers, etc. which allows collaboration on the go. However, sometimes these apps suffer from the size of little screens and lack of visibility or not being able to display multiple things at the same time. The before mentioned devices such as the Google Jamboard, Microsoft Surface Hub, and the Cognitive Whiteboard by Ricoh and IBM Watson might be able to make the agenda of a meeting permanently visible, display slides and have an electronic whiteboard at the same time which might bring online visual collaboration to a new level.

We are curious to see not only the size of the screens and the resolution of displays but more specifically how these devices enhance existing visual tools such as the ones we present in this chapter as well as what new types of visual tools and new interactions possibilities they will offer to bring the way we collaborate and share and create knowledge to a new level.

References

Bresciani, S., & Comi, A. (2017). Facilitating culturally diverse groups with visual templates in collaborative systems: Increasing structuration to improve precision. *Cross Cultural and Strategic Management, 24*(1), 78–98.

Bresciani S., & Eppler M. J. (2009). The benefits of synchronous collaborative information visualization: Evidence from an experimental evaluation. *IEEE transactions on visualization and computer graphics, 15*(6).

Clark, H. H., & Brennan, S. E. (1991). Grounding in communication. Perspectives on Socially Shared Cognition; American Psychological Association. J. M. L. a. S. D. T. E. L.B. Resnick 127–149.

DeSanctis, G., & Poole, M. S. (1994). Capturing the complexity in advanced technology use: Adaptive structuration theory. *Organization Science, 5*(2), 121–147.

Eppler, M. J., Hoffmann, M. H., & Kernbach, S. (2015, July). Navicons for Collaboration-Navigating and Augmenting Discussions through Visual Annotations. In *Information Visualisation (iV), 2015 19th International Conference on* (pp. 386–391). IEEE.

Eppler, M. J., & Kernbach, S. (2016). Dynagrams: Enhancing design thinking through dynamic diagrams. *Design Studies*, 47, 91–117.

Kernbach, S. (2015). The facilitative power of visual artifacts for knowledge sharing in client-consultant interactions. *Academy of Management Proceedings*, *2015*(1):14578–14578.

Kernbach, S., Bresciani, S., & Eppler, M. J. (2015). Slip-sliding-away: A review of the literature on the constraining qualities of PowerPoint. *Business and Professional Communication Quarterly*, *78*(3), 292–313.

Kernbach, S., Eppler, M. J., & Bresciani, S. (2015). The use of visualization in the communication of business strategies: An experimental evaluation. *International Journal of Business Communication*, *52*(2), 164–187.

Mengis, J., & Eppler, M. J. (2006). Seeing versus arguing the moderating role of collaborative visualization in team knowledge integration. *J Universal Knowledge Manage.*, *1*(3), 151–162.

Paivio, A. (1991). Dual coding theory: Retrospect and current status. *Canadian Journal of Psychology, 45*(3), 255–287.

Suthers, D. D. (2001). Towards a systematic study of representational guidance for collaborative learing discourse. *Journal of Universal Computer Science, 7*(3), 254–277.

Tversky, B. (2005). Visuospatial reasoning. *The Cambridge handbook of thinking and reasoning* 209–240.

Author Biographies

Sebastian Kernbach, Dr. is Post-Doctoral Researcher, Executive Consultant and Trainer at the University of St. Gallen (Switzerland). His research focuses on knowledge visualization, visual education and visual storytelling. He is the Founder of the Visual Collaboration Lab (www.visualcollaborationlab.org), which helps individuals, teams and organizations to have more meaningful and memorable interactions combining insights from creativity, education, and psychology. He works with organizations such as Axa, European Central Bank, Fraunhofer, Hilti and others. He is Guest Professor at the Central University of Finance and Economics in Beijing and the African Doctoral Academy at Stellenbosch University.

Sabrina Bresciani, Dr. is an Assistant Professor at the University of St. Gallen (Switzerland), where she conducts research on the topics of intercultural management communication and visualization for organizational communication. She has been a researcher at Harvard University (U.S.), the University of Cambridge (U.K.) and the National University of Singapore. She is currently a visiting Professor at the University of the Pacific (Peru), at the Central University of Finance and Economics (China) and at the University of Lugano (Switzerland). She is the Founder and President of Kolours, a non-profit social enterprise supporting disadvantaged people in India through design. https://sabrinabresciani.com/

How to Achieve Better Knowledge Utilization with Knowledge Externalization Mechanisms in Social Intranets

Vanessa Bachmaier and Isabella Seeber

Abstract Organizations rely on employees to externalize their tacit knowledge in order to effectively conduct knowledge-intensive work processes. Tacit knowledge externalization is particularly important in times of digital transformation where product and service innovation cycles become shorter and require creative, quick decision-making. Our understanding of mechanisms that constitute tacit knowledge externalization and how this relates to knowledge use is, however, limited. This paper contributes towards closing this gap by testing whether the suggested mechanisms of content generation, storytelling, organizational communication, professional collaboration, and practice demonstration are associated with knowledge use when supported by corporate social media. The implications for research and practice are discussed.

Keywords Tacit knowledge externalization · Corporate social media Hospitality industry · Instrument development · Instrument validation

1 Introduction

Social media is argued to play an important role in times of digital transformation as it changes how people live, work, and communicate (Panahi et al. 2016). Social media provides organizations with improved decision-making capabilities with data derived from multiple social media sources, which is estimated to make businesses more productive (Alberghini et al. 2014). This is especially true for the hospitality industry (Sigala and Chalkiti 2007). For example, customer reviews on Yelp, TripAdvisor and other social platforms have proven relevant resources for hotels and other businesses in this industry to understand customer satisfaction and drive

V. Bachmaier
Innsbruck, Austria

I. Seeber (✉)
University of Innsbruck, Innsbruck, Austria
e-mail: isabella.seeber@uibk.ac.at

K. North et al. (eds.), *Knowledge Management in Digital Change*, Progress in IS,
https://doi.org/10.1007/978-3-319-73546-7_9

business value (Pantano et al. 2017). But can social media do more? There exists increasing effort to deduce concrete shortcomings in guest experiences from reviews in order to improve the relevant service functions and train employees accordingly (Sigala and Chalkiti 2007). For many work processes in hotels, e.g., guest services and the handling of complaints, work-related know-how must be shared among employees to provide guests with the unique services and experiences (Sigala and Chalkiti 2007). Hotel employees that share their personal knowledge acquired from guest interactions are able to generate process-, product- and marketing-innovations (Nieves et al. 2014). Hence, it is essential for businesses to provide employees with rich social interaction opportunities so that they can articulate their personal tacit knowledge in order to create understandable, usable knowledge that can be stored and made available for organizational members for increased sustained competitive advantage (Wagner et al. 2014). Tacit knowledge externalization (TKE) represents one of those knowledge-intensive processes that is fostered by corporate social media (CSM). TKE describes activities that employees perform to convert their difficult to verbalize and low on tacitness into explicit, well-articulated knowledge (Nonaka 1994; Nonaka and Konno 1998; Nonaka et al. 2000). CSM tools, such as wikis, blogs, and social networking sites (SNS) facilitate those TKE activities (Standing and Kiniti 2011; Zaffar and Ghazawneh 2012). Several studies (Lee et al. 2010; Panahi et al. 2013; Wagner et al. 2014) have argued that the open, participatory nature of CSM enables employees to engage in rich TKE activities, such as free-form authoring, coauthoring and coediting content, commenting contributions, and sharing multimedia. More recently, research on TKE activities suggested that there exist latent mechanisms that explain how these activities foster TKE. Thus far, they identified content generation, storytelling, organizational communication, professional collaboration, and practice demonstration (Kosonen and Kianto 2009; Kiniti and Standing 2013; Panahi et al. 2013; Bachmaier 2017). As these important mechanisms are relatively new, research has not yet confirmed whether these mechanisms cause similar effects on CSM. In this respect, one crucial effect is the positive influence that TKE has on knowledge utilization. Research in the offline-context could confirm this effect and argue for its overall performance and innovation (Nonaka and Konno 1998; Nonaka et al. 2000; Popadiuk and Choo 2006; Sigala and Chalkiti 2007). Yet, there exists little understanding if TKE on CSM is able to achieve similar positive effects.

This paper contributes towards closing this gap. The goal is to show which TKE mechanisms have the strongest effects on knowledge utilization when supported by CSM. Only by understanding how TKE manifests on CSM, we can effectively design support for employees and drive business functions through better trained employees. We administered a survey and analyzed responses from 381 employees from the hospitality industry using the CSM platform hotelkit. We find that TKE on CSM has strong positive effects on knowledge utilization and that storytelling and practice demonstration are the most powerful mechanisms for that. All other mechanisms showed significant positive effects but were weaker in their impact. We discuss our findings with a special focus on how practitioners can make use of the TKE survey instrument.

2 Theoretical Background

2.1 Tacit Knowledge Externalization

Knowledge theories have recognized two major types of organizational knowledge: tacit and explicit (Byosiere and Luethge 2008). Nonaka and Takeuchi (1995) conceptualize those two spheres of knowledge as the extreme ends of a tacit–explicit continuum along which knowledge is created. *Explicit* knowledge refers to knowledge that can be translated into formal, systematic language and, as such, is relatively easy to recognize and share in manuals and guidelines, to name two common examples (Nonaka 1994; Nonaka and Takeuchi 1995; Nonaka et al. 2000). In contrast, *tacit knowledge* refers to highly personal knowledge deeply rooted in both action and commitment in a specific context and therefore difficult to formalize and express. Two dimensions of tacit knowledge can be differentiated: the cognitive, which consists of mental models, schemata, beliefs, and values deeply engrained in the human mind, and the technical, which embodies the types of crafts or individual skills commonly called know-how (Nonaka 1994; Nonaka and Takeuchi 1995).

Such knowledge held by employees must be passed along to other organizational members to increase organizational performance and ultimately develop competitive advantage (Nonaka et al. 2000; Cabrera and Cabrera 2002). Consequently, the knowledge-based view of the firm (Kogut and Zander 1992; Grant 1996) prioritizes the effective management of knowledge flows such as knowledge creation and transfer (Cabrera and Cabrera 2005). While knowledge transfer focuses on the transmission and absorption of both, tacit and explicit knowledge (Cabrera and Cabrera 2002), knowledge creation focuses on the conversion of tacit into explicit knowledge and vice versa (Nonaka 1994). In this study, we focus on tacit into explicit knowledge conversions, i.e. tacit knowledge externalization. Hence, while in knowledge transfer, both, the availability and usability of tacit and explicit knowledge are required (Davenport and Prusak 1998), TKE merely requires the availability of explicit knowledge created from tacit knowledge.

Theories of organizational knowledge creation derive from the belief that individuals, groups, and organizations can create knowledge together (Byosiere and Luethge 2008). Nonaka (1994) refers to this interaction in his seminal work on the SECI model that describes the spiraling process of organizational knowledge creation. The model illustrates four sequential modes of the knowledge creation process—namely, socialization, externalization (or TKE), combination, and internalization (Nonaka 1994; Nonaka and Takeuchi 1995; Nonaka et al. 2000)—each of which is characterized by different activities within an organization (Byosiere and Luethge 2008). First, socialization involves sharing tacit knowledge among individuals by converting it into new tacit knowledge without articulating it (Nonaka and Takeuchi 1995). For example, mentoring and apprenticeships instruct tacitly through observation and imitation (Chou et al. 2005). Second, externalization, or TKE, entails articulating tacit knowledge acquired from moments of socialization into codified, explicit concepts, as in the act of writing instruction manuals (Nonaka et al. 2000). Third, combination involves the conversion of explicit knowledge into more systematic sets of explicit knowledge (Nonaka and Takeuchi 1995)—for instance, by integrating information extracted from databases and creating new explicit knowledge (Chou et al. 2005). Fourth and lastly, internalization refers to the embodiment of explicit knowledge into tacit knowledge, as in the act of applying knowledge learnt from manuals (Byosiere and Luethge 2008).

Nonaka et al. (2000) have suggested that externalization occurs when individuals use discursive consciousness as a means to rationalize and articulate the world around them. It is through reflection that words develop into phrases and, in turn, crystallized concepts (Nonaka and Takeuchi 1995). Employees articulate their tacit knowledge through face-to-face dialogue, the textualization of documents, or via ICT (Suppiah and Singh Sandhu 2011; Haag and Duan 2012). For example, TKE occurs when an employee holds a face-to-face dialogue with colleagues to communicate an idea or, based on his or her work-related experiences, describes work processes in writing (Nonaka and Konno 1998; Ray 2014) on documents or ICT.

It is heavily discussed, however, if and to what extent tacit knowledge can be externalized in general and on ICT in particular (Panahi et al. 2013). Although a strand of research (Haldin-Herrgard 2000; Johannessen et al. 2001; Hislop 2002) argues that tacit knowledge is not accessible for consciousness and therefore unable to be articulated, another strand (Kogut and Zander 1992; Jasimuddin et al. 2005; Chennamaneni and Teng 2011; Dinur 2011; Panahi et al. 2013) claims that at least parts of tacit knowledge can become conscious and thus can be extracted and articulated, particularly with the support of innovative ICT, such as social media. The latter strand moreover claims that tacit knowledge can be classified as low, medium, or high based on its degree of tacitness. While highly tacit knowledge (e.g., intuitions and mental models) is hardly articulable and therefore less likely to be externalized, low and medium tacit knowledge (e.g., ideas and lessons learned) can easily be externalized using social media (Lopez-Nicolas and Soto-Acosta 2010; Panahi et al. 2013). In that light, this paper adopts the latter position and focuses primarily on types of tacit knowledge that can be externalized and shared through various CSM-based activities.

2.2 *Tacit Knowledge Externalization on Corporate Social Media*

In contrast to public social media (e.g., Facebook and Wikipedia), CSM such as corporate wikis, blogs, and social networking sites, some of which are shown in Fig. 1, are restricted to company-internal use. Arguably, the dynamic CSM activate employees' TKE by allowing social interaction and user-generated content even across spatiotemporal barriers; today, employees can easily generate work-related content and share multimedia on firm-internal dynamic, interactive blogs,

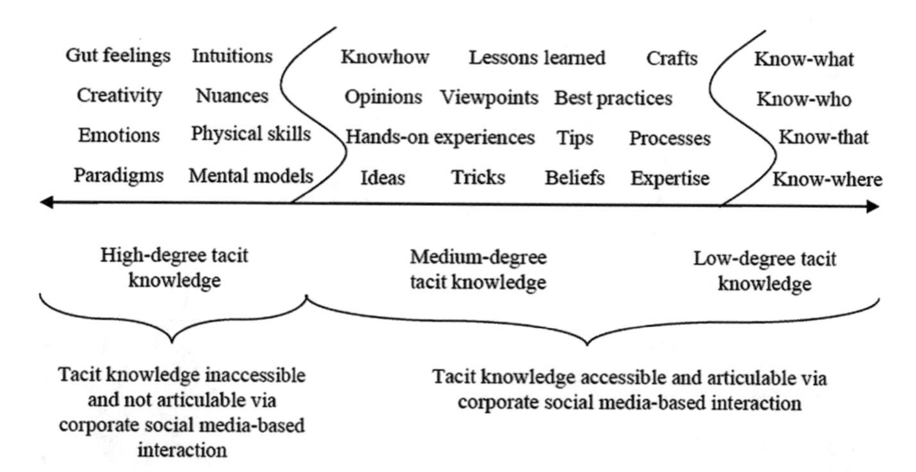

Fig. 1 Degrees of tacit knowledge articulable via corporate social media (Ambrosini and Bowman 2001; Panahi et al. 2016)

microblogs, wikis, and social networking sites. Previously, static forms of ICT (e.g., email, telephone) were unable to support such activities (Panahi et al. 2013).

Several conceptual and broadly qualitative studies (Kosonen and Kianto 2009; Lee et al. 2010; Standing and Kiniti 2011; Panahi et al. 2013; Wagner et al. 2014) have explored employees' activities on CSM during TKE. For example, in their conceptual study on CSM for knowledge conversion, Wagner et al. (2014) argued that employees externalize tacit knowledge on blogs and wikis by authoring and editing work-related content. In support, Lee et al. (2010) analyzed corporate blogging practices of 500 organizations and found that hundreds of employees externalized their work-related knowledge by sharing personal experiences and success stories on corporate blogs. Furthermore, Panahi et al. (2016) in their study on physicians' knowledge sharing on CSM discovered that SNS allowed discussing work-related topics with colleagues, answering colleagues' work-related questions, and making suggestions for work. Some studies have gone further, and by building on Nonaka's (1994) work, they have qualitatively explored mechanisms of TKE on CSM. For example, in their study on wikis, Kiniti and Standing (2013) found that employees express tacit knowledge through collaboration with colleagues in authoring and editing content. Similarly, Kosonen and Kianto (2009) conducted group interviews to examine how employees used wikis in order to manage information and found that they explicated tacit knowledge on wikis as part of their professional collaboration with employees. In their literature review on social media for knowledge sharing, Panahi et al. (2013) reported that professional collaboration, storytelling, and organizational communication all describe mechanisms through which individuals externalize their tacit knowledge on social media. Figure 2 integrates findings of the abovementioned research and summarizes how the externalization of tacit knowledge is understood in this study. The dashed rectangle highlights the traceable CSM-enabled activities of employees to externalize their

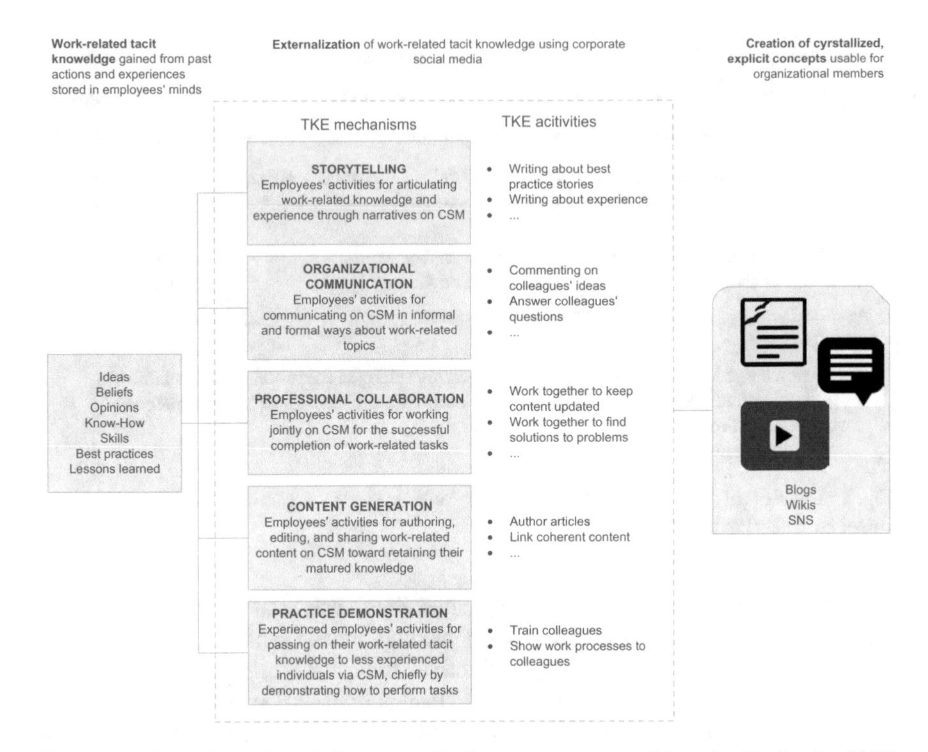

Fig. 2 Employees' tacit knowledge externalization on corporate social media (Bachmaier 2017)

knowledge. The activities differ with respect to the purpose they fulfil. Based on this, one can cluster these activities into mechanisms such as storytelling, professional collaboration, organizational communication, content generation, and practice demonstration (Kosonen and Kianto 2009; Kiniti and Standing 2013; Panahi et al. 2013; Bachmaier 2017). Thus far, research has mainly focused on the final presentation format such as wiki, blog and SNS technologies but overlooked how these stocks of knowledge come about (Panahi et al. 2013; Wagner et al. 2014) also argue that studies in the field are principally conceptual and qualitative, indicating a lack of quantifiable, empirical evidence to support their arguments. Based on their calls for quantitative work we aim to fill this gap.

3 Method

We administered an online survey to all active users of the corporate social intranet hotelkit (www.hotelkit.net) working in the service sector. Launched in 2012, "hotelkit" is a cloud-based corporate social media platform especially designed for hotels. It features several types of CSM, including a news blog, an idea blog, a

SNS, and a wiki. The primary purpose of "hotelkit" is to support service employees—it counts more than 300 hotels in Western Europe as clients—in communicating about and collaborating on firm-related topics. The study was announced to hotel managers in October 2015, by the owner of hotelkit himself, through a message posted on all hotelkit licenses' message board. Two weeks later, a link to the online survey was sent out per e-mail to 4231 hotel employees using hotelkit, excluding pilot test participants. Two reminder e-mails followed within the next two weeks. The online survey was completed by 416 hotelkit users. We excluded 22 cases because respondents declared themselves as hotelkit non-users. An additional 13 cases were excluded in which the participants were apprentices with only limited access to hotelkit, i.e. rights to consume but not to produce content, hence not enabled to externalize tacit knowledge. Therefore, we analysed a total of 382 cases. Participants were predominantly women, between 19 and 30 years old, mostly high-school graduates holding a professional education degree in tourism, and were by occupation lower-management employees. Most respondents worked in organizations with 40 or fewer employees.

3.1 Measurement

Tacit knowledge externalization is conceptualized as reflective–formative in its measurement.[1] Items can be found in Table 1 and are modeled as reflective indicators of the lower-order latent constructs. Lower-order latent constructs represent the five dimensions modelled as formative indicators of the higher-order construct TKE. To measure the higher-order latent construct, we followed the indicator-reuse approach recommended by Lohmöller (1989). The approach was deemed suitable since each latent construct was measured with at least one indicator (Ringle et al. 2012) and each lower order construct had approximately the same number of items (Becker et al. 2012).

Knowledge use refers to the degree to which externalized knowledge is used by employees in performing their work (Chen and Hung 2010). Knowledge use was measured with a 3-item instrument on a 7-point Likert scale adopted from Chen and Hung (2010), which entailed "I often use some kind of knowledge that I get from our virtual community to solve problems at work," "I frequently use some kind of knowledge that I get from our virtual community to improve my professional knowledge in my area of expertise," and "I regularly use some kind of knowledge that I get from our virtual community to handle challenges and changes in the work in the future." To adjust the instrument for the CSM context, the term "virtual community" was replaced with "hotelkit".

[1]The survey construct is self-developed and details on the instrument development and validation process can be requested from the authors.

Table 1 Overview of TKE items

Higher-order construct code	Indicator code	Indicator
		On hotelkit,
Content Generation (CG)	TK01_03	I author articles
	TK01_04	I edit articles
	TK01_05	I attach self-authored documents to content
	TK01_06	I edit documents that are attached to content
	TK01_07	I link coherent content
Storytelling (ST)	TK02_01	I write about my experiences at work
	TK02_02	I write about specific happenings at work
	TK02_03	I write about specific situations at work
	TK02_04	I write about achievements at work
	TK02_05	I write about problem cases at work
	TK02_06	I write about best practice examples at work
	TK02_07	I write about the emergence of work-related ideas in specific situations
Organizational Communication (OC)	TK03_01	I communicate with colleagues
	TK03_02	I answer colleagues' questions
	TK03_03	I comment on colleagues' contributions
	TK03_06	I exchange information with colleagues
Professional Collaboration (PC)	TK04_01	I work together with colleagues to generate content
	TK04_02	I work together with colleagues to keep content updated
	TK04_03	I work together with colleagues to generate work-related ideas
		I work together with colleagues to find solutions to problems
	TK04_04	I work together with colleagues to plan projects
		I work together with colleagues to coordinate tasks
	TK04_05	I advise colleagues in work-related matters
	TK04_06	I guide colleagues in work-related matters
		I train colleagues
Practice Demonstration (PD)	TK05_01	I show work processes to colleagues
	TK05_02	I show colleagues how to act in specific work-related situations
	TK05_03	I support colleagues in performing their tasks
	TK05_04	
	TK05_05	
	TK05_06	

3.2 Reliability and Validity

As a first step, missing data, outliers, and the normality of dimensions were assessed using IBM SPSS Statistics 21. We identified one extreme outlier that was excluded from further analysis. Whereas the values of univariate skewness ranged between −2.067 and 0.427, the values of univariate kurtosis ranged between −1.509 and 3.647. The Shapiro–Wilk test confirmed that the dimensions did not follow normal distribution (Hair Jr et al. 2016), which posed implications for confirmatory factor analysis. Given the non-normality of the underlying data distribution, we relied on a component-based approach—namely, partial least squares (PLS) (Urbach and Ahlemann 2010). The significance of indicators and paths was assessed by using the bootstrap resampling procedure (Efron and Tibshirani 1993) with 5000 resamples.

Following the validation guidelines of Straub et al. (2004) and Lewis et al. (2005), we assessed reliability, convergent validity, discriminant validity. For tacit knowledge externalization we had to assess reliability and validity first for the lower-order and then for the second-order constructs. The evaluation of lower-order latent constructs was deemed satisfactory (see Table 2 in Appendix). To establish reliability, we assessed Cronbach's Alpha. All 5 mechanisms reached a Cronbach's Alpha between 0.884 and 0.931, which exceeds the recommended threshold of 0.7 (Tenenhaus et al. 2005). Convergent validity was assessed with Average Variance Extracted (AVE). The AVE for the lower-order constructs ranged from 0.692 to 0.743. A construct's AVE should be at least 0.50 (Fornell and Larcker 1981) and hence all mechanism were deemed to satisfy convergent validity. Discriminant validity was assessed with the heterotrait–monotrait (HTMT) criterion. The HTMT of correlations should not exceed 0.9 (Henseler et al. 2014). The values of the HTMT criterion ranged from 0.435 to 0.703. The evaluation of the higher-order latent construct followed. Because of its formative nature, we tested *convergent validity* of the lower-order latent construct to the higher-order construct by using Edwards' adequacy coefficient (Edwards 2001). Edwards (2001) suggests that the value of the adequacy coefficient should exceed 0.50, which was the case with 0.626. Since lower-order latent constructs are modelled formatively to the higher-order latent construct, they should be conceptually distinct and not collinear (MacKenzie et al. 2011). As such, the *discriminant validity* of formative measurements needs to be tested by assessing the variance inflation factor (VIF), which should be smaller than 5 (Ringle et al. 2012). The VIF of the lower-order latent constructs ranged from 1.659 to 2.395, meaning that the conditions for discriminant validity of the higher-order latent construct were met.

For knowledge use, validity and reliability tests were deemed satisfactory, with a composite reliability of 0.949, Cronbach's alpha of 0.919, lowest factor loading of 0.923, AVE of 0.860, and square root of AVE in excess of all of its interconstruct correlations.

Table 2 Test for reliability and convergent validity

	CG	ST	OC	PC	PD	KU
Cronbach's alpha	0.888	0.931	0.884	0.918	0.931	0.919
Composite reliability	0.918	0.945	0.920	0.936	0.946	0.949
Average variance extracted	0.692	0.709	0.741	0.709	0.743	0.860

Taken together, all reliability and validity tests showed satisfying values and therefore we proceeded with estimating the structural model in PLS-SEM.

4 Results

Nomological validity exists when a construct measures the theoretical effect on antecedents or consequences, if not both (Lewis et al. 2005). Our analysis proceeded by including knowledge use as an additional hypothesized construct in the structural model (Freeze and Raschke 2007). As Fig. 3 shows, the theoretical relationship between TKE on CSM platforms and knowledge use has been suggested by several studies, since CSM platforms make employees' externalized tacit knowledge persistently visible and available (Leonardi et al. 2013; Rode 2016).

We next investigated the structural model and assessed the significance of paths by using the bootstrap resampling procedure (Efron and Tibshirani 1993) with 5000 resamples. We found that TKE had a significant effect on knowledge use ($\beta = 0.667$, $p < 0.001$), with an explained variance of $R^2 = 0.445$. This means that knowledge utilization can be facilitated with TKE mechanisms. The investigation into the single

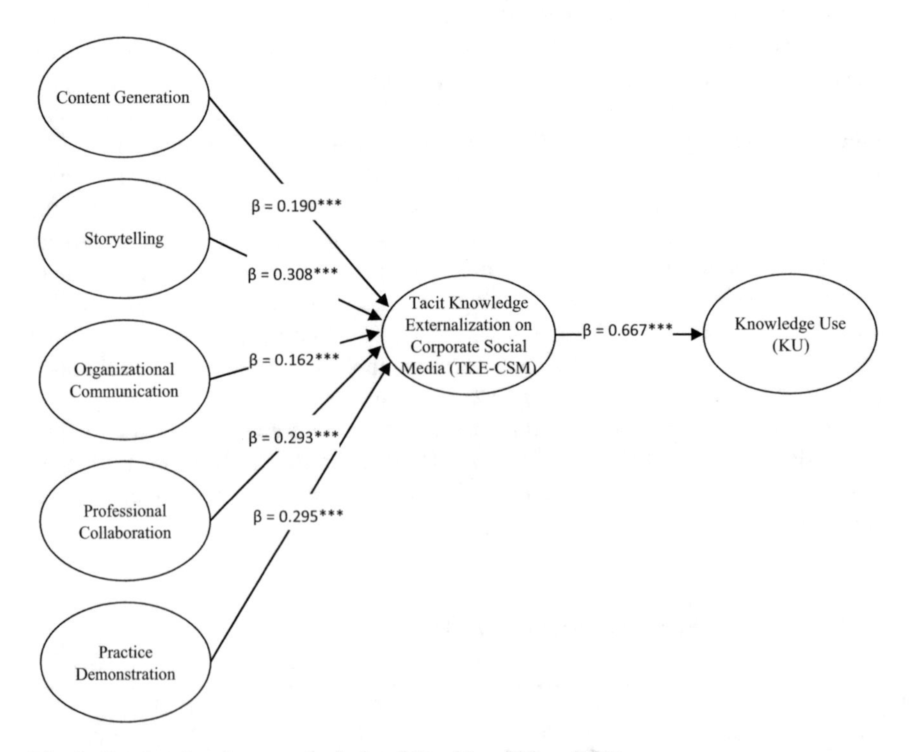

Fig. 3 Results of testing nomological validity. *Note* ***$p < 0.001$

effect of each mechanism on TKE reveals that while all mechanisms significantly relate to the externalization of knowledge, their intensity differs. The strongest effect had the mechanisms storytelling ($\beta = 0.308$, $p < 0.001$), professional collaboration ($\beta = 0.293$, $p < 0.001$), and practice demonstration ($\beta = 0.295$, $p < 0.001$). The effects of content generation ($\beta = 0.190$, $p < 0.001$) and organizational communication ($\beta = 0.162$, $p < 0.001$) were not as strong but still highly significant.

5 Discussion and Limitations

The objective of this study was to measure to what extend employees externalize their tacit knowledge on CSM and use it later on. We found that employees can make use of five mechanisms: content generation, storytelling, organizational communication, professional collaboration, and practice demonstration. Our findings additionally confirm the positive relationship between employees' TKE on CSM platforms and knowledge use.

The finding that practice demonstration has the highest path coefficient suggests that employees primarily use CSM platforms to demonstrate their practical skills. CSM functionalities, including multimedia sharing (e.g., pictures, videos, and audio) (Panahi et al. 2013), can support this dimension of TKE. Until the emergence of CSM platforms and other Web 2.0 tools, practice demonstration was considered to be possible only through face-to-face interaction (Panahi et al. 2013). Owing to CSM, employees have found new ways to externalize tacit knowledge in practice demonstration (Nonaka 1994; von Krogh 1998; Haldin-Herrgard 2000; Swap et al. 2001). Our findings suggest that employees engage in CSM-supported TKE activities such as guiding and mentoring colleagues, demonstrating work practices, and supporting colleagues in performing their tasks in specific situations.

Professional collaboration was found to be the second highest enabler of TKE on CSM platforms. The importance of this dimension supports previous findings that face-to-face communication is no longer the principal way of collaborating on work-related projects and themes (Panahi et al. 2013). Instead, employees assimilate CSM into their daily operations in order to professionally and asynchronously collaborate across geographical distances (Wakefield et al. 2008). Our findings suggest that employees could use CSM to collaborate to develop ideas, find solutions, and manage projects towards externalizing tacit knowledge.

Our findings furthermore suggest that storytelling, which describes the sharing of experiences, is a way for employees to externalize their tacit knowledge on CSM platforms. Digital storytelling was identified as a mechanism for externalizing tacit knowledge in past research (Panahi et al. 2013). According to our findings, CSM provide employees with rich spaces to engage in activities, such as to write about work-related experiences, happenings, and situations, yet also to tell stories about best practices and successes. Specifically best practices and success stories represent a crucial source of tacit knowledge by which other employees can improve their work practices toward increasing firm performance (Hisyam Selamat and Choudrie 2004).

Organizational communication was found to be another way for employees to externalize tacit knowledge on CSM platforms. By providing various spaces for open and free-form communication, CSM enabled employees to communicate on and exchange work-related topics (Panahi et al. 2013). They enable TKE activities such as commenting and providing immediate feedback on peers' contributions and engaging in discussions by answering colleagues' questions and making suggestions.

Content generation represents another mechanism that allows employees to externalize their tacit knowledge on CSM platforms. Relevant CSM functionalities support activities such as authoring and editing of articles and sharing and editing files. This is in line with prior quantitative, empirical studies that suggest that support for connecting activities are more important for organizational knowledge creation than support for collecting mechanisms (Kaschig et al. 2016).

The findings should be interpreted in the light of some limitations. First, although the instrument was developed for potential application in various industries, it was validated in the hospitality industry. Although the empirical results are consistent with the theoretical reasoning, cross-validating the instrument in other industries would further define its applicability in other industries (Jimenez-Castillo and Sanchez-Perez 2013). Therefore, to underpin the claim that the instrument has external validity, we suggest replication studies in other industries (e.g., manufacturing) and cultural settings (Flynn and Pearcy 2001). Second, data on dependent and independent variables was collected as part of one survey and hence there exists a risk of common method bias. We used Harman's one-factor test, to see if a single factor can account for a majority of covariance in the independent and dependent variables (Podsakoff et al. 2003). There was no single factor as a result of the principal component analysis that explained a majority of the covariance and hence common method variance is considered low. Third, response bias may have had an effect, given the cognitive and socio-historical background of the respondents—for example, whether they consider CSM to be useful for their work. We mitigated this risk by considering only those responses from employees that used the CSM and therefore should find it useful for some work-related matter. Fourth, this study deals with tacit knowledge externalization, the process of which is personal, context-specific, complex, and affected by surrounding conditions (Panahi et al. 2013). Thus, some insights gained might be specific to context and not directly transferable to other settings.

6 From Research to Practice

6.1 The TKE-CSM Instrument as an Instrument for Human Resource Managers

With this study, we provide practitioners with a 28-item instrument (see Table 1) that can easily be administered to measure the degree of employees' TKE on CSM. This contribution should allow human resource managers to support and motivate

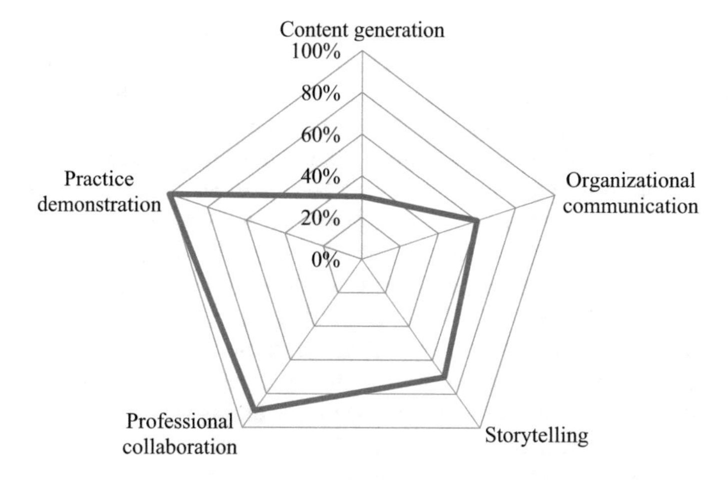

Fig. 4 Spider diagram of hotel employees' degree of tacit knowledge externalization on corporate social media

employees, teams, and departments. For example, on the basis of some fictive survey responses, Fig. 4 shows the degree of a hotel employees' TKE on hotelkit who predominantly externalize their tacit knowledge through practice demonstration (100%) and professional collaboration (90%).

These employees further externalize their tacit knowledge through storytelling (70%) and organizational communication (60%) and, to a lesser degree, content generation (30%). In that sense, human resource managers could offer training to relevant employees, teams, and departments to increase the degree of content generation and motivate their sustained efforts in improving the respective dimension of TKE-CSM with rewards.

It might be sufficient for a company to perform well in 3 out of 5 TKE mechanisms. However, since the aim of this chapter is to show how companies could use the results of a TKE-CSM survey, we suggest that the results of a survey could support companies in identifying not well performing dimensions and trigger them to develop strategies to increase those dimensions. Moreover, a social intranet is a rather expensive investment for companies. Accordingly, having capable employees use the full functionalities, spaces and possibilities of a CSM for externalizing tacit knowledge shall be the chief target of companies investing in social intranets in order to fully exploit the investment.

6.2 *The TKE-CSM Instrument as an Instrument for CSM Managers*

Based on our assessment of employees' TKE on CSM (Fig. 4), CSM managers and designers can indeed improve CSM. It is known that professional collaboration occurs mostly on wikis (Pei Lyn Grace 2009), in storytelling on blogs (Martin-Niemi and Greatbanks 2010), through organizational communication on social networking sites (Kaplan and Haenlein 2010), via practice demonstration through videoconferencing or multimedia sharing (Okumus 2013), and with content generation on all types of CSM through authoring, editing, and commenting (Wagner et al. 2014). Given that content generation in our survey has the weakest degree of employees' TKE, CSM managers can explore the corporate wiki, blog, and social networking site in order to identify room for improvement (e.g., the design of the authoring, editing, and commenting functions). CSM functionalities can then be redesigned to increase employees' degrees of content generation. The results of the assessment of employees' degree of TKE in Fig. 4 furthermore show that they actively engage in practice demonstration, meaning that CSM managers can install multimedia-sharing tools on all CSM (e.g., wikis, blogs, and social networking sites) to allow employees to demonstrate their practical skills wherever they want.

Organizations can also assess the overall usefulness of their CSM in supporting their employees' TKE by using the instrument developed in this study. Depending on the context and purposes of organizations, the usefulness of CSM could vary, however.

6.3 Opening the Black Box of TKE for Hotel Managers

Regarding the sector examined in our study (i.e., hospitality), the knowledge obtained from results could especially promote a deeper understanding of hotel managers in harnessing the power of their employees' tacit knowledge. The knowledge-intensive processes in the hospitality sector still represent a widely neglected field in CSM-mediated TKE research (Sigala and Chalkiti 2007). Although it has gained worldwide importance due to its rapid growth, the awareness of how the sector manages employees' tacit knowledge remains in its infancy (Sigala and Chalkiti 2007). In that regard, our study shows that the operationalization of employees' TKE on CSM makes the concept more tangible for the hospitality sector and provides opportunities for hotel managers to understand, measure, and manage their employees' TKE. In fact, many items of the TKE construct (e.g., comment on colleagues' contributions or write about problem cases at work) could be operationalized within the CSM to train "artificial intelligence" components of the CSM with TKE activities. This could help to build a smart knowledge environment that cognitively assists service employees to better attend to customer needs and solve customer problems.

References

Alberghini, E., Cricelli, L., & Grimaldi, M. (2014). A methodology to manage and monitor social media inside a company: A case study. *Journal of Knowledge Management, 18*, 255–277.

Ambrosini, V., & Bowman, C. (2001). Tacit knowledge: Some suggestions for operationalization. *Journal of Management Studies, 38*(6), 811–829.

Bachmaier, V. (2017). Examining the role of corporate social media platforms for tacit knowledge externalization in the hospitality industry. Doctoral Thesis, University of Innsbruck.

Becker, J.-M., Klein, K., & Wetzels, M. (2012). Hierarchical Latent Variable Models in PLS-SEM: Guidelines for Using Reflective-Formative Type Models. *Long Range Planning, 45*(5–6), 359–394.

Byosiere, P., & Luethge, D. J. (2008). Knowledge domains and knowledge conversion: an empirical investigation. *Journal of Knowledge Management, 12*(2), 67–78.

Cabrera, A., & Cabrera, E. F. (2002). Knowledge-Sharing Dilemmas. *Organization Studies, 23*(5), 687–710.

Cabrera, E. F., & Cabrera, A. (2005). Fostering knowledge sharing through people management practices. *The International Journal of Human Resource Management, 16*(5), 720–735.

Chen, C.-J., & Hung, S.-W. (2010). To give or to receive? Factors influencing members' knowledge sharing and community promotion in professional virtual communities. *Information and Management, 47*(4), 226–236.

Chennamaneni, A., & Teng, J. T. (2011). An Integrated Framework for Effective Tacit Knowledge Transfer. AMCIS.

Chou, T.-C., Chang, P.-L., Tsai, C.-T., & Cheng, Y.-P. (2005). Internal learning climate, knowledge management process and perceived knowledge management satisfaction. *Journal of Information Science, 31*(4), 283–296.

Davenport, T., & Prusak, L. (1998). Learn how valuable knowledge is acquired, created, bought and bartered. *The Australian Library Journal, 47*(3), 268–272.

Dinur, A. (2011). Tacit knowledge taxonomy and transfer: Case-based research. *Journal of Behavioral and Applied Management, 12*(3), 246.

Edwards, J. R. (2001). Multidimensional Constructs in Organizational Behavior Research: An Integrative Analytical Framework. *Organizational Research Methods, 4*(2), 144–192.

Efron, B., & Tibshirani, R. J. (1993). *An Introduction to the Bootstrap*. Boston: MA, Springer, US.

Flynn, L. R., & Pearcy, D. (2001). Four subtle sins in scale development: some suggestions for strengthening the current paradigm. *International Journal of Market Research, 43*(4), 409.

Fornell, C., & Larcker, D. F. (1981). Evaluating structural equation models with unobservable variables and measurement error. *Journal of Marketing Research,* 39–50.

Freeze, R. D., & Raschke, R. L. (2007). An Assessment of Formative and Reflective Constructs in IS Research. In *ECIS*.

Grant, R. M. (1996). Toward a knowledge-based theory of the firm: Knowledge-based Theory of the Firm. *Strategic Management Journal, 17*(S2), 109–122.

Haag, M., & Duan, Y. (2012). Understanding Personal Knowledge Development in Online Learning Environments: An Instrument for Measuring Externalisation, Combination and Internalisation. *Electronic Journal of Knowledge Management, 10*(1).

Hair, J. F. Jr., Hult, G. T. M., Ringle, C., & Sarstedt, M. (2016). *A primer on partial least squares structural equation modeling (PLS-SEM).* Sage Publications.

Haldin-Herrgard, T. (2000). Difficulties in diffusion of tacit knowledge in organizations. *Journal of Intellectual Capital, 1*(4), 357–365.

Henseler, J. J., Ringle, C. M. M., & Sarstedt, M. (2014). A new criterion for assessing discriminant validity in variance-based structural equation modeling. *Journal of the Academy of Marketing Science, 43*(1), 115–135.

Hislop, D. (2002). Mission impossible? Communicating and sharing knowledge via information technology. *Journal of Information Technology, 17*(3), 165–177.

Hisyam, S. M., & Choudrie, J. (2004). The diffusion of tacit knowledge and its implications on information systems: The role of meta-abilities. *Journal of Knowledge Management, 8*(2), 128–139.

Jasimuddin, S. M., Klein, J. H., & Connell, C. (2005). The paradox of using tacit and explicit knowledge: Strategies to face dilemmas. *Management Decision, 43*(1), 102–112.

Jimenez-Castillo, D., & Sanchez-Perez, M. (2013). Integrated market-related internal communication: development of the construct. *International Journal of Market Research, 55*(4), 563.

Johannessen, J.-A., Olaisen, J., & Olsen, B. (2001). Mismanagement of tacit knowledge: the importance of tacit knowledge, the danger of information technology, and what to do about it. *International Journal of Information Management, 21*(1), 3–20.

Kaplan, A. M., & Haenlein, M. (2010). Users of the world, unite! The challenges and opportunities of Social Media. *Business Horizons, 53*(1), 59–68.

Kaschig, A., Maier, R., & Sandow, A. (2016). The effects of collecting and connecting activities on knowledge creation in organizations. *The Journal of Strategic Information Systems, 25*(4), 243–258.

Kiniti, S., & Standing, C. (2013). Wikis as knowledge management systems: Issues and challenges. *Journal of Systems and Information Technology, 15*(2), 189–201.

Kogut, B., & Zander, U. (1992). Knowledge of the firm, combinative capabilities, and the replication of technology. *Organization Science, 3*(3), 383–397.

Kosonen, M., & Kianto, A. (2009). Applying wikis to managing knowledge-A socio-technical approach. *Knowledge and Process Management, 16*(1), 23–29.

Lee, H. W., Lim, K. Y., & Grabowski, B. L. (2010). Improving self-regulation, learning strategy use, and achievement with metacognitive feedback. *Educational Technology Research and Development, 58*(6), 629–648.

Leonardi, P. M., Huysman, M., & Steinfield, C. (2013). Enterprise Social Media: Definition, History, and Prospects for the Study of Social Technologies in Organizations. *Journal of Computer-Mediated Communication, 19*(1), 1–19.

Lewis, B. R., Templeton, G. F., & Byrd, T. A. (2005). A methodology for construct development in MIS research. *European Journal of Information Systems, 14*(4), 388–400.

Lohmöller, J.-B. (1989). *Latent Variable Path Modeling with Partial Least Squares*. Heidelberg: Physica-Verlag HD.

Lopez-Nicolas, C., & Soto-Acosta, P. (2010). Analyzing ICT adoption and use effects on knowledge creation: An empirical investigation in SMEs. *International Journal of Information Management, 30*(6), 521–528.

Mackenzie, S. B., Podsakoff, P. M., & Podsakoff, N. P. (2011). Construct measurement and validation procedures in MIS and behavioral research: Integrating new and existing techniques. *MIS Quarterly, 35*(2), 293–334.

Martin-Niemi, F., & Greatbanks, R. (2010). The ba of blogs: Enabling conditions for knowledge conversion in blog communities. *Vine, 40*(1), 7–23.

Nieves, J., Quintana, A., & Osorio, J. (2014). Knowledge-based resources and innovation in the hotel industry. *International Journal of Hospitality Management, 38,* 65–73.

Nonaka, I. (1994). A dynamic theory of organizational knowledge creation. *Organization Science, 5*(1), 14–37.

Nonaka, I., & Konno, N. (1998). The Concept of "Ba": Building a Foundation for Knowledge Creation. *California Management Review, 40*(3), 40–54.

Nonaka, I., & Takeuchi, H. (1995). *The knowledge-creating company: How Japanese companies create the dynamics of innovation.* Oxford university press.

Nonaka, I., Toyama, R., & Konno, N. (2000). SECI, Ba and Leadership: a Unified Model of Dynamic Knowledge Creation. *Long Range Planning, 33*(1), 5–34.

Okumus, F. (2013). Facilitating knowledge management through information technology in hospitality organizations. *Journal of Hospitality and Tourism Technology, 4*(1), 64–80.

Panahi, S., Watson, J., & Partridge, H. (2013). Towards tacit knowledge sharing over social web tools. *Journal of Knowledge Management, 17*(3), 379–397.

Panahi, S., Watson, J., & Partridge, H. (2016). Information encountering on social media and tacit knowledge sharing. *Journal of Information Science, 42*(4), 539–550.

Pantano, E., Priporas, C.-V., & Stylos, N. (2017). 'You will like it!'using open data to predict tourists' response to a tourist attraction. *Tourism Management, 60,* 430–438.

Pei Lyn Grace, T. (2009). Wikis as a knowledge management tool. *Journal of Knowledge Management, 13*(4), 64–74.

Podsakoff, P. M., Mackenzie, S. B., Lee, J.-Y., & Podsakoff, N. P. (2003). Common method biases in behavioral research: a critical review of the literature and recommended remedies. *Journal of Applied Psychology, 88*(5), 879.

Popadiuk, S., & Choo, C. W. (2006). Innovation and knowledge creation: How are these concepts related? *International Journal of Information Management, 26*(4), 302–312.

Ray, D. (2014). Overcoming cross-cultural barriers to knowledge management using social media. *Journal of Enterprise Information Management, 27*(1), 45–55.

Ringle, C. M., Sarstedt, M., & Straub, D. (2012). A critical look at the use of PLS-SEM in MIS Quarterly.

Rode, H. (2016). To share or not to share: The effects of extrinsic and intrinsic motivations on knowledge-sharing in enterprise social media platforms. *Journal of Information Technology, 31*(2), 152–165.

Sigala, M., & Chalkiti, K. (2007). Improving performance through tacit knowledge externalisation and utilisation: Preliminary findings from Greek hotels. *International Journal of Productivity and Performance Management, 56*(5/6), 456–483.

Standing, C., & Kiniti, S. (2011). How can organizations use wikis for innovation? *Technovation, 31*(7), 287–295.

Straub, D., Boudreau, M.-C., & Gefen, D. (2004). Validation guidelines for IS positivist research. *The Communications of the Association for Information Systems, 13*(1), 63.

Suppiah, V., & Singh, Sandhu M. (2011). Organisational culture's influence on tacit knowledge-sharing behaviour. *Journal of Knowledge Management, 15*(3), 462–477.

Swap, W., Leonard, D., & MimiShields, A. L. (2001). Using mentoring and storytelling to transfer knowledge in the workplace. *Journal of Management Information Systems, 18*(1), 95–114.

Tenenhaus, M., Vinzi, V. E., Chatelin, Y.-M., & Lauro, C. (2005). PLS path modeling. *Computational Statistics & Data Analysis, 48*(1), 159–205.

Urbach, N., & Ahlemann, F. (2010). Structural equation modeling in information systems research using partial least squares. JITTA: *Journal of Information Technology Theory and Application, 11*(2), 5.

Von Krogh, G. (1998). Care in Knowledge Creation. *California Management Review, 40*(3), 133–153.

Wagner, D., Vollmar, G., & Wagner, H.-T. (2014). The impact of information technology on knowledge creation: An affordance approach to social media. *Journal of Enterprise Information Management, 27*(1), 31–44.

Wakefield, R. L., Leidner, D. E., & Garrison, G. (2008). Research note—a model of conflict, leadership, and performance in virtual teams. *Information Systems Research, 19*(4), 434–455.

Zaffar, F. O., & Ghazawneh, A. (2012). Knowledge sharing and collaboration through social media–the case of IBM. In *Proceedings of the 7th Mediterranean Conference on Information Systems, MCIS.*

Author Biographies

Vanessa Bachmaier, Dr., received her Ph.D. from the University of Innsbruck in Austria. She had earned significant work experience prior to her studies in companies such as booking.com. After her PhD she returned to practice and heads the knowledge management department at Skidata. Her research covers knowledge management on corporate social media and has been published at international conferences.

Isabella Seeber, Dr., is a Post Doc at the University of Innsbruck in Austria where she also earned her Ph.D. Her research focuses on team and crowd-based idea convergence, feedback in innovation contests, and knowledge externalisation on social media. Her research has been published at many international conferences and in journals such as Journal of Management Information Systems, Computers in Human Behaviour and Group Decision and Negotiation.

Balancing Knowledge Protection and Sharing to Create Digital Innovations

Stefan Thalmann and Ilona Ilvonen

Abstract The creation of digital innovations requires active participation and knowledge sharing on behalf of all collaboration partners in inter-organisational settings. However, while the participants collaborate, they also have their own interests and as they are competitors in many cases, they have to protect their competitive knowledge. Collaboration thus requires balancing of knowledge sharing and protection on both the organizational and individual level. This paper reviews literature from several domains to assess how the balancing act is scoped and what kind of measures to achieve this balance prior research has identified. The balancing act is examined on the channel, partner and artefact levels. The paper identifies the balancing act as decisions made over the course of the collaboration both by the organizations as a whole, and by individuals on concrete knowledge artefacts in their daily work. Implications from the point of view of creating digital innovation are presented.

Keywords Knowledge protection · Knowledge sharing · Inter-organizational collaboration · Digital innovation · Balancing act · Risks and benefits

1 Motivation

New ways of combining digital and physical innovations, as well as intensified inter-organizational collaborations, create new challenges to the protection of organizational knowledge. Whereas traditional innovations mainly depend on

---

S. Thalmann (✉)
Pro2Future GmbH and Know-Center GmbH, Graz, Austria
e-mail: stefan.thalmann@tugraz.at

S. Thalmann
Institute for Interactive Systems and Data Science, Graz University of Technology, Graz, Austria

I. Ilvonen
NOVI Research Center, Tampere University of Technology, Tampere, Finland
e-mail: Ilona.ilvonen@tut.fi

K. North et al. (eds.), *Knowledge Management in Digital Change*, Progress in IS,
https://doi.org/10.1007/978-3-319-73546-7_10

physical artefacts, digital innovations predominantly rely on innovative ideas and knowledge (Yoo et al. 2012). An essential characteristic of digital innovations is that they are created in a collaborative manner, involving business partners, as well as customers (Yoo et al. 2012). The required heterogeneity of knowledge and knowledge sources of digital innovations especially demands inter-organizational collaboration (Yoo et al. 2012). The increased complexity of the innovation process leads to a geographical dispersion and a distribution across multiple organizations (Von Hippel 2009). However, research shows that participation in inter-organizational knowledge development processes bears knowledge protection risks (Loebbecke et al. 2016). Recent studies indicate that balancing knowledge sharing and protection is a challenge for organizations, especially in network settings (Hernandez et al. 2015; Pahnke et al. 2015).

Due to high demand for inter-organizational knowledge exchange during the creation and adaption of digital innovations, organizations have to join networks with competitors. In such networks they have to share knowledge to get appropriate feedback and simultaneously protect their crucial knowledge for misappropriation by other network members (Trkman and Desouza 2012). Hence, they want to find a satisfactory mixture between sharing and protection (Loebbecke et al. 2016).

This mixture is also of high interest for network management. If network members are very protective and aim at skimming knowledge, the knowledge sharing in the network is low and the network is less attractive for participants (Manhart et al. 2015). Hence, network management is concerned with facilitating sharing. But does more sharing always mean less protection and vice versa? Literature frequently considers knowledge protection as a barrier for knowledge sharing and as its counterpart (Mazloomi Khamseh and Jolly 2008). However, current research shows that effective knowledge protection enhances knowledge sharing, as well as team coordination and performance (Lee et al. 2015). These contradictory results underline the importance for research on the relationship between knowledge sharing and knowledge protection. To get an overview about the different research streams on balancing knowledge sharing and protection, we performed a literature review.

2 Background

Management literature widely acknowledges the importance of knowledge (Conner and Prahalad 1996; Grant 1996a) and the knowledge-based view of the firm (Grant 1996b; Sveiby 2001). In his knowledge-creation theory, Nonaka views the organization as a knowledge-creating entity and argues that not only knowledge but specifically the capability to create, share, and utilize knowledge is the most important source of a firm's competitive advantage (Nonaka 1991, 1994; Nonaka and Toyama 2003).

The barriers for knowledge sharing have received a lot of attention (Riege 2005). The first perspective considers the characteristics of knowledge hampering sharing.

In this regard, scholars discuss the causal ambiguity of knowledge (Szulanski 1996a, b) and the difficulty of transferring tacit knowledge (Haldin-Herrgard 2000) as barriers. The second perspective focuses on the characteristics of the people involved in knowledge sharing. Scholars argue that the sender's lack of motivation is a barrier resulting from differences in status, culture or space (Zimmermann and Ravishankar 2014), lack of time (O'Dell and Grayson 1998), lack of trust (Pinjani and Palvia 2013), or a culture of hoarding knowledge (Sveiby and Simons 2002). Concerning the knowledge receiver's characteristics, barriers include a low absorptive capacity (Szulanski 1996a; Zimmermann and Ravishankar 2014), limited previous knowledge (Hansen 1999; Szulanski 2000; Reagans and McEvily 2003), weak social ties (Szulanski 1996a), a high geographical (Haldin-Herrgard 2000; Yoo and Kanawattanachai 2001), cultural (Leyland 2006; Pawlowski 2008), or cognitive distance (Cramton 2001), or different terminologies (Szulanski 1996b; Reagans and McEvily 2003). The third perspective considers the organizational setting and includes barriers such as a lack of rewards (Bock et al. 2005), lack of corporate culture (McDermott and O'dell 2001), lack of managerial leadership (David and Fahey 2000), reluctance to use IT systems (Riege 2005), or lack of technical support (Hendriks 1999).

Knowledge protection and, thus, the perspective that less knowledge sharing can be better for organizations is rarely mentioned in this stream of research so far (Ahmad et al. 2014). However, a few examples are mentioned in literature: Trkman and Desouza (2012) as well as Olander et al. (2014) argue that knowledge sharing is not "risk free" and that the value of knowledge sharing can be overshadowed by the damage caused by lost knowledge. Ahmad et al. (2014) formulates this risk more alarming by concluding that knowledge leakage, with potentially devastating consequences, could occur on a number of fronts. Hence, sharing the "wrong" knowledge could reduce the firm's rent-generating potential (Von Krogh 2012) and the openness for knowledge sharing depends on the industry sector and the level of openness for sharing is circumstance-specific (Erickson and Rothberg 2009). As a consequence of this Bogers (2011) suggested a "knowledge exchange strategy" to specify the knowledge sharing behaviour with external partners. Taking this gap into account, the goal of our work is to provide a better understanding of the counterforces of knowledge sharing to gain a more differentiated view towards barriers of knowledge sharing.

Organizations need to build protective capabilities to ensure that the relevant knowledge stays within their boundaries while they engage in inter-organizational knowledge sharing (Von Krogh 2012). As such, knowledge protection focuses on (1) the prevention of unwanted knowledge spillovers (Ahmad et al. 2014), (2) the reduction of knowledge visibility (Lee et al. 2007), and (3) knowledge loss (Jennex and Durcikova 2013). We define knowledge protection as the collection of the formal practices that organizations enforce and the informal practices that individuals perform to prevent unwanted disclosure, spillover, or loss of knowledge.

To gain the greatest benefits, organizations have to exchange knowledge with external parties, whilst, at the same time, they have to protect themselves against knowledge appropriation by externals (Jordan and Lowe 2004). Thereby, vulnerabilities for knowledge outflows are to a certain degree accepted as the price for generating access to valuable knowledge (Sofka et al. 2014) or for successfully executing joint tasks (Hamel 1991). However, finding a balance between knowledge protection and sharing is considered as especially challenging for firms for several reasons: (1) The protection of explicit knowledge remains hard to achieve as property rights are very costly to write and enforce (Chan and Lee 2011), leading to firms be more reluctant towards the sharing of explicit knowledge. (2) Although security literature provides approaches towards awareness trainings as well as access and authorization schemes, this does not fully cover the question how to protect knowledge in people's brains (Desouza 2006). Reasons for this are ill-defined decision rights and ownership of the knowledge (Grant 1996c) and that it is difficult for employees as well as for the organizations itself to know which knowledge is critical for the organization (Trkman and Desouza 2012). (3) Tacit knowledge is sticky and complex (Nelson and Winter 1982), and cannot be easily codified, articulated and is not visible when observed (Nonaka and Takeuchi 1995) which hinders the transfer on the one hand, but makes the planning and enforcement of knowledge protection challenging on the other hand. However, organizations have to make their tacit knowledge explicit to make use of it (Coff et al. 2006). These issues hamper the management of sharing and protection of knowledge.

The relationship between knowledge protection and knowledge sharing has received little attention to date (Ahmad et al. 2014). The few works to have looked at this relationship include Trkman and Desouza (2012), as well as Olander et al. (2014), who argue that knowledge sharing is not "risk-free" and that the damage caused by lost knowledge can overshadow the value of knowledge sharing. Further the general assumption is that the more protective a partner is the lower is the level of knowledge sharing (Simonin 2004), as a more protective behaviour reduces the motivation of the knowledge sharing partners to contribute (Norman 2004; Sazali et al. 2010). Further, the sharing literature recommends that the more knowledge is shared with a knowledge transfer partner the higher is the motivation for the partner to contribute, and the higher is the amount of knowledge received from the partner (de Faria and Sofka 2010).

3 Procedure

This paper presents a structured literature review according to Webster and Watson (Webster and Watson 2002a). The review was undertaken in three stages: (1) identifying the relevant literature, (2) structuring the review, and (3) contributing to theory.

In stage (1), a full review of top journals of the base domains was conducted. The review was conducted in spring 2017, and focused on articles published between 2005 and 2016. Selection of journals was based on their rankings, if available (Azar and Brock 2008; Crossan and Apaydin 2010; Serenko and Bontis 2013). The review was complemented with backward and forward searches of highly cited and relevant articles. To identify potentially relevant papers, the building-blocks approach (Rowley and Slack 2004) was applied, transforming relevant concepts into search statements and extending the statements by using synonyms and related terms. For example, articles that focus on knowledge about protection instead of protection of knowledge were excluded from the in-depth analysis.

In stage (2), a concept matrix (Webster and Watson 2002b) that identifies the main elements of analysis was developed. One column each was defined to cover how the papers consider the balancing of knowledge sharing and protection, its scope and application. The matrix was iteratively refined and extended (Webster and Watson 2002b) with new insights emerging from the literature.

In stage (3), the goal was to get a deeper understanding on balancing knowledge sharing and protection that incorporates the specifics of the identified base domains. Patterns within and across the base domains were identified using the concept matrix. The goal was to synthesize the findings from the base domains and align the findings with the research question. Thereby, we analysed the papers according to the three dimensions proposed by (Manhart et al. 2015): knowledge artefact, sharing partner and communication channel.

4 Discussion of Results

4.1 Theoretical Foundations for Knowledge Protection

Many of the reviewed papers do not note a specific theoretical framework for examining knowledge protection. The strategic management and knowledge management literature very much rests on the foundation of the resource based view and knowledge based view (Grant 1996b) of the firm, although these are not often mentioned in the papers. These views, however, reason the starting point of the whole need for sharing and protecting knowledge: knowledge is an important organizational resource. Among the theories that are used to explain the protection measures are transaction cost economics (e.g. Mazloomi Khamseh and Jolly 2008; Trkman and Desouza 2012), control theory (e.g. Loebbecke et al. 2016), social exchange theory (e.g. Fauchart and Von Hippel 2008), self-regulation theory (Jarvenpaa and Majchrzak 2016) and institutional theory (e.g. Di Stefano et al. 2014). The theoretical approaches to the topic can thus be varied, and emphasize the practical and cross-disciplinary nature of the sharing-protection balancing act.

4.2 Perspectives to Knowledge Protection

The scope of balancing of knowledge sharing and protection requires first examination of the scope of protection and scope of sharing, as the scope of balancing is a combination of these two. Balancing is hard to do, if the knowledge protection efforts are completely separate from the knowledge sharing activity, e.g. knowledge protection efforts are concentrated on by security professionals or the IT department, whereas knowledge sharing is the business of knowledge managers or HR. As written in the background section, knowledge sharing efforts need to overcome barriers, and the same barriers can work as protection mechanisms in case of unwanted sharing takes place (Marabelli and Newell 2012). When knowledge protection and sharing are approached simultaneously by the same group of people, the complementarity of efforts is possible to achieve (Jennex and Durcikova 2014).

The need to balance knowledge sharing and protection primary occurs in inter-organizational settings, as an effect of competitive forces (Loebbecke et al. 2016). Hence, we focus on inter-organisational settings in the following.

First, balancing of knowledge sharing and protection can be investigated from the perspective of the individual or the organisation. Many of the sharing and protection decisions are in the end made by individuals in every-day communication situations (Jarvenpaa and Majchrzak 2010; Trkman and Desouza 2012; Jarvenpaa and Majchrzak 2016). However, individuals have roles in organizations, and they are influenced by organisational policies that play a part in their decisions about knowledge sharing or protection (Daghfous et al. 2013). For example, decisions about collaborations with other organisations as part of strategic alliances or memberships in professional networks are taken by organisations and such decisions influence the decision space of individuals.

According to (Manhart et al. 2015), knowledge protection activities can (1) focus on restricting the sharing within a certain communication channel, i.e. participate in a knowledge sharing network, (2) focus on restricting the sharing with specific sharing partners, i.e. share only with trusted peers, or (3) focus on restricting the sharing of concrete knowledge artefacts, i.e. knowledge related to a certain topic. In the following (see Table 1), we will use these three perspectives on knowledge protection to structure and discuss our results according the following dimensions:

- The need to share knowledge. Why is knowledge sharing needed?
- The need to protect knowledge. Why should knowledge be protected?
- The temporal dimension of the decision. When and how often should the balancing take place?
- The act of balancing. How to contrast benefits and risks? What benefits and risks should be contrasted?
- Balancing measures. How to enforce the balancing decision?

4.3 Channel

With a channel or forum, we mean different kinds of networks of organizations that are formed on digital platforms, or use digital channels for their communication. The digital platform can be used for one-to-one communication (e.g. messenger

Table 1 The scope and approaches for balancing knowledge sharing and protection

Contrast scope	Need to share knowledge	Need to protect knowledge	Time-frame	The act of balancing	Balancing measures
Channel	Keep in contact with a community and get updates on recent developments. Jointly co-develop and share knowledge, which is needed. Ease of use or popularity of channel	Make sure that critical knowledge is not misused, technical vulnerabilities of channel, limitation of the number of channels to enable control	Especially beginning of collaboration, in change situations	Evaluate the competences of the network members, the intensity/openness of sharing and collaboration and thus the possible knowledge gains and contrast with the competitive situation with members, the sharing culture of the channel, features to protect knowledge in the channel	Policies or informal controls such as norms and values
Partner	Necessity to collaborate to ensure a good mix of competences to jointly co-create knowledge	Make sure that the co-created knowledge can be appropriated and that knowledge shared is not misused	Beginning of collaboration, re-evaluation in change situations	Evaluate the existing knowledge and competencies of the partner to judge on the expected benefits arising from the collaboration and contrast with the risk factors (i.e. competitive position in the market), the trust or possibilities to observe the behavior of the partner	Contract design in strategic alliances, IPR regime, partner selection and sourcing decisions, relational governance and trust
Artefact	Share knowledge to facilitate creation of new knowledge	Preventing unwanted spillover or misuse of knowledge; preventing losing exclusiveness of knowledge	In concrete sharing situations (short term); across collaboration life-cycle, short term and long term perspective	Contrast the expected benefits from the knowledge gained in the sharing with the expected risks of knowledge exposure Due to the concrete relationship to a knowledge artefact the risk/benefit analysis can be made case by case	Timing of sharing, secrecy, recognition of important knowledge, ensuring inflow-outflow analysis, relational governance, trust, contracts and IPR

chats, private slack messages, etc.) or cross-organizational group communications (Facebook, Yammer, Slack etc.) between actors, when the channel is just that, a channel for communication that could take place also elsewhere. On the other hand, the channel or platform may be one that launches the collaboration altogether (entering a GrabCAD challenge, participating a developer forum). The decision of entering a specific channel is made on the organizational level, taking into account what kind of other organizations and actors are involved with the channel.

Mostly the articles that were reviewed for this study do not address the perspective of digital channels or their role in balancing knowledge sharing and protection. When digital communication platforms were discussed, they were seen as one context or a proxy for communication (Marabelli and Newell 2012; Loebbecke et al. 2016) or as a source of risk (Jennex and Durcikova 2014; Sarigianni et al. 2016), or pointed out that this perspective should be emphasized more (Ilvonen et al. 2015; Manhart and Thalmann 2015). The perspectives of this section have thus been sought also from other literature, since we consider it important for understanding balancing, even if the reviewed literature does not reflect this importance.

The need to share knowledge over a particular channel originates from the need to participate in a community, get up to date on what other members of the network are doing, and engage in communication that establishes trust within the network. Due to the concentration of knowledge hubs, shorter innovation cycles and the increasing complexity of knowledge, organisations face serious challenges in creating and absorbing the required knowledge on their own. Hence, organization have to engage in inter-organizational networks to stay competitive (Trkman and Desouza 2012; Schäper and Thalmann 2015). Hence, joining a communication channel and thus the community that uses it is an important decision to get updates on recent developments, to absorb and develop critical knowledge.

A communication channel connects a community and thus spreads shared knowledge to many peers, creating a need to protect the knowledge (Manhart et al. 2015). On one hand participation on a channel enhances the chances for getting feedback and to learn, but on the other hand this causes a risk of unwanted and especially not recognized knowledge spillovers (Castellaneta et al. 2016). The latter point is a particular risk in networks as not all communication partners are known (Hernandez et al. 2015) and their behaviour is difficult to observe (Di Stefano et al. 2014). As a consequence, informal measures such as trust or relational governance, and also formal measures such as contracts or NDAs, are difficult to apply (Di Stefano et al. 2014; Pahnke et al. 2015).

The time-frame of decisions about channels are twofold: The decision for joining a communication channel and thus for participating in a community has a strategic long-term character and is executed occasionally (Giarratana and Mariani 2014). In contrast, individual members decide whether to use a specific channel on a regular basis.

The balancing act of sharing and protecting knowledge in regard of a specific channel is centred on finding the benefits and risks related to that channel

(Ilvonen et al. 2015). For the benefits, the competences of the network members, the intensity/openness of sharing and the intensity of collaboration are evaluated to judge on possible knowledge gains (Toh and Polidoro 2013). For the risks, especially the competitive situation with members (Holmes et al. 2016), the sharing culture of the channel (Di Stefano et al. 2014) and the possibilities to protect knowledge in the channel (Manhart et al. 2015) are considered. Benefits and risks are contrasted having the perspective of a long-term collaboration in mind. In such a perspective, also gains from reabsorbing knowledge from others (over a longer period of time) can be taken into account (Alnuaimi and George 2015).

Measures that can help in this balancing act are, for example, establishing norms and values in networks (Di Stefano et al. 2014) or agreeing on technical features to build trusted subgroups (Manhart et al. 2015). Policies that restrict the access to communication channels (Sarigianni et al. 2016) may also be of help in the balancing. The measures aim to maximize the benefits sought from the channel and minimize the risks linked to the use of the channel.

The questions that are answered after evaluating the risks and benefits and considering the available balancing measures is: Do we enter this forum or network? Do we use this channel? After answering this question the organization can then go further into the partner and artefact aspects of deciding with whom and what knowledge is to be shared.

4.4 Partner

On the partner level, organizations evaluate whether to collaborate with a specific organization or not, and negotiate agreements about the terms and boundaries of their collaboration (e.g. Toh and Polidoro 2013; Zanarone et al. 2015). The decision about establishing partnerships are done on the organizational level, although again, individuals carry out the communication and take the partnership to practice. In addition to entering a loosely tied network of organizations that is formed by a digital channel, an organization can establish stronger partnerships with some of the collaboration partners for creating digital innovations, e.g. strategic alliances (Lin et al. 2012). For example participating on a GrabCAD challenge (Ilvonen et al. 2015) may lead into a joint product development initiative with some of the challenge participants.

The need to share knowledge between partners spurs from the reason the partnership is established: the partnering organizations bring into the mix their competences, that are used to jointly co-create new knowledge and to co-create digital innovations (Yoo et al. 2012). In order to create innovations together, the organizations need to share knowledge related to the innovation process (e.g. Lin et al. 2012; Loebbecke et al. 2016). The reason for collaboration may be to expand the market in order to have more to compete over, or to create a completely new market altogether. With digital innovation, the aim may be one of these, or both.

The need to protect knowledge in the partnership level stems from the need or organizations to protect and enhance their competitive position (e.g. Kale et al. 2000; Jean et al. 2014). Although establishing the collaboration necessary to create new innovations requires knowledge sharing (Trkman and Desouza 2012; Zanarone et al. 2015; Loebbecke et al. 2016) the flip side of this sharing is the risk of unintentional spillover of knowledge (Alnuaimi and George 2013, 2015), that benefits the collaboration partners more than the organization where the knowledge originated from (Alnuaimi and George 2013, 2015). Knowledge protection is needed thus to make most of the results of the innovation process (Olander et al. 2015).

Regarding the time frame, the partner scope of balancing activity is especially emphasized at the beginning of collaboration or an innovation project (Sazali et al. 2010; Holmes et al. 2016). Before engaging in deeper collaboration and knowledge sharing, the IPR regime should be sorted out (Bou-Llusar and Segarra-Cipre´s 2006; Nandkumar and Srikanth 2015). The agreements can have a long-term perspective, which also necessitates re-evaluations when there are changes in the way the partners collaborate (i.e. changes in the type, quantity and direction of knowledge sharing between the partners triggered, for example, by the different life-cycle stages of the co-created innovation) (Loebbecke et al. 2016).

For the balancing act that is required in partnering decisions the benefits of collaborating with a specific organization are weighed against the risks the collaboration carries (Norman 2002; Becerra et al. 2008). Evaluating the benefits includes evaluating what the partnering organization has to offer, what kind of competences and knowledge they have and how well they are able to share it to collaborators. The risk side of this is the readiness of a partnering organization to exploit knowledge that they gain in exchange, and consequently their ability to weaken the competitive position of the parent organization (Zanarone et al. 2015). These risks are weighed against the benefits and the potential outcomes as a whole are assessed (Giarratana and Mariani 2014).

The measures that an organization can take in balancing the benefits with the risks is the application of a careful IPR regime (Nandkumar and Srikanth 2015; Zanarone et al. 2015) with the partners. Careful consideration of who to partner with in the first place is essential (Toh and Polidoro 2013). Mutual trust among the partners needs to be in place for the partnership to be formed, but the trustworthiness of the partner regarding different knowledge artefacts still needs to be evaluated after the partnering decision has been made (Becerra et al. 2008; Olander et al. 2015). Careful selection decisions regarding what sort of IPR regime is to be followed, if IPR can be enforced in the country, and who from the collaborating organizations will participate are done, and these decisions have long-lasting effects on the innovation process (Nandkumar and Srikanth 2015). How well these decisions are communicated across the organizations, will then play a role in the artefact scope decisions that are made over time.

Key question: Do I share knowledge with this partner? How to define the scope of collaboration with the partners? What are suitable protection measures with this partner?

4.5 Artefact

An important scope in which knowledge sharing and protection needs to be done is the perspective of artefact: a certain piece of knowledge. Although visible in knowledge protection literature (e.g. Loebbecke et al. 2016) the common distinction between tacit and explicit knowledge (e.g. Nonaka and Toyama 2003) does not necessarily address the needs of knowledge protection efforts. Strategic management literature (e.g. Reitzig and Puranam 2009 or Ceccagnoli 2009) emphasize patentability of knowledge as a measure for need of protection: If knowledge is patentable, it has the potential to generate value for the organization, and hence it should be protected. While tacitness and explicitness help in identifying where the knowledge resides and how it is transferred and shared, the competitive value or importance of knowledge determines whether it should be protected, regardless of its type (Thalmann et al. 2014). The recognition of the importance and value of knowledge (Ilvonen et al. 2016) is hence the first step toward finding a balance between sharing and protecting that piece of knowledge.

The individual knowledge artefacts need to be shared between individual people in the daily operations to co-create digital innovations (Yoo et al. 2010). Although depending on the channel, the sharing can be multilateral and reach many people at the same time, the knowledge artefact originates from one person, who makes the decision to share the knowledge. As a precondition to the individual level knowledge sharing, the organization level decision of entering a specific channel or sharing with certain partner have been made. On the level of knowledge artefacts individual people take these decisions to practice: An organizational policy to share less valuable knowledge, costly-to-imitate knowledge, old knowledge whose risk of leakage is less destructive (Khamseh and Jolly 2014) is carried out by individuals over and over again in operative knowledge sharing situations. Current research also showed the importance for well-defined knowledge protection policies defining the knowledge boundaries of organisations (Lee et al. 2015).

On the artefact level, the timing of sharing knowledge (Alnuaimi and George 2015), the sequencing of sharing (Moschini and Yerokhin 2008), the use of secrecy (Castellaneta et al. 2016), or hiding of details (Manhart et al. 2015) can be used as effective protection measures.

The consideration of time frame of the balancing decision from the point of view of the knowledge artefact is very wide. A decision may have both short term and long-term implications, and these need to be taken into account when the balancing is done. For example, sharing a piece of knowledge of a certain technology component that is being developed may grant an organization access to a network and give some knowledge of other components in exchange. In the short term, the outcome of sharing is thus positive. However, in addition to this short-term perspective the individual that shares the knowledge artefact needs to evaluate, whether sharing this knowledge will result in losing competitive advantage. Although predicting future outcomes may be a gamble, considering the temporal dimension of sharing knowledge artefacts is better than to ignore them (Alnuaimi and George 2015).

As knowledge is socially constructed (Marabelli and Newell 2012) the knowledge sharing across organizational boundaries creates new knowledge that is embedded in the network of people from different organizations. Although knowledge may be sticky and transferring it may take effort (Trkman and Desouza 2012) there is always knowledge that is fairly easily shared and transferred to others so that collaboration is possible (Loebbecke et al. 2016). The balancing act requires contrasting the benefits of sharing the knowledge artefact with the risk, while considering both short-term and long-term effects of the decision. The contrasting elements here are not only the role of the knowledge artefact for the competitive position of the organization, but also what may be gained in return (Zanarone et al. 2015; Loebbecke et al. 2016). From physics we learn that every action has an equal and opposite reaction. This fundamental law, however, does not necessarily apply to collaboration between organizations. Careful consideration is needed to determine and predict what kind of reaction sharing a particular knowledge artefact at a specific time may result in.

The knowledge artefact needs to drive the balancing measures that are taken (Loebbecke et al. 2016). Balancing of knowledge sharing and protection done by individuals can be characterized by using "cat's whiskers" or an art (Jarvenpaa and Majchrzak 2016). This means that an individual needs to be aware of and keep in mind the contracts the organizations has done with the collaboration partners (Zanarone et al. 2015), and remember the organization level instructions about knowledge sharing and withholding (Manhart and Thalmann 2015) while engaging in collaboration and knowledge exchange with people from other organizations.

Key question answered for balancing regarding the knowledge artefact is: Should I share this piece of knowledge with this particular knowledge sharing partner in this particular situation and channel? In other words, the artefact level is the one that in the end carries out the balancing between knowledge protection and sharing.

5 Conclusions

The above discussion on the perspectives to balancing knowledge sharing and protection show that the balancing act is not trivial, nor does it happen only at a certain point of time at the management of a digital innovation endeavour. Our review shows that there is literature, which addresses the need to find balance between sharing and protection, although most of the papers describe the challenge, and empirically testing the ideas is left in the "avenues for future research" sections of the papers. The varied theoretical approaches of the reviewed papers show that there are many approaches as to how this empirical examination can be done, but no approach specifically focuses on digital innovations so far.

In our review, we showed that the balancing happens on three levels of detail: (1) the decision about using certain communication channels, (2) the decision to share/collaborate with certain partners and (3) the decision to share a certain knowledge artefacts. Thereby, it turned out that balancing on all three levels require a careful consideration of the benefits and risks of sharing or not sharing. For creating digital innovations, all three levels are relevant as well. The decision about participation in open-innovation platforms or knowledge exchange networks are such a strategic decision on the channel level. In addition to a careful assessment of risks and benefits, also effective legal and contractual measures are needed to successfully exploit the co-created digital innovation. For the decision to collaborate with a partner, research findings from the literature on strategic alliances seem also applicable here. However, it should be noted that creating digital innovation typically requires several different organisations (Yoo et al. 2010). On the artefact, level and thus on the level, individuals are deciding, clear policies and rules of behaviour are needed. On the one hand, to prevent unwanted spill-overs but on the other hand also to push the knowledge sharing needed for creating digital innovations.

Future research should specifically focus on the practices applied in spaces where digital innovations are co-created and also think about formal and informal measures to balance knowledge sharing and protection in such spaces. Such measures are critical as they ensure a lively contribution needed for co-creating digital innovations.

Acknowledgements The Know-Center and Pro2Future are funded within the Austrian COMET Program—Competence Centers for Excellent Technologies—under the auspices of the Austrian Federal Ministry of Transport, Innovation and Technology, the Austrian Federal Ministry of Economy, Family and Youth and by the State of Styria. COMET is managed by the Austrian Research Promotion Agency FFG.

This research is partly funded by the Finnish Foundation for Economic Education for the second author.

References

Ahmad, A., Bosua, R., & Scheepers, R. (2014). Protecting organizational competitive advantage: A knowledge leakage perspective. *Computers & Security, 42,* 27–39.

Alnuaimi, T., & George, G. (2013). The retrieval of knowledge after spillovers. In *Academy of management proceedings*. Briarcliff Manor: Academy of Management.

Alnuaimi, T., & George, G. (2015). Appropriability and the retrieval of knowledge after spillovers. *Strategic Management Journal*.

Azar, O. H., & Brock, D. M. (2008). A citation-Based ranking of strategic management journals. *Journal of Economics & Management Strategy, 17*(3), 781–802.

Becerra, M., Lunnan, R., & Huemer, L. (2008). Trustworthiness, risk, and the transfer of tacit and explicit knowledge between alliance partners. *Journal of Management Studies, 45*(4), 691–713.

Bock, G.-W., Zmud, R. W., Kim, Y.-G., & Lee, J.-N. (2005). Behavioral intention formation in knowledge sharing: Examining the roles of extrinsic motivators, social-psychological forces, and organizational climate. *MIS Quarterly*, 87–111.

Bogers, M. (2011). The open innovation paradox: Knowledge sharing and protection in R&D collaborations. *European Journal of Innovation Management, 14*(1), 93–117.

Bou-Llusar, J. C., & Segarra-Cipre´S, M. (2006). Strategic knowledge transfer and its implications for competitive advantage: An integrative conceptual framework. *Journal of Knowledge Management, 10*(4), 100–112.
Castellaneta, F., Conti, R., & Kacperczyk, A. (2016). Money secrets: How does trade secret legal protection affect firm market value? Evidence from the uniform trade secret act. *Strategic Management Journal.*
Ceccagnoli, M. (2009). Appropriability, preemption, and firm performance. *Strategic Management Journal, 30*(1), 81–98.
Chan, P. C. W., & Lee, W. B. (2011). Knowledge audit with intellectual capital in the quality management process: An empirical study in an electronics company. *The Electronic Journal of Knowledge Management, 9*(2), 98–116.
Coff, R. W., Coff, D. C., & Eastvold, R. (2006). The knowledge-leveraging paradox: How to achieve scale without making knowledge imitable. *Academy of Management Review, 31*(2), 452–465.
Conner, K. R., & Prahalad, C. K. (1996). A resource-based theory of the firm: Knowledge versus opportunism. *Organization Science, 7*(5), 477–501.
Cramton, C. D. (2001). The mutual knowledge problem and its consequences for dispersed collaboration. *Organization Science, 12*(3), 346–371.
Crossan, M. M., & Apaydin, M. (2010). A multi-dimensional framework of organizational innovation: A systematic review of the literature. *Journal of Management Studies, 47*(6), 1154–1191.
Daghfous, A., Belkhodja, O., & Angell, L. C. (2013). Understanding and managing knowledge loss. *Journal of Knowledge Management, 17*(5), 639–660.
David, W., & Fahey, L. (2000). Diagnosing cultural barriers to knowledge management. *The Academy of Management Executive, 14*(4), 113–127.
De Faria, P., & Sofka, W. (2010). Knowledge protection strategies of multinational firms—A cross-country comparison. *Research Policy, 39*(7), 956–968.
Desouza, K. C. (2006). Knowledge security: An interesting research space. *Journal of Information Science and Technology, 3*(1).
Di Stefano, G., King, A. A., & Verona, G. (2014). Kitchen confidential? Norms for the use of transferred knowledge in gourmet cuisine. *Strategic Management Journal, 35*(11), 1645–1670.
Erickson, G. S., & Rothberg, H. N. (2009). Intellectual capital in business-to-business markets. *Industrial Marketing Management, 38*(2), 159–165.
Fauchart, E., & Von Hippel, E. (2008). Norms-based intellectual property systems: The case of French chefs. *Organization Science, 19*(2), 187–201.
Giarratana, M. S., & Mariani, M. (2014). The relationship between knowledge sourcing and fear of imitation. *Strategic Management Journal, 35*(8), 1144–1163.
Grant, R. M. (1996a). Prospering in dynamically-competitive environments: Organizational capability as knowledge integration. *Organization Science, 7*(4), 375–387.
Grant, R. M. (1996b). Toward a knowledge-based theory of the firm. *Strategic Management Journal, 17*(2), 109–122.
Grant, R. M. (1996c). Toward a knowledge-based theory of the firm. *Strategic Management Journal, 17*(S2), 109–122.
Haldin-Herrgard, T. (2000). Difficulties in diffusion of tacit knowledge in organizations. *Journal of Intellectual Capital, 1*(4), 357–365.
Hamel, G. (1991). Competition for competence and inter-partner learning within international strategic alliances. *Strategic Management Journal, 12*(4), 83–103.
Hansen, M. T. (1999). The search-transfer problem: The role of weak ties in sharing knowledge across organization subunits. *Administrative Science Quarterly, 44*(1), 82–111.
Hendriks, P. (1999). Why share knowledge? The influence of ICT on the motivation for knowledge sharing. *Knowledge and Process Management, 6*(2), 91.
Hernandez, E., Sanders, W. G., & Tuschke, A. (2015). Network defense: Pruning, grafting, and closing to prevent leakage of strategic knowledge to rivals. *Academy of Management Journal, 58*(7), 1233–1260.

Holmes, R. M., Li, H., Hitt, M. A., Deghetto, K., & Sutton, T. (2016). The effects of location and MNC attributes on MNCs' establishment of foreign R&D Centers: Evidence from China. *Long Range Planning, 49*(5), 594–613.

Ilvonen, I., Alanne, A., Helander, N., & Väyrynen, H. (2016). Knowledge sharing and knowledge security in Finnish companies. In *2016 49th Hawaii International Conference on System Sciences (HICSS)*. IEEE.

Ilvonen, I., Jussila, J., Kärkkäinen, H., & Päivärinta, T. (2015). Knowledge security risk management in contemporary companies–Toward a proactive approach. In *2015 48th Hawaii International Conference on System Sciences (HICSS)*. IEEE.

Jarvenpaa, S., & Majchrzak, A. (2016). Interactive self-regulatory theory for sharing and protecting in inter-organizational collaborations. *Academy of Management Review, 41*(1), 9–27.

Jarvenpaa, S. L., & Majchrzak, A. (2010). Research commentary-vigilant interaction in knowledge collaboration: Challenges of online user participation under ambivalence. *Information Systems Research, 21*(4), 773–784.

Jean, R. J., Sinkovics, R. R., & Hiebaum, T. P. (2014). The effects of supplier involvement and knowledge protection on product innovation in customer–supplier relationships: A study of global automotive suppliers in China. *Journal of Product Innovation Management, 31*(1), 98–113.

Jennex, M., & Durcikova, A., (2013). Assessing knowledge loss risk. In *International Conference on System Sciences, HICSS46, Hawaii 46th Hawaii*. IEEE Computer Society.

Jennex, M., & Durcikova, A. (2014). Integrating IS security with knowledge management: Are we doing enough? *International Journal of Knowledge Management (IJKM), 10*(2), 1–12.

Jordan, J., & Lowe, J. (2004). Protecting strategic knowledge: insights from collaborative agreements in the aerospace sector. *Technology Analysis & Strategic Management, 16*(2), 241–259.

Kale, P., Singh, H., & Perlmutter, H. (2000). Learning and protection of proprietary assets in strategic alliances: Building relational capital. *Strategic Management Journal, 21*(3), 217–237.

Khamseh, H. M., & Jolly, D. (2014). Knowledge transfer in alliances: The moderating role of the alliance type. *Knowledge Management Research & Practice, 12*(4), 409–420.

Lee, J., Min, J., & Lee, H. (2015). *Setting a knowledge boundary for enhancing work coordination and team performance: Knowledge protection regulation across teams*. Fort Worth, USA: ICIS.

Lee, S. C., Chang, S. N., Liu, C. Y., & Yang, J. (2007). The effect of knowledge protection, knowledge ambiguity, and relational capital on alliance performance. *Knowledge and Process Management, 14*(1), 58–69.

Leyland, M. L. (2006). The role of culture on knowledge transfer: The case of the multinational corporation. *The Learning Organization, 13*(3), 257–275.

Lin, C., Wu, Y. J., Chang, C., Wang, W., & Lee, C. Y. (2012). The alliance innovation performance of R&D alliances-the absorptive capacity perspective. *Technovation, 32*(5), 282–292.

Loebbecke, C., Van Fenema, P. C., & Powell, P. (2016). Managing inter-organizational knowledge sharing. *The Journal of Strategic Information Systems, 25*(1), 4–14.

Manhart, M., & Thalmann, S. (2015). Protecting organizational knowledge: A structured literature review. *Journal of Knowledge Management, 19*(2), 190–211.

Manhart, M., Thalmann, S., & Maier, R. (2015). The Ends of Knowledge Sharing in Networks: Using Information Technology to Start Knowledge Protection. In *23rd European Conference on Information Systems (ECIS)*. Münster, Germany.

Marabelli, M., & Newell, S. (2012). Knowledge risks in organizational networks: The practice perspective. *The Journal of Strategic Information Systems, 21*(1), 18–30.

Mazloomi Khamseh, H., & Jolly, D. R. (2008). Knowledge transfer in alliances: Determinant factors. *Journal of Knowledge Management, 12*(1), 37–50.

Mcdermott, R., & O'dell, C. (2001). Overcoming cultural barriers to sharing knowledge. *Journal of Knowledge Management, 5*(1), 76–85.

Moschini, G., & Yerokhin, O. (2008). Patents, research exemption, and the incentive for sequential innovation. *Journal of Economics & Management Strategy, 17*(2), 379–412.
Nandkumar, A., & Srikanth, K. (2015). Right person in the right place: How the host country IPR influences the distribution of inventors in offshore R&D projects of multinational enterprises. *Strategic Management Journal*.
Nelson, R. R., & Winter, S. G. (1982). *An evolutionary theory of economic change*. Cambridge: Harvard University Press.
Nonaka, I. (1991). The knowledge-creating company. *Harvard Business Review, 69*(6), 96–104.
Nonaka, I. (1994). A dynamic theory of organizational knowledge creation. *Organization Science, 5*(1), 14–37.
Nonaka, I., & Takeuchi, H. (1995). *The knowledge-creating company*. New York: Oxford University Press.
Nonaka, I., & Toyama, R. (2003). The knowledge-creating theory revisited: Knowledge creation as a synthesizing process. *Knowledge Management Research & Practice, 1*(1), 2–10.
Norman, P. M. (2002). Protecting knowledge in strategic alliances: Resource and relational characteristics. *The Journal of High Technology Management Research, 13*(2), 177–202.
Norman, P. M. (2004). Knowledge acquisition, knowledge loss, and satisfaction in high technology alliances. *Journal of Business Research, 57*(6), 610–619.
O'dell, C., & Grayson, C. J. (1998). If only we knew what we know. *California Management Review, 40*(3), 154–174.
Olander, H., Vanhala, M., & Hurmelinna-Laukkanen, P. (2014). Reasons for choosing mechanisms to protect knowledge and innovations. *Management Decision, 52*(2), 207–229.
Olander, H., Vanhala, M., Hurmelinna Laukkanen, P., & Blomqvist, K. (2015). HR related knowledge protection and innovation performance: The moderating effect of trust. *Knowledge and Process Management*.
Pahnke, E., Mcdonald, R., Wang, D., & Hallen, B. (2015). Exposed: Venture capital, competitor ties, and entrepreneurial innovation. *Academy of Management Journal, 58*(5), 1334–1360.
Pawlowski, J. M. (2008). Culture profiles: Facilitating global learning and knowledge sharing. In *16th International Conference on Computers in Education (ICCE)*. Taipei, Taiwan.
Pinjani, P., & Palvia, P. (2013). Trust and knowledge sharing in diverse global virtual teams. *Information & Management, 50*(4), 144–153.
Reagans, R., & Mcevily, B. (2003). Network structure and knowledge transfer: The effects of cohesion and range. *Administrative Science Quarterly, 48*(2), 240–267.
Reitzig, M., & Puranam, P. (2009). Value appropriation as an organizational capability: The case of IP protection through patents. *Strategic Management Journal, 30*(7), 765–789.
Riege, A. (2005). Three-dozen knowledge-sharing barriers managers must consider. *Journal of Knowledge Management, 9*(3), 18–35.
Rowley, J., & Slack, F. (2004). Conducting a literature review. *Management Research News, 27* (6), 31–39.
Sarigianni, C., Thalmann, S., & Manhart, M. (2016). Protecting knowledge in the financial sector: An analysis of knowledge risks arising from social media. In *49th Hawaii International Conference on System Sciences (HICSS), 2016*. IEEE.
Sazali, A., Raduan, C., Jegak, U., & Haslinda, A. (2010). The effects of partner protectiveness and transfer capacity on degree of inter-firm technology transfer in international joint ventures. *International Journal of Economics and Management, 4*(2), 334–349.
Schäper, S., & Thalmann, S. (2015). Addressing challenges for informal learning in networks of organizations. In *23rd European Conference on Information Systems (ECIS)*. Münster, Germany.
Serenko, A., & Bontis, N. (2013). Global ranking of knowledge management and intellectual capital academic journals: 2013 update. *Journal of Knowledge Management, 17*(2), 307–326.

Simonin, B. L. (2004). An empirical investigation of the process of knowledge transfer in international strategic alliances. *Journal of International Business Studies, 35*(5), 407–427.
Sofka, W., Shehu, E., & De Faria, P. (2014). Multinational subsidiary knowledge protection-do mandates and clusters matter? *Research Policy, 43*(8), 1320–1333.
Sveiby, K.-E. (2001). A knowledge-based theory of the firm to guide in strategy formulation. *Journal of Intellectual Capital, 2*(4), 344–358.
Sveiby, K.-E., & Simons, R. (2002). Collaborative climate and effectiveness of knowledge work-an empirical study. *Journal of Knowledge Management, 6*(5), 420–433.
Szulanski, G. (1996a). Exploring internal stickiness: Impediments to the transfer of best practice within the firm. *Strategic Management Journal, 17*(S2), 27–43.
Szulanski, G. (1996b). Exploring internal stickiness: Impediments to the transfer of best practice within the firm. *Strategic Management Journal 17*(Winter Special Issue), 27–43.
Szulanski, G. (2000). The process of knowledge transfer: A diachronic analysis of stickiness. *Organizational Behavior and Human Decision Processes, 82*(1), 9–27.
Thalmann, S., Manhart, M., Ceravolo, P., & Azzini, A. (2014). An integrated risk management framework: Measuring the success of organizational knowledge protection. *International Journal of Knowledge Management, 10*(2), 28–42.
Toh, P. K., & Polidoro, F. (2013). A competition-based explanation of collaborative invention within the firm. *Strategic Management Journal, 34*(10), 1186–1208.
Trkman, P., & Desouza, K. C. (2012). Knowledge risks in organizational networks: An exploratory framework. *The Journal of Strategic Information Systems, 21*(1), 1–17.
Von Hippel, E. (2009). Democratizing innovation: The evolving phenomenon of user innovation. *International Journal of Innovation Science, 1*(1), 29–40.
Von Krogh, G. (2012). How does social software change knowledge management? Toward a strategic research agenda. *The Journal of Strategic Information Systems, 21*(2), 154–164.
Webster, J., & Watson, R. T. (2002a). Analyzing the past to prepare for the future: Writing a literature review. *Management Information Systems Quarterly, 26*(2).
Webster, J., & Watson, R. T. (2002b). Analyzing the past to prepare for the future: Writing a literature review. *MIS Quarterly*, xiii–xxiii.
Yoo, Y., Boland, R. J., Jr., Lyytinen, K., & Majchrzak, A. (2012). Organizing for innovation in the digitized world. *Organization Science, 23*(5), 1398–1408.
Yoo, Y., Henfridsson, O., & Lyytinen, K. (2010). Research commentary-The new organizing logic of digital innovation: An agenda for information systems research. *Information Systems Research, 21*(4), 724–735.
Yoo, Y., & Kanawattanachai, P. (2001). Developments of transactive memory systems and collective mind in virtual teams. *The International Journal of Organizational Analysis, 9*(2), 187–208.
Zanarone, G., Lo, D., & Madsen, T. L. (2015). The double edged effect of knowledge acquisition: How contracts safeguard pre existing resources. *Strategic Management Journal.*
Zimmermann, A., & Ravishankar, M. (2014). Knowledge transfer in IT offshoring relationships: The roles of social capital, efficacy and outcome expectations. *Information Systems Journal, 24*(2), 167–202.

Author Biographies

Stefan Thalmann, Dr., is Assistant Professor at the Graz University of Technology in Austria, AREA Manager for cognitive decision support at the Pro2Future GmbH and senior researcher at the Know-Center GmbH. His research interests are in knowledge protection, knowledge management, technology-enhanced learning and predictive maintenance.

Ilona Ilvonen, D.Sc.(Tech), is a Post-Doctoral Researcher and University Teacher at Tampere University of Technology in Finland. Her dissertation on the topic of knowledge security was accepted in 2013, and since she has continued to pursue her research interest on understanding knowledge sharing and protection. She has published numerous conference articles on the information security, knowledge management and information systems fields, and considers cross-disciplinary research to be essential for reaching understanding of complex phenomena such as knowledge protection.

Localizing Knowledge in Networks of SMEs—Implication of Proximities on the IT Support

Stefan Thalmann and Stephan Schäper

Abstract The concentration of knowledge development around the economy's big players and into few regions leads to rising inequalities of knowledge distribution. Due to shorter innovation cycles, more and more knowledge is ephemeral. To stay competitive, both trends force organizations to absorb increasingly more distant knowledge faster and with less opportunities of reuse. This situation is particularly challenging for small and medium-sized enterprises (SMEs) with their limited resources. Joining networks focused on the acquisition of external knowledge and is one promising solution for SMEs. So far, there is little research on strategies that facilitate localization of knowledge, particularly in networks of SMEs. In this paper, therefore, we first identified the phases of localizing external knowledge, followed by an investigation on the role of proximities during the localization process and the potential for supportive IT.

1 Introduction

The concentration of knowledge development around big players and areas with more mature research infrastructure, changes the conditions under which smaller and lower-income competitors access world-class knowledge[1]. This trend is underlined by a current OECD study, stating that over 33% of R&D and around 25% of skilled employment occurs in the top 10% of the OECD regions, coming to

[1]OECD, OECD Science, Technology and Industry Outlook 2014, OECD Publishing, 2014

S. Thalmann (✉)
Know-Center GmbH and Pro2Future GmbH, Graz, Austria
e-mail: stefan.thalmann@tugraz.at

S. Thalmann
Institute for Interactive Systems and Data Science,
Graz University of Technology, Graz, Austria

S. Schäper
Julius Blum GmbH, Höchst, Austria

K. North et al. (eds.), *Knowledge Management in Digital Change*, Progress in IS,
https://doi.org/10.1007/978-3-319-73546-7_11

the conclusion that the strongest interactions between stakeholders take place within a radius of approximately 200 km.[2] This high level of concentration tends to increase knowledge inequalities, leading to "islands of excellence", which concentrate high-performance innovators that co-exist with groups of poorly performing companies.[3]

Thus, organizations have to assimilate external knowledge from more places, for more complex outputs, and originating from multiple sectors, locations, and cultural settings (Malecki 2010). These conditions hinder knowledge flows as they make knowledge assimilation more difficult. To overcome these barriers, organizations have to spend more efforts on assimilating and synthesizing such distant knowledge by relating it to their context (Malecki 2010).

Due to shorter innovation cycles, knowledge nowadays becomes more and more dynamic, short-cycled, and therefore ephemeral (Salovaara and Tuunainen 2015). It has been observed that globalization is accompanied by an increasing production of ephemeral knowledge in the form of coordination standards, including those between organizations (Torre and Rallet 2005). This trend of short-cycled, ephemeral knowledge is particularly challenging for SMEs with their limited resources and dependency on exploiting external knowledge sources in order to stay competitive (Egbu et al. 2005). To cope with this challenge, organizations rely more and more on networks, facilitating the absorption of knowledge (Arora and Gambardella 1990; Hagedoorn 1995; Gulati 1999).

Networks are faced with this demand from current and potential participants, and are seeking new opportunities to support their members in acquiring this crucial external knowledge from globally distributed sources. The most promising way to support knowledge acquisition is the adaptation of knowledge, taking the contextual requirements of the members into account (Thalmann 2014). Hence, subgroups with high proximities are an important pre-condition for knowledge sharing, knowledge transfer, and technology acquisition (Gertler 1995). The important question for networks in this regard is: Which proximities are relevant for the localization of knowledge?

Thus, the concept of proximity is seen as crucial to cope with the dynamics of knowledge (Torre and Rallet 2005). The concept of proximities has become a popular and powerful theoretical basis for approaching mechanisms behind networks' emergence, evolution, and structural changes (Cantner and Graf 2006; Balland 2012). Networks, therefore, need support for (1) offering the right capabilities to localize knowledge and (2) building groups of members with high proximities.

Within our study, we investigate the process of localizing knowledge in networks of SMEs. We are particularly investigating the role of proximities during the localization as well as the potential for supportive IT.

---

[2]OECD, Regions and Innovation: Collaborating across Borders, OECD Reviews of Regional Innovation, OECD Publishing, 2013.

[3]OECD, OECD Science, Technology and Industry Outlook 2014, OECD Publishing, 2014.

2 Background

From an organizational theorist's view, knowledge is embedded in, and constructed from and through social relationships and interactions (Nonaka 1991; Blackler 1995). Hence, knowledge is embedded into a context (Alavi and Leidner 2001), which needs to be considered while sharing knowledge(Argote and Ingram 2000). In the literature, ways to support knowledge sharing and facilitate knowledge flows by context-aware systems are discussed (e.g. Lum and Lau 2002; Zimmermann et al. 2005; Williams 2007). However, so far, the literature has focused on long-term stable knowledge and neglected ephemeral knowledge (Salovaara and Tuunainen 2013, 2015).

In contrast to the more stable kernel knowledge, characterized by its sustainability and reusability (Leseure and Brookes 2004), the so-called ephemeral knowledge tends to be useful for distinct projects or cases and a defined timeframe, but lacks in being useful again in other scenarios or later in time (Leseure and Brookes 2004; Salovaara and Tuunainen 2015). Examples include knowledge about technologies becoming obsolete with the release of new, replacing technologies or knowledge about guidelines and standards also being outdated with the release of new, adopted standards or guidelines. Thus, ephemeral knowledge cannot solely be viewed as a static container managed via traditional knowledge management approaches such as permanent repositories of knowledge (Salovaara and Tuunainen 2013), but rather more dynamic and collaborative processes are needed to assimilate this kind of knowledge (Salovaara and Tuunainen 2015). The marginal cost of absorbing knowledge rises with the distance and lower levels of contextual proximity (Audretsch 1998), and more acquisition barriers can occur due to the missing localization of knowledge (Howells 2002).

Thus, organizations participating in networks having higher knowledge identification, assimilation, and transformation capabilities are more likely to successfully absorb external knowledge (Cheng et al. 2014). The successful transfer of knowledge through networks requires a "common stock of knowledge" and a shared system of meaning (Kogut and Zander 1992), particularly to interpret and apply the newly acquired knowledge correctly (Howells 2002). This requires localized social capital in terms of shared norms as well as common knowledge and networks, and thus limits the size of the social system in which knowledge can be assimilated (Laursen et al. 2012). Hence, forming small subgroups, with a great match of common prior knowledge and values, facilitates the reception of knowledge (Tortoriello et al. 2012).

Similar firms located in a region share a common set of values and knowledge, forming a cultural environment (Dahl and Pedersen 2004), where successful knowledge spill-overs are more likely within a similar local context (Audretsch and Lehmann 2005). Geographical agglomerations have a territorial configuration comprising general formal constraints, communal regimes, or a common climate of understanding and trust, which most likely enhances the localization process

(Maskell 2001). Further, benefits from spillovers decline with distance as well (Keller 2004). Particularly relevant for localizing knowledge are shared values, meanings, and understandings, specifically territorially embedded and tacit knowledge, as well as institutional structures (Hudson 1999). The local context is a critical lens through which organizations perceive their environment (Fuellhart 1999).

As a sufficient quantity of internal competences is necessary to absorb knowledge produced in other regions (Autant-Bernard et al. 2013) and as it is difficult for SMEs to hold the critical quantity of internal competences, it can be assumed that the absorption of non-localized knowledge is challenging for SMEs. Thus, the local context can be hypothesized to be a particularly important factor mediating the ability of small firms to capture relevant information from the environment due to their limited resources (Donckels and Lambrecht 1997).

3 Procedure

The primary goal of our study is to investigate which proximities are relevant for the localization of external knowledge in networks of SMEs, and how networks can improve their localization processes.

Based on this goal, we will answer the following research question: How can networks of organizations make use of proximities to enhance the localization of knowledge? To answer this, it is first necessary to identify the current process of localizing external knowledge in networks. Further, we will investigate which proximities are relevant while performing this process. By doing so, we expect to gather not only insights about the proximities, but also about ways in which networks should tackle the localization and particularly the implications of proximities on the IT support for this purpose.

We considered semi-structured interviews in networks of SMEs sufficient to answer our research question. A semi-structured interview was selected because the varied professional, educational, and personal histories of the sample group would be highly likely to hinder the use of a standardized interview approach (Louise Barriball and While 1994). In order to explore the respondents' opinions, the possibility to clarify interesting and relevant issues, foster information completeness, and explore sensitive topics within each interview turned out as the main advantages (Louise Barriball and While 1994) during the investigation.

In total, we led 53 interviews in eleven networks of organizations, between January and October 2014. We conducted the interviews in the scope of a research project, which focuses on IT support for informal learning and knowledge sharing in networks of SMEs. The networks and key informants were selected based on convenience sampling and the networks are affiliated to the research project. We organized our study in two phases. The interviews were held in German

(Construction) and English (Healthcare). We asked the individuals to describe their personal behavior regarding localization of external knowledge and the social relationships of acting as representatives of their organization. Further, they reflected about their learning and knowledge sharing behavior. Thus, our unit of analysis is the individual itself, however, acting in an organization that is part of one or many networks.

First, we interviewed eleven key informants occupying a central management role in one of the eleven networks. The interviews took approximately 2 h each and were conducted face-to-face. The goal was to get an initial overview of the networks and to identify promising candidates for the subsequent informant interviews. We approached five SME networks in Germany and six in the United Kingdom and each key informant represented one. We selected networks in which informal learning in the workplace is important and IT is already used for informal learning. Second, we performed 42 informant interviews with members identified by the key informants. We conducted the informant interviews via telephone, and they took approximately one hour each. Eight interviewees had less than five years and 45 had more than five years of working experience, indicating that most interviewees had profound working experience. Table 1 provides a description of the investigated networks (sector, number of member organizations) and the number of performed interviews.

The audio-recorded interviews were transcribed verbatim and cleansed afterwards. The data analysis of the transcripts was then done by applying an informed inductive coding procedure based on Mayring (2014), carried out via Atlas.ti. Based on these first insights, we developed 16 codes for our three dimensions of analysis (proximities, localization of knowledge, networks) including coding rules. The whole data analysis process was accompanied by multiple meetings where (1) the meanings of the codes were clarified and discussed, and (2) initial findings were discussed and continuously challenged.

Table 1 Network overview

ID	Sector	Member orgs.	Number of interviews
N1	Construction	130	6
N2	Construction	30	6
N3	Construction	92	5
N4	Construction	270	6
N5	Construction	~1600	6
N6	Healthcare	41	2
N7	Healthcare	27	5
N8	Healthcare	538	6
N9	Healthcare	50	2
N10	Healthcare	~2600	4
N11	Healthcare	150	5

4 Findings

4.1 The Localization Process

We recognized four phases for the process of localizing knowledge in networks of SMEs:

Initiation of the Localization

One important trigger of localizing external knowledge is the need to acquire new knowledge to solve work problems. A new work context, new materials, or new machines characterize the need for knowledge acquisition. Further, in cases where new knowledge is available or employees become aware of new knowledge, the localization process is initiated. Examples include newly published guidelines or standards or newly available products. In these cases, the person becoming aware of the new knowledge initiates a discussion in his personal network and his own organization. Later, a request to the network level is initiated, in case the application of new knowledge is vague or interdependencies are not evaluated.

Here, the backing from the network is considered crucial, as one interviewee explained [N5-a]: "If a new standard is released, it is crucial to identify the relevant aspects. In most cases, only 10% of this long document are relevant for us. […] We need safety and reliability to avoid claims for compensation if we interpreted details of the new standard wrongly." Another interviewee pointed out [N9-a]: "All the guidelines need to be interpreted, and that's why I've talked about mind lines, which is our interpretation of the guidelines, and there's a lot of work." Further, a senior member of the network initiates a request in case they did not find a suitable solution in the local network, and he considers the usefulness and relevance for a potential solution for other network members to be high.

Evaluation and Problem Solving

After escalating the problem or the need for specifying the application of new knowledge, the network management forwards the issue to domain experts or discusses it in meetings. One network member said [N7-a]: "One member might volunteer to sit and read it and go through it with the GP and then send it around and then you can decide whether to implement it or not." Further, the evaluation of the issue in a larger group can also involve other subgroups within the network. Here, the chairperson of a regional network said [N5-a]: "First I contacted the chairpersons of the other member networks and they discussed this in their local meetings. Then they reported their point of view and possible findings in the next meeting." In one network, such local meetings are institutionalized as topic specific subgroups with the local focus [N1-e]: "Local aspects are considered in the regional subgroups. […] where people talk about how things work in their organizations and how they have solved a specific problem […] that is very important because [legal bodies] have no idea how businesses are running."

If needed, the networks investigate new solutions by contacting experts, as one interviewee said [N1-b]: "Firstly, we try to evaluate the issue internally, however in the extreme case we involve external experts." Another interviewee said [N3-a]: "We have two domain experts. They can provide a solution within 36 h." We also found that evaluation and problem solving can be an iterative process, especially in the network where subgroups serve to evaluate the applicability of new knowledge. Here, the results can also be escalated back to those who introduced a new work context, new materials, or new machines and a new iteration of this phase starts again.

Formalization

The networks need to consolidate the outcome of the prior phase before application. One interviewee explained [N2-a]: "The challenge is to aggregate the knowledge and to formalize the different insights in a way that we can distribute it in a formalized and quality-proven way. This means that we have knowledge on a good level, which is checked and evaluated, to ensure that we avoid mistakes." The network management is responsible for the formalization of knowledge, supported by internal as well as external domain experts. Frequently mentioned in this context are lawyers to evaluate the legal consequences. One interviewee said [N1-b]: "I always recommend one solution in an information letter and the balancing of yes or no is included." Finally, network members need a clear recommendation to reduce their concerns. Thus, the final goal of the formalization process is a piece of formalized knowledge, which is rigorously evaluated, for which the network takes the responsibility and gives clear advice as to how the knowledge can be applied in an appropriate way.

Distribution

After formalizing the knowledge, it is adapted to the target groups. One interviewee said [N3-a]: "We had a new wood protection standard last year. In this case, we had to adapt the details accordingly. […] we try to focus on the crucial aspects and we say pay attention to this and that aspect […] because you cannot prepare everything." After adopting it, the network distributes the knowledge via newsletters, info mails, presentations, or formal training sessions. One interviewee explained [N1-b]: "New results, which need to be distributed quickly, are sent via e-mail newsletter, and we also have a portal with all attachments."

4.2 *Proximities*

We found that proximities between network members play an important role during the described process of localizing knowledge. In the following, we investigate how the different proximities influence the localization process.

Cognitive Proximity

The cognitive proximity describes the similarities of the personal or organizational knowledge base needed to learn and innovate (Boschma 2005).

Cognitive proximity affects the localization of knowledge as it lowers the barriers for knowledge transfer. One network member said [N2-b]: "The experts simply have a common stack of knowledge, they know what not to do and it's not necessary to explain things over and over again." Further, the cognitive proximity is not necessarily restricted to regions as one interviewee pointed out that [N11-c]: "There're a few people in different parts of the country who kind of know what's going on in their areas and are connected." The cognitive proximity is also one important criterion for forming subgroups to develop and evaluate knowledge. Here, a common knowledge base with diverse, but complementary capabilities could be one outcome.

Different personal networks can have different purposes. One interviewee said [N1-c]: "A 1A network is a network with decision-makers and 1B or 2 networks are networks to get information from a very specific community." Here, decision-makers or people from one specific community have a high cognitive proximity in one specific area of relevance. Thus, the member classifies the networks according to the proximity and the potential goal in the localization process. Thereby, it appears that the group size effects the knowledge exchange and particularly the homogeneity of the group as an interviewee explained [N1-e]: "The larger the group, the more difficult it is to discuss topics, and if you have a small and stable group you can easily refer to previously discussed aspects." However, one interviewee explained some downsides of groups with a high cognitive proximity [N2-b]: "If you want to find new solutions this [homogenous group] can also be an obstacle."

Summing up, the cognitive proximity seems to be one important dimension in selecting knowledge transfer partners and forming personal help and advice networks. However, during the localizing process, the cognitive proximity is not related to the entire knowledge base. Rather, the cognitive proximity relates only to those parts of the knowledge base, needed to localize the knowledge. The consequence is a multitude of goal-oriented and overlapping help and advice networks.

Organizational Proximity

The organizational proximity describes the similarities in organizational arrangements and their economic and financial (inter)dependencies, as well as norms and values (Boschma 2005).

The degree of organizational proximity is high within one industry sector, which is beneficial for the adaption of knowledge by associations. Here, one interviewee working in an industry association said [N1-d]: "The administration knows nothing about the organizational procedures in our companies. They issue regulations without considering the impacts. Here the industry association comes into play and intervenes. We ask our members and forward a proposal […] In the best case, they adapt the regulation." The advantage there is that knowledge, which is subject to regulations, could be adapted to organizational characteristics, considering their high organizational proximity.

The communication rules, as well as the agreement on common goals or the scope of the network, are important so that people feel comfortable. Such rules can be enforced by an administrator [N4-a]: "Our administrator is very rigid. He reads a lot and boots out people or deletes member profiles if necessary." However, the norms and values can also be a consensus in the group and members can be selected according to their conformance to the norms and values [N4-e]: "I invited people who are good and who share the same philosophy."

In addition, rules of cooperating and working together are important. If complementary network members start cooperating across professional boards, they define rules for their work together [N4-e]: "We formed a group of craftsman with its own name, and we have a consensus how we collaborate." The members of the group not only work effectively together as they have established routines, but they also share knowledge and jointly develop and evaluate knowledge in the group.

The organizational proximity provides stable conditions that facilitate learning and innovation, but it is challenging to find ways to consider the explicit aspects of institutional arrangements. During the localization, the organizational proximity is particularly relevant for translating the new knowledge into work practices fitting to the existing organizational settings.

Social Proximity

The social proximity is related to social ties and relationships based on trust, friendship, kinship, and joint experience between individuals (Boschma 2005).

Trust-based social relationships facilitate the exchange of tacit knowledge. One interviewee described a network with a high social proximity [N2-b]: "We have a humane, open, and friendly way of exchanging knowledge in our group. I think you have the feeling of belonging to a group where idealism, ecological construction, and improving our world plays an important role, to a certain extent." He continued explaining the impact on knowledge sharing [N2-b]: "[…] this creates a basis of trust and mutual appreciation which reduces rivalry and barriers of contribution." The trust and the low barriers of contribution are particularly important while discussing and evaluating new knowledge, or while finding solutions. One member said [N11-c]: "Because I know him, I will email him about things and get sort of perspectives from him but that bypasses any formal process."

Another interviewee explained that in anonymous forums with a low social proximity [N2-b]: "people act more carefully as self-expression plays a role […] and people can easily skim knowledge without contributing." Here, it seems that a low social proximity in electronic communication is particularly limiting the knowledge transfer. One interviewee noted, for electronic communication, that [N5-d] "it is very challenging to transfer the message if your communication partner is unknown." Thus, knowledge about the personalities needs to be collected, and social ties need to be built, to allow an effective knowledge exchange [N4-a]: "You start carefully, and then you become more confident, and finally you get a lot of requests. For me, it is very important to know how to interact with each other, and then you often read the same names, and then you know what to say to whom or how to say something."

Summing up, the social proximity is perceived as very important while evaluating a problem, creating shared understanding, and finding new solutions. It seems that a high social proximity leads to a more open communication by lowering the barriers for contribution and active engagement in the previous activities of the localized learning process.

Geographic Proximity

The geographic proximity defines the spatial and physical distance between actors (Boschma 2005).

There is knowledge which can only be applied in certain regions, as one interviewee pointed out [N6-a]: "We've got a very different population. It's probably not going to work for us." One interviewee highlighted [N1-d]: "For example, we have sea ports in northern Germany. In southern Germany, you have only river ports and not such big ports. That implies specific regional challenges in connecting these ports to the infrastructure and so forth. That's a topic for people from the north, in which only very few people from Bavaria are interested in." The motivation to share also depends on the locality, as one interviewee pointed out [N11-e]: "Look, I just want to share my experience with my local friends." Here, the geographic proximity is relevant for forming local subgroups in which new knowledge is evaluated.

One interviewee explained that it is also relevant for online networks to break through the anonymity [N4-e]: "There are many things which are not nice, resulting from the virtuality in which we act. [...] The background is to establish teams of craftsmen or communication teams which can realize solutions in the region." In addition to the common regional knowledge, geographic proximity reduces the efforts for physical meetings. Further, the regional identity seems to be important for forming a regional subgroup [N2-d]: "We are all from southern Germany. We like each other, and we all have the same interests."

Summing up, it seems that the geographic proximity plays a complementary role in building and strengthening social, organizational, professional, and cognitive proximity. Geographic identity and the possibility to arrange physical meetings more easily are important while discussing, evaluating, and distributing new knowledge. Further, we observed that networks mostly organize their structures according to the geographic dimension.

Professional Proximity

The professional proximity is related to a group of people having a common professional background and a shared professional identity (Schamp et al. 2004). Professional identity describes group interactions in the workplace, focusing on how people compare and differentiate themselves from other professional groups (Adams et al. 2006).

Sharing the same professional background is important for trusting network members, as one interviewee stated [N4-a]: "Definitely, not only formality, it is the profession. Well, as I am a construction expert, I am in construction-related networks. I am not joining an electrical industry network for example." One interviewee highlights the vision and the professional identity in his online community, which makes a difference [N5-e]: "There are also hobby-communities, but we have a real vision and truly a profession." We also came across that networks with a high professional identify differentiate themselves sometimes by stigmatization [N2-e]: "Well, I believe, between architects and workers, there are tensions. The architects are stuck with their profession and think the workers do not know it better. And of course, the workers think the architects cannot accept their opinion. In this branch, that is still the case."

Taking such differences between communities into account is crucial for the success of the localization. Especially, the acceptance of localized knowledge can be reduced if it is adopted for the wrong professional group. In this regard, it is also important to consider that a specific language characterizes such networks [N10-b]: "You need to think what is it that I can and can't say which will be in keeping with my professional status as a doctor." The importance of having the right language for the professional group is highlighted by one interviewee [N4-b]: "You do not understand the lawyer jargon […] therefore I try to adopt the content that fits to the professional community." Further, a high professional proximity increases the trust within the involved members and lowers the barrier of absorbing knowledge.

Subgroups with a high professional proximity are used to work out solutions before they are disseminated within the network [N7-b]: "Um, there were three of us, three practice managers that worked on that and we shared it each, we checked sort of the timelines, that things were right, that we'd interpreted it correctly because different people interpret different guidelines differently and then once we were happy with the final structure and how it looked then we shared it out to all the Practice Managers." Thereby, final decisions are made by accepted senior members of the network [N9-a]: "I or my colleagues will meet as many of the team members as possible, but particularly the senior clinical and hopefully the senior nurse will always[…]give the binding advice."

However, the exchange between networks with a high professional proximity is also considered crucial to check the applicability of knowledge [N7-a]: "We feel that we GPs really need a voice [in the management] board, so that they don't roll out these programs without thinking how doable is it on the ground." He further points out [N7-a]: "We've got a lot to offer, we really do have a lot to offer the Board." However, hierarchical perspectives linked to professional identities can hinder this fruitful exchange.

Summing up, the professional proximity seems to be an important dimension for localizing knowledge in networks. A high professional proximity can speed up the localization of external knowledge by making use of the professional expertise. Therefore, the members' professional identity plays an important role for facilitating the collaboration and finding solutions. Further, having knowledge adopted to professions lowers the barrier of knowledge absorption. Nevertheless, focusing on networks with a high professional proximity can also harm the exchange between professions in a supply chain.

5 Discussion

SMEs seek support from their network to assimilate external knowledge. Due to the decontextualized nature of global knowledge and their limited capacities, this assimilation is challenging for SMEs. However, SMEs need this external knowledge to be compliant and, thus, to reduce the risk of negative consequences of non-compliance. Further, this external knowledge is also important to gain competitive advantage. Therefore, speeding up the localization process and ensuring a high-quality outcome are crucial for networks. The members expect the network management to be in charge of a fast and high-quality localization.

The networks can use their strong domain knowledge to localize knowledge. Networks provide knowledge-pools (Meyer and Skak 2002), and, thus, their main advantage is to increase the efficiency of accessing external knowledge, leading to a better application of it (Grant and Baden-Fuller 2004). Therefore, networks can use their strong knowledge about their members to prepare content that fits to specific (sub)groups. Putting the lens of our discussion in the scope of the explained proximities, it seems promising that those proximities are affecting the network by understanding the members' needs, relationships, and expectations.

We argue that proximities between network members play an important role not only during the process of localizing knowledge, but also for the networks in building up their structure, subgroups, and IT infrastructure for their members. Having a geographically wide-spread community with very low social and cognitive proximities will lead to different requirements than a homogenous, regionally tied group. Research argues for the positive effect of IT, as IT can support and cultivate knowledge synergies by creating electronic networks and facilitating cross-firm socialization for knowledge integration (Tippins and Sohi 2003).

Within the scope of IT, we will further discuss the implications of the different proximities in the different phases of the localization process on IT. Figure 1 provides an overview of our implications for IT capabilities. As people switch back and forth between different networks, the integration of different IT used in different (sub)networks is important.

Initiation of the Localization

In the initiation phase, proximities can be used to evaluate the target group and to decide to what extent the new knowledge needs to be adapted. SMEs rely on subgroups to pre-filter the knowledge they are interested in, saving time and enhancing the applicability of the identified knowledge.

For networks, quickly identifying knowledge relevant for (some or all) members of the network is challenging. Therefore, the network must recognize potential

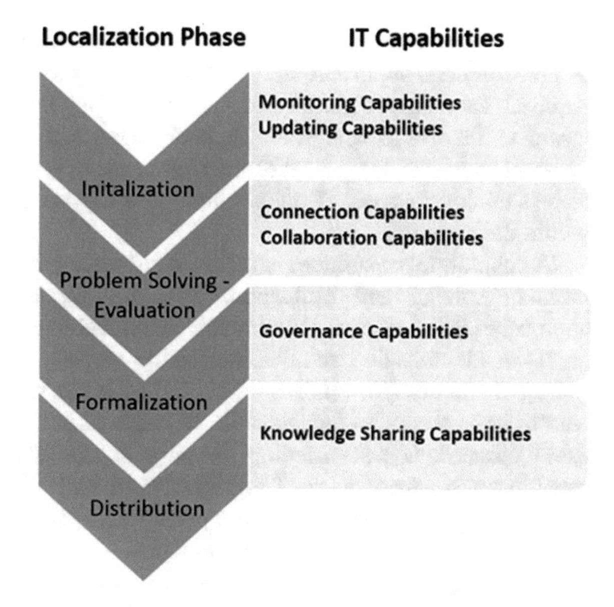

Fig. 1 Proposed IT capabilities

target groups as well as subgroups responsible for escalating observations. In this regard, IT solutions, which provide monitoring capabilities for knowledge identification and pre-filtering, are needed, and can be combined with higher updating capabilities to inform members about new knowledge. Therefore, the network should provide IT services such as monitoring services, push notifications to update their members, or RSS aggregation services supporting their members in gaining access to new knowledge, even if members are geographically dispersed. Receiving quick updates via RSS feeds helps network members to synthesize and share knowledge from multiple sources (Shneiderman 2007).

Evaluation and Problem Solving

We discovered that trust in groups with a high social and professional proximity facilitates the willingness to contribute. Networks, thus, need to find suitable subgroups for evaluating and solving problems. Thereby, the composition in terms of proximities is crucial to allow quick, target group oriented, and high-quality work.

The network should thereby provide its members connection capabilities to support them in being linked through IT. If people are able to more quickly identify or create suitable subnetworks, it will lead to compositions with higher proximities. IT systems, in this regard, should foster the formation of such subgroups and collaboration spaces. User profiles in a social network show potential, as these web-based services allow individuals to construct profiles, create a list of other users, as well as view and share their list of connections (Ellison 2007). We argue that these effects have a positive impact on the outcome of this phase.

We also showed challenges between different subgroups, which could harm the identification and evaluation of solutions, e.g. between subgroups with high professional proximity. Whereas a high proximity helps to localize knowledge and to find solutions applicable in the local context, cross-community collaboration is required for many cases, as one interviewee pointed out [N8-a]: "trying to get, trying to involve people who we know have a different view. So you do get different views of things rather than just relying on the people who, potentially always think the same." IT offers solutions to support people in those environments within the network.

Collaboration capabilities should be enhanced by the networks to facilitate problem solving and evaluation. Research argues for collaborative tagging approaches that formalize the communication and knowledge exchanges, leading to an easier identification and communication of people from another or within the same community (Ravenscroft et al. 2012). Further, we identified that networks tend to build their structure according to regions, as a high geographical proximity eases discussions and meetings, leading to the need for higher collaboration capabilities. If networks use IT for virtual communication and collaboration such as Skype or Google Docs to connect people with a lower geographical proximity, networks can lower their regional limitation.

Therefore, research argues to provide a so-called social knowledge environment, using social software tools systematically for knowledge management, allowing shared online collaboration spaces as well as productivity tools and other business applications (Pawlowski et al. 2014).

Formalization

Rigorously evaluating the localized knowledge is crucial before distribution. First, network members' trust in the quality of knowledge and second, limitations in the evaluation could cause damage claims. Therefore, experts having substantial knowledge of or being a member of the target group should perform the evaluation of the knowledge. Networks could use proximities to define and form these subgroups. We showed that senior members of subgroups, in particular, should be involved to ensure the acceptance of the prepared knowledge. Management realizes the formalization in our networks, having heterogeneous groups with a very strong professional identity that are distributed geographically. Thus, networks need the capabilities to ensure that the quality of the localized knowledge is evaluated. We argue in this regard for the provision of governance capabilities.

If the network offers solutions that provide the needed expertise, people with substantial knowledge can be linked to this phase more reliably. From a network side, IT services that support the preparation of the content presentation and formalization for each subgroup would be appreciated. Research showed that modern, IT-based governance systems are more and more based on developing policies and roles, which focus on the benefits of the collective rather than maximizing the interests of specific individuals or sub-communities (Huang et al. 2015). A rather simple tool providing those capabilities could be a wiki, supporting easy, collaborative editing of online content and enables simple, distributed, and traceable changes (Razmerita et al. 2014).

Distribution

Networks currently distribute the formalized knowledge via multiple channels, and in multiple formats, but lack a strategy to ensure the target group fit. As the main disadvantage of electronic communication regarding distribution, we identified that a low social proximity tends to increase the barrier of knowledge sharing. Therefore, we argue that knowledge sharing capabilities can support the members by addressing the drawbacks and helping to choose the right content, in the right format, for the right target group. Current research argues that online platforms like Twitter, weblogs, or social network sites like LinkedIn are effective platforms for knowledge distribution within and across organizations as they can facilitate expert and expertise locating, socializing, reaching out, and horizon broadening (Jarrahi and Sawyer 2013).

Thus, weblogs are already used widely to publish personalized knowledge, by having features such as personal editorship, a hyperlinked post structure, frequent updates, and free public access to the contents and archives (Yu et al. 2010). Such features seem promising to increase personalization and traceability especially for informal learning in networks (Schäper and Thalmann, 2015), and also challenges of low geographic proximities can be addressed. Research stated that explicit knowledge is shared effectively as documents or texts, whereas the distribution of tacit knowledge tends to be easier through pictures, videos, and audios (Malhotra et al. 2005).

6 Conclusion and Outlook

We identified that the process of localizing external knowledge is crucial in networks of SMEs to ensure the absorption of external knowledge of the member SMEs. Research so far discussed the great importance of networks for the knowledge absorption of SMEs. However, the crucial process of localizing external knowledge for the member SMEs received less attention so far. The major contribution of this paper is describing this crucial process, considering the perspectives of proximities based on interviews in eleven networks of SMEs. Further, we discussed how networks could use the proximity perspective to support their members' absorption process. We also discovered that the increasing importance of ephemeral knowledge will further raise the need to localize this knowledge and particularly to perform this localization quickly and with a high-quality output.

Our interviewees reported about the need for communication with some people having high and other people having low proximities. They need a group of people with high proximities for effective knowledge sharing and development, but they also need to ensure the collaboration along the supply chain as well.

We identified that people are more likely to listen to and share with people that are similar in terms of the proximities. Thus, we consider the concept of the dimensions of proximity to be useful to support workers, by recommending people for particular tasks or situations, and, therefore, to localize knowledge. Network managers should consider the different proximities while defining target groups for external knowledge and building subgroups. As SMEs are facing limited resources, offering infrastructures and IT services tends to be in the responsibility of networks to drive this process forward.

In terms of generalizability, we do not claim generalization due to our small sample. Nevertheless, we selected eleven networks with overlapping perspectives between them, showing the broad relevance. Our findings offer an initial glance at challenges for localizing external knowledge in SME networks as well as the rising opportunities for IT support provided by the network. In future research, we plan to focus on the requirements of ephemeral knowledge more explicitly. Further, we plan to investigate how IT can facilitate localization by designing supportive IT systems in a particular network.

Acknowledgements The Know-Center and Pro2Future are funded within the Austrian COMET Program—Competence Centers for Excellent Technologies—under the auspices of the Austrian Federal Ministry of Transport, Innovation and Technology, the Austrian Federal Ministry of Economy, Family and Youth and by the State of Styria. COMET is managed by the Austrian Research Promotion Agency FFG.

References

Adams, K., Hean, S., Sturgis, P., & Clark, J. M. (2006). Investigating the factors influencing professional identity of first-year health and social care students. *Learning in Health and Social Care, 5*(2), 55–68.

Alavi, M., & Leidner, D. E. (2001). Knowledge management and knowledge management systems: Conceptual foundations and research issues. *MIS Quarterly*, 107–136.

Argote, L., & Ingram, P. (2000). Knowledge transfer: A basis for competitive advantage in firms. *Organizational Behavior and Human Decision Processes, 82*(1), 150–169.

Arora, A., & Gambardella, A. (1990). Complementarity and external linkages: The strategies of the large firms in biotechnology. *The Journal of Industrial Economics*, 361–379.

Audretsch, B. (1998). Agglomeration and the location of innovative activity. *Oxford Review of Economic Policy, 14*(2), 18–29.

Audretsch, D. B., & Lehmann, E. E. (2005). Does the knowledge spillover theory of entrepreneurship hold for regions? *Research Policy, 34*(8), 1191–1202.

Autant-Bernard, C., Fadairo, M., & Massard, N. (2013). Knowledge diffusion and innovation policies within the European regions: Challenges based on recent empirical evidence. *Research Policy, 42*(1), 196–210.

Balland, P.-A. (2012). Proximity and the evolution of collaboration networks: Evidence from research and development projects within the global navigation satellite system (GNSS) industry. *Regional Studies, 46*(6), 741–756.

Blackler, F. (1995). Knowledge, knowledge work and organizations: An overview and interpretation. *Organization Studies, 16*(6), 1021–1046.

Boschma, R. (2005). Proximity and innovation: A critical assessment. *Regional Studies, 39*(1), 61–74.

Cantner, U., & Graf, H. (2006). The network of innovators in Jena: An application of social network analysis. *Research Policy, 35*(4), 463–480.

Cheng, H., Niu, M.-S., & Niu, K.-H. (2014). Industrial cluster involvement, organizational learning, and organizational adaptation: An exploratory study in high technology industrial districts. *Journal of Knowledge Management, 18*(5), 971–990.

Dahl, M. S., & Pedersen, C. Ø. (2004). Knowledge flows through informal contacts in industrial clusters: Myth or reality? *Research Policy, 33*(10), 1673–1686.

Donckels, R., & Lambrecht, J. (1997). The network position of small businesses: An explanatory model. *Journal of Small Business Management, 35*(2), 13.

Egbu, C. O., Hari, S., & Renukappa, S. H. (2005). Knowledge management for sustainable competitiveness in small and medium surveying practices. *Structural Survey, 23*(1), 7–21.

Ellison, N. B. (2007). Social network sites: Definition, history, and scholarship. *Journal of Computer-Mediated Communication, 13*(1), 210–230.

Fuellhart, K. (1999). Localization and the use of information sources: The case of the carpet industry. *European Urban and Regional Studies, 6*(1), 39–58.

Gertler, M. S. (1995). "Being there": Proximity, organization, and culture in the development and adoption of advanced manufacturing technologies. *Economic Geography, 71*(1), 1–26.

Grant, R. M., & Baden-Fuller, C. (2004). A knowledge accessing theory of strategic alliances. *Journal of Management Studies, 41*(1), 61–84.

Gulati, R. (1999). Network location and learning: The influence of network resources and firm capabilities on alliance formation. *Strategic Management Journal, 20*(5), 397–420.

Hagedoorn, J. (1995). Strategic technology partnering during the 1980s: Trends, networks and corporate patterns in non-core technologies. *Research Policy, 24*(2), 207–231.

Howells, J. R. (2002). Tacit knowledge, innovation and economic geography. *Urban studies, 39* (5–6), 871–884.

Huang, J., Baptista, J., & Newell, S. (2015). Communicational ambidexterity as a new capability to manage social media communication within organizations. *The Journal of Strategic Information Systems, 24*(2), 49–64.

Hudson, R. (1999). 'The learning economy, the learning firm and the learning region' a sympathetic critique of the limits to learning. *European Urban and Regional Studies, 6*(1), 59–72.

Jarrahi, M. H., & Sawyer, S. (2013). Social technologies, informal knowledge practices, and the enterprise. *Journal of Organizational Computing and Electronic Commerce, 23*(1–2), 110–137.

Keller, W. (2004). International technology diffusion. *Journal of Economic Literature, 42*(3), 752–782.

Kogut, B., & Zander, U. (1992). Knowledge of the firm, combinative capabilities, and the replication of technology. *Organization Science, 3*(3), 383–397.

Laursen, K., Masciarelli, F., & Prencipe, A. (2012). Regions matter: How localized social capital affects innovation and external knowledge acquisition. *Organization Science, 23*(1), 177–193.

Leseure, M. J., & Brookes, N. J. (2004). Knowledge management benchmarks for project management. *Journal of Knowledge Management, 8*(1), 103–116.

Louise Barriball, K., & While, A. (1994). Collecting data using a semi-structured interview: A discussion paper. *Journal of Advanced Nursing, 19*(2), 328–335.

Lum, W. Y., & Lau, F. C. (2002). A context-aware decision engine for content adaptation. *IEEE Pervasive Computing, 1*(3), 41–49.

Malecki, E. J. (2010). Global knowledge and creativity: New challenges for firms and regions. *Regional Studies, 44*(8), 1033–1052.

Malhotra, A., Gosain, S., & Sawy, O. A. E. (2005). Absorptive capacity configurations in supply chains: gearing for partner-enabled market knowledge creation. *MIS Quarterly*, 145–187.

Maskell, P. (2001). Towards a knowledge-based theory of the geographical cluster. *Industrial and Corporate Change, 10*(4), 921–943.

Mayring, P. (2014). Qualitative content analysis: Theoretical foundation, basic procedures and software solution.

Meyer, K., & Skak, A. (2002). Networks, serendipity and SME entry into Eastern Europe. *European Management Journal, 20*(2), 179–188.

Nonaka, I. (1991). The knowledge-creating company Harvard business review November-December.

Pawlowski, J. M., Bick, M., Peinl, R., Thalmann, S., Maier, R., Hetmank, L., et al. (2014). Social knowledge environments. *Business & Information Systems Engineering, 6*(2), 81–88.
Ravenscroft, A., Schmidt, A., Cook, J., & Bradley, C. (2012). Designing social media for informal learning and knowledge maturing in the digital workplace. *Journal of Computer Assisted Learning, 28*(3), 235–249.
Razmerita, L., Kirchner, K., & Nabeth, T. (2014). Social media in organizations: Leveraging personal and collective knowledge processes. *Journal of Organizational Computing and Electronic Commerce, 24*(1), 74–93.
Salovaara, A., & Tuunainen, V. K. (2013). Software developers' online chat as an intra-firm mechanism for sharing ephemeral knowledge.
Salovaara, A., & Tuunainen, V. K. (2015). Mediated sharing as software developers' strategy to manage ephemeral knowledge. ECIS.
Schäper, S., & Thalmann, S. (2015). Addressing challenges for informal learning in networks of organizations. In ECIS.
Schamp, E. W., Rentmeister, B., & Lo, V. (2004). Dimensions of proximity in knowledge-based networks: The cases of investment banking and automobile design. *European Planning Studies, 12*(5), 607–624.
Shneiderman, B. (2007). Creativity support tools: Accelerating discovery and innovation. *Communications of the ACM, 50*(12), 20–32.
Thalmann, S. (2014). Adaptation criteria for the personalised delivery of learning materials: A multi-stage empirical investigation. *Australasian Journal of Educational Technology, 30*(1).
Tippins, M. J., & Sohi, R. S. (2003). IT competency and firm performance: Is organizational learning a missing link? *Strategic Management Journal, 24*(8), 745–761.
Torre, A., & Rallet, A. (2005). Proximity and localization. *Regional Studies, 39*(1), 47–59.
Tortoriello, M., Reagans, R., & McEvily, B. (2012). Bridging the knowledge gap: The influence of strong ties, network cohesion, and network range on the transfer of knowledge between organizational units. *Organization Science, 23*(4), 1024–1039.
Williams, C. (2007). Transfer in context: Replication and adaptation in knowledge transfer relationships. *Strategic Management Journal, 28*(9), 867–889.
Yu, T.-K., Lu, L.-C., & Liu, T.-F. (2010). Exploring factors that influence knowledge sharing behavior via weblogs. *Computers in Human Behavior, 26*(1), 32–41.
Zimmermann, A., Specht, M., & Lorenz, A. (2005). Personalization and context management. *User Modeling and User-Adapted Interaction, 15*(3), 275–302.

Author Biographies

Stefan Thalmann, Dr. is Assistant Professor at the Graz University of Technology in Austria, AREA Manager for cognitive decision support at the Pro2Future GmbH and senior researcher at the Know-Center GmbH. His research interests are in knowledge protection, knowledge management, technology-enhanced learning and predictive maintenance.

Stephan Schäper works as an international IT Project Manager focusing on SAP rollouts and migration projects mainly in South East Asia and Asia Pacific for an international manufacturer of furniture fittings. His interests are in technology-enhanced methods for knowledge management and sharing to enhance the localization of knowledge within networks of one or multiple enterprises.

Part III
Leading and Learning 4.0

Digital Leadership

Thorsten Petry

Abstract We are living in a complex environment with dynamic and fundamental changes. A core aspect of these changes is the exponential development of new technologies. In addition to new competencies and the dynamic capability to adapt the competencies within a company, this dynamic and complex environment also leads to new leadership challenges. In an age of acceleration, managers have to juggle with different options and be agile. A pragmatic test, measure and learn approach is often more successful than very detailed analysis and long-term planning. In addition, single managers are often overstrained in such an environment. Therefore, leadership in the digital economy needs to be more decentralized and should use the collective competence and intelligence in the company. This article describes the characteristics of leadership in the digital economy as well as some adequate leadership tools. However, the article ends with a "but", i.e. leaders should not push too hard and dump all traditional management tools. A successful leadership will typically require some kind of ambidexterity—efficient business execution and agile business adaption.

1 Introduction

As the results from IBM's global CEO-studies show, technological factors are the number one reason for business transformations in companies worldwide (IBM 2015). Other studies come to comparable results (e.g. Accenture and EIU 2014; Petry et al. 2015). Most companies are already faced with **huge technological changes** and expect even more challenges in the near future. Based on a McKinsey study digitalization has only begun to transform the economic performance of companies. On average, industries are less than 40% digitalized (Bughin et al. 2017). The current impact of the digitalization differs by industry and function,

T. Petry (✉)
RheinMain University of Applied Sciences, Wiesbaden, Germany
e-mail: thorsten.petry@hs-rm.de

K. North et al. (eds.), *Knowledge Management in Digital Change*, Progress in IS,
https://doi.org/10.1007/978-3-319-73546-7_12

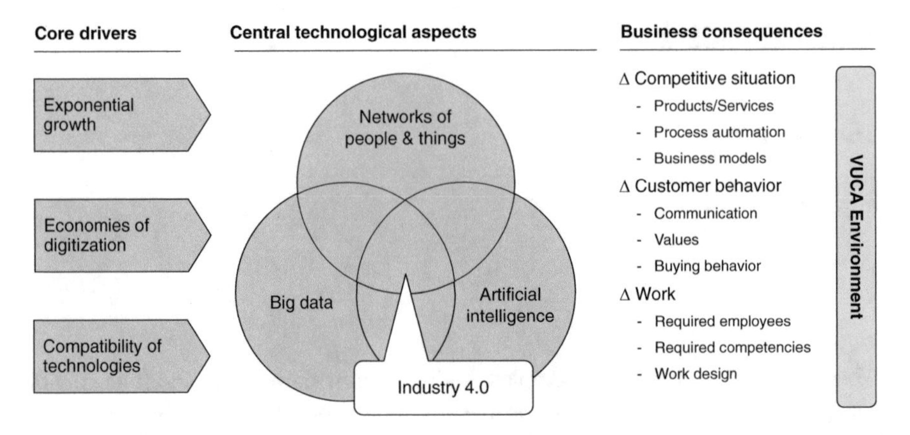

Fig. 1 Core aspects of digitalization (based on Petry 2016)

but in the end all industries and companies will be influenced significantly (BCG 2015; PWC 2014).

However, what does digitalization actually mean? In order to understand the **core aspects of the digitalization** the following model (see Fig. 1), that is based on the work of Erik Brynjolfsson and Andrew McAfee at the MIT in Boston, is quite helpful.

Based on their analysis Brynjolfsson and McAfee (2014) highlight three core digitalization drivers, that affect all kind of digital technologies:

1. Exponential growth of digital technologies, i.e. continuous doubling of performance in equal intervals
2. Economies of digitization, i.e. nearly zero marginal costs of digital products
3. Compatibility of different technologies, i.e. different innovations are supporting each other's value

The central technological aspects are the connection of people and things via the internet and cloud technology. Social media platforms like facebook, twitter, linkedin, blogs etc. are establishing huge personnel networks; the same approach could be used with companies (social collaboration platforms). The internet of things is setting up networks of machines, wearables, products etc. Based on these networks a tremendous amount of data is produced. This big data can be used (in real-time) for data analytics and business predictions. In order to do this, artificial intelligence and cognitive computing functionalities are getting more and more important. The industry 4.0 concept is at the center of these three technological aspects.

These technological developments have a significant impact on the competitive landscape, e.g. new business models, new competitors, new products and services, higher process automation etc. based on new technological possibilities. Many business ecosystems are in a fundamental transformation process (e.g. automotive industry). This is closely linked to modifications in the customer behavior, i.e.

immediate, open and many-to-many communication (e.g. via social media), changed values (e.g. sharing economy) and different buying behavior (e.g. Amazon, Check24). Finally, the technological developments also affect the amount of required employees (e.g. Frey and Osborne 2013), the required competencies of employees (e.g. data analytics) and managers (discussed in this article) as well as the way how work is organized (e.g. new work, Petry 2016).

Considering all these aspects and the tremendous amount and speed of change, the digital economy could be defined as a **VUCA environment**. The acronym "VUCA" is coming from US-American military jargon and is representing the characteristics volatility, uncertainty, complexity and ambiguity (Bennett and Lemoine 2014).

- Volatility = frequent and strong changes
- Uncertainty = unclear situations, lack of predictability
- Complexity = interdependence of multiple elements, cause-and-effect chain unclear
- Ambiguity = inconsistent and contradictory environment, cause-and-effect confusion

In such a VUCA environment, business developments are often not foreseeable or predictable and therefore detailed analysis and planning gets quite difficult. Therefore, some successful leadership approaches of the past are not working any more.

2 Digital Leadership Characteristics

In the VUCA environment of the digital age, leadership requirements are changing. Therefore, a lot of managers see the need for a paradigm shift in leadership (Kellerman 2012; Hlupic 2014). This is the main result of an in-depth interview study of 400 German managers. 77% of the participants expect such a paradigm shift (INQA 2014). Hamel (2009), ranked the world's most influential business thinker by the Wall Street Journal, is talking of a "management revolution that is likely to be as profound and unsettling as the one that gave birth to the modern industrial age. … [T]his transformation will radically reshape the nature of work, boundaries of the enterprise, and the responsibilities of business leaders." This means digital leadership is nothing that is just done by one person, e.g. a chief digital officer (CDO), but needs to be done by every manager.

Excursus: **Chief Digital Officer (CDO)**

More and more companies are implementing the role of a chief digital officer (CDO) – either on company level and/or in different business units. Although the specific role differs a lot and there is not one commonly accepted role description, a CDO could be generally defined as the person that is responsible for driving and coordinating the overall digital transformation in the company or business unit. This should include strategic, organizational, cultural and technological aspects. Therefore, the CDO is a general management – not just IT – position that should be closely linked to the CEO. It is expected, that this role will vanish over the time, as digital will be more and more part of "normal" business. Therefore, it is a transformation manager role.

All managers—although different in magnitude—need to adopt there leadership style to the VUCA environment of the digital age. But, what are the characteristics of this new, digital leadership?

A volatile, uncertain, complex and ambiguous environment requires a flexible approach with fast (re)actions. In the digital economy, managers have to juggle with different options. A pragmatic test, measure and learn approach is often more successful than very detailed analysis and long-term planning. A leader has to define a (rough) direction, think in different scenarios, maintain several options, realize weak signals, experiment with ideas and learn very fast from success and failure. All this could be described as **agile leadership**.

It is also important to realize that a leader cannot know everything. Individuals are overstrained in a VUCA environment. It is impudent to centrally control and steer such a complex system. Because of that, leadership must be more decentralized and shared. Leaders need to use the collective intelligence in the company (**participative leadership**, Pearce and Conger 2003; Greenleaf 2002). Similar to the transformative leadership approach (Bass 1985), managers should create conditions in which intrinsic motivated knowledge workers can bring in their experiences and competencies and fulfill their specific tasks. Self-organization and self-management within communities are getting more and more important and managers need to become more like community managers and coaches.

A core requirement for more participation, self-organization and self-management are strong networks. The individual knowledge workers need to be linked to each other. Leaders need to be networkers, they have to support the connection of internal as well as external competences (**networking leadership**). As the US-American materials science company W. L. Gore highlights, a major task of leadership is to "maximize opportunities for personal interactions" (Hamel 2007).

In addition, leaders in the digital economy need to lead openly, i.e. open communication, give and receive feedback openly, be open for criticism. Digital leadership is **open leadership** (Li 2010; Petry 2014). Unfortunately, this is quite hard for somebody who learned over years and decades, that knowledge is power, that trust is good, but control is better, and that the important decisions are done behind closed doors. Based on a study by Hays and Pierre Audoin Consultants (2015) thinking in separated functional silos is the most often reason for failure in the digital transformation.

To sum it up so far: The volatility, uncertainty, complexity and ambiguity (VUCA) of the digital economy requires

- the connection of, not only data and machines, but also knowledge workers,
- an open communication and open access to information,
- the participative use of individual and collective intelligence,
- and an agile thinking and behavior.

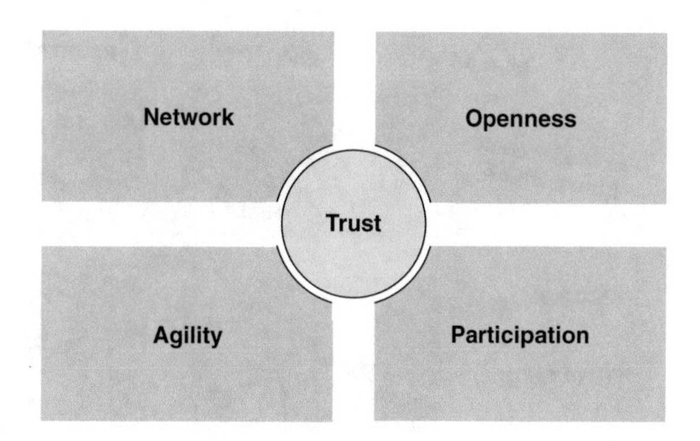

Fig. 2 NOPA+ model (based on Petry 2016)

In order to lead in an agile way, participative, networked and openly, managers need to trust their colleagues and employees (**trust-based leadership**). Without trust in their competences as well as their motivation, they won't let go. But digital leadership requires to "let go" of traditional command-and-control approaches.

The five characteristics of network, openness, participation, agility plus trust form the so called **NOPA+ model** of digital leadership (based on Petry 2014 and Buhse 2014). As explained, leadership needs to be more networked, open, participative and agile—based on trust (Fig. 2).

3 Digital Leadership Tools

After knowing these digital leadership characteristics, the next question is how to lead in a NOPA+ way? Which tools and approaches could help to bring digital leadership to live?

There are several tools that could support a NOPA+ leadership approach (for details see Petry 2016). Figure 3 shows some of these tools and visualizes the focused leadership characteristic. In order to support an internal network of the company's knowledge workers, a social collaboration/Enterprise 2.0 platform (e.g. IBM Connections, Yammer, Jive) could be very powerful (McAfee 2006; Petry and Schreckenbach 2015). Such Enterprise 2.0 tools and services use different social media features, e.g. social bookmarking and linking, tagging, rating, user commenting, user discussions, user content generation or syndication via RSS feeds. These **networking tools** encourage open communication and collaboration. However, even in digital times, physical meetings and networking is still very important. Digital permanence does not replace physical presence. Therefore Axel-Springer, a leading German digital publishing company, is using a lot of analog meet and greet formats, e.g. blind lunch or learning lunch (Burr 2016).

These networking tools also support openness. Another approach that is focused on openness and networking is working/leading out loud (Pearce 2013; Stepper 2015). At companies like Bosch (Connect), Audi (team) or Deutsche Bank

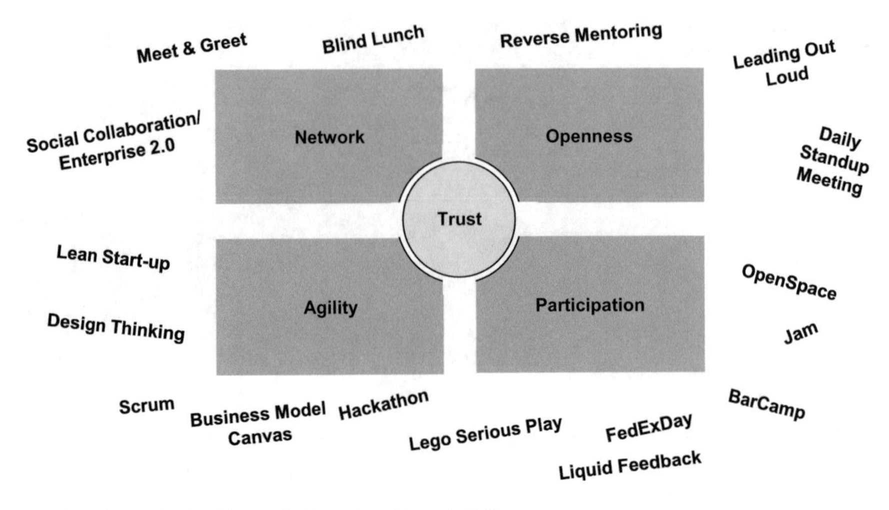

Fig. 3 Digital leadership tools (based on Petry 2016)

(Ask Me Anything) leaders are using executive blogs, podcasts and other **open communication formats** to regularly show what they are working on and to explain their leadership tasks, thinking and decisions (Dückert 2016). This transparency helps to build trust and is the basis for participation in leadership issues.

In order to use the competence and collective intelligence in the company, **participative workshop approaches** are helpful. An analog OpenSpace or a digital Jam for example could help to develop ideas in a collaborative way by using the competencies of a large amount of knowledge workers. A BarCamp could be used in order to develop more specific concepts in a participative way. If you want to develop prototypes, a Hackathon, a FedExDay or a Lego Serious Play workshop are tools worth to be considered. In addition, LiquidFeedback could be a very interesting approach to share leadership and come up with better decisions, based on the use of all available knowledge and the collective intelligence within the company.

Building and testing prototypes is also a core aspect of **agile management approaches**. Although there are some differences between for example Scrum, Lean Start-up or Design Thinking, all these agile management approaches share the same core characteristics. They are focused on agility, based on networking, openness and collaboration/participation. Starting point should be the real customer need (functionality, design). The process follows the "develop—try—fail—retry—fail again—retry—succeed" logic. Core elements are teamwork, focus, time-boxing, visualizing, prototyping, experimentation, failure tolerance, early and regular feedback as well as iterations.

To sum it up: There are several tools that could support NOPA+ leadership. So leaders can use a pre-defined digital leadership toolbox, but it should be adapted to the specific company and situation. A core aim of the most tools is to stimulate an open and participative discussion to quickly come up with new ideas.

4 Leadership Ambidexterity

This article on how to lead in digital times ends with a "but". Yes, leadership needs to be more networked, open, participative, agile and trust-based, but leaders should not push too hard and dump all traditional management tools. A successful leadership typically requires some kind of **ambidexterity**—efficient business execution and agile business adaption (Gibson and Birkinshaw 2004). As studies show, digitization is not erasing hierarchy totally (Petry and Schreckenbach 2015). Therefore, the leadership pendulum (Fig. 4) should not swing too far. In most cases leadership could not be fully socialized. Extreme self-management could work in small teams and specific situations, but it will most probably not be the best fit for bigger companies.

The German Bosch group, a leading global supplier of technology and services, emphasizes that they are seeking a combination of their traditional "German industry excellence" with more "Silicon Valley agility" (Bosch 2013). While some areas like production have to preserve a focus on efficient business execution based on standardized processes, stability and specialization, other areas like product development or customer service have to be more agile. Therefore the NOPA+ model will not fitting equally well in all areas.

Bosch is talking of effective leadership as the **mastery of slide control** (Fig. 5). Leaders have to find the right adjustment of this slide control for a specific unit, team or entity (e.g. production vs. product development). In addition leaders need to be able to switch their leadership style (e.g. head of production is in parallel leading a team to develop new production processes).

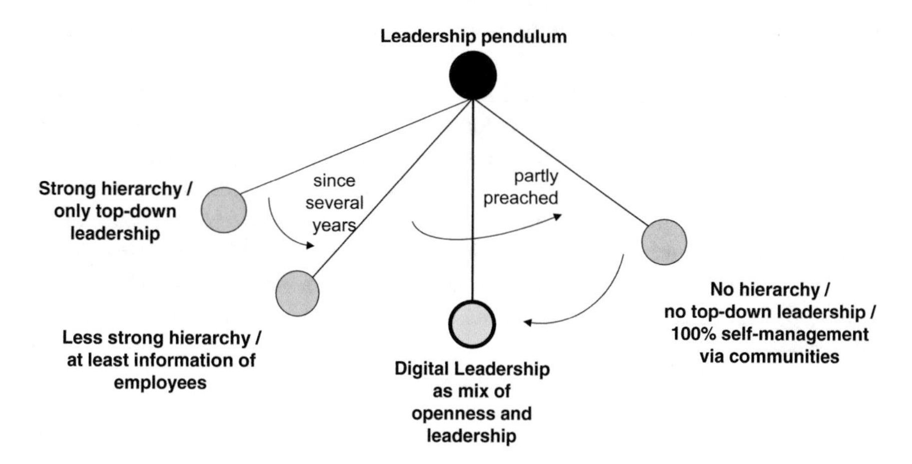

Fig. 4 Leadership pendulum (based on Petry 2016)

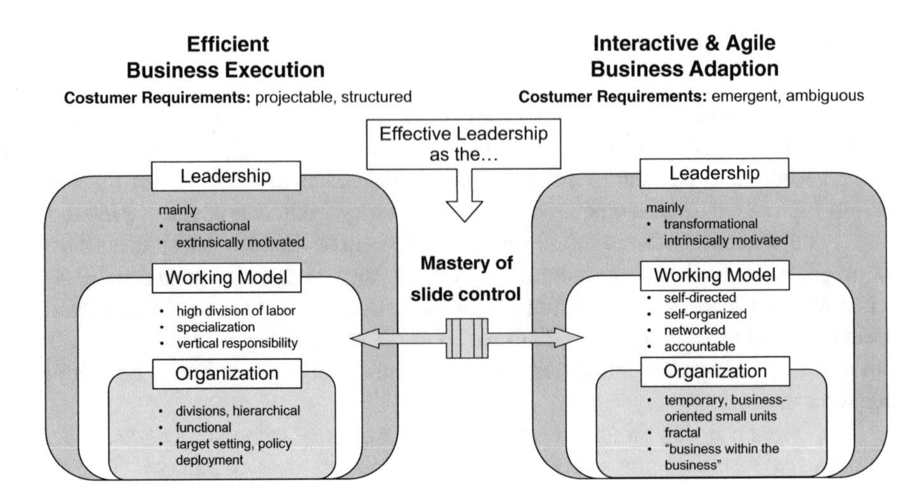

Fig. 5 Effective leadership as the mastery of slide control (based on Bosch 2013)

5 Conclusion

We are living in a complex environment with dynamic and fundamental changes. A core aspect of these changes is the exponential development of new technologies that lead to a VUCA environment. This environment also leads to new leadership challenges. Leadership needs to be more networked, open, participative, agile and trust-based. This implies that in a time of distributed information and knowledge also leadership needs to be distributed and shared. Decisions have to be taken more decentralized, and in case of central decisions, the collective intelligence within the company should be used.

There are several tools that could support such a NOPA+ leadership (e.g. networking tools, open communication formats, participative workshop methods or agile management approaches). So leaders can use a pre-defined digital leadership toolbox, but it should be adapted to the specific company and situation. In addition, leaders should not push too hard and dump all traditional management tools. A successful leadership typically requires some kind of ambidexterity—efficient business execution and agile business adaption. Leadership could not be fully socialized, leaders are still needed in the digital economy. But the leadership style will change.

Literature

Accenture, & EIU. (2014). *Digital double-down: How far will leaders leap ahead?* Whitepaper.
Bass, B. M. (1985). *Leadership and performance beyond expectations*. New York.
BCG. (2015). *How to jump-start a digital transformation*. Whitepaper.

Bennett, N., & Lemoine, G. J. (2014). What VUCA really means for you. *Harvard Business Review, 1,* 27.

Bosch. (2013). *Enabling Enterprise 2.0*. EFQM Good Practice Competition.

Brynjolfsson, E., & McAfee, A. (2014). The second machine age: An industrial revolution powered by digital technologies. *Digital Transformation Review, 5,* 12–17.

Bughin, J., LaBerge, L., & Mellbye, A. (2017). The case for digital reinvention. *McKinsey Quarterly, 2,* 1–15.

Buhse, W. (2014). *Management bei Internet*. Kulmbach.

Burr, J. (2016). Management der Digitalen Transformation bei Axel Springer. In T. Petry (Ed.), *Digital Leadership—Erfolgreiches Führen in Zeiten der Digital Economy*. Freiburg et al., 341–353.

Dückert, S. (2016). Leitbild der digitalen Führungskraft. In Petry, T. (Ed.). *Digital Leadership—Erfolgreiches Führen in Zeiten der Digital Economy*. Freiburg et al., 115–125.

Frey, C. B., & Osborne, M. A. (2013). *The future of employment: How susceptible are jobs to computerisation?* Working Paper Oxford University.

Gibson, C. B., & Birkinshaw, J. (2004). The antecedents, consequences, and mediating role of organizational ambidexterity. *Academy of Management Journal, 2,* 209–226.

Greenleaf, K. (2002). *Servant leadership: A journey into the nature of legitimate power and greatness*. New Jersey: Paulist Press.

Hamel, G. (2007). *The future of management*. Boston.

Hamel, G. (2009). *Appraisal in foreword*. McAfee, A.: Enterprise 2.0, Boston.

Hays, & Pierre Audoin Consultants. (2015). *Von starren Prozessen zu agilen Projekten: Unternehmen in der digitalen Transformation*. Whitepaper.

Hlupic, V. (2014): *The Management Shift: How to harness the power of people and transform your organization for sustainable success*. London.

IBM. (2015). *Redefining competition: Insights from the Global C-suite Study—The CEO perspective*. CEO Study 2015.

INQA. (2014). *Führungskultur im Wandel*. Berlin.

Kellerman, B. (2012). *The end of leadership*. New York.

Li, C. (2010). *Open leadership: How social technology can transform the way you lead*. San Francisco.

McAfee, A. (2006). Enterprise 2.0: The dawn of emergent collaboration. *MIT Sloan Management Review, 3,* 20–28.

Pearce, C. L., & Conger, J. A. (2003): *Shared leadership—Reframing the hows and whys of leadership*. Thousand Oaks et al.

Pearce, T. (2013). *Leading out loud: A guide for engaging others in creating the future*. Jossey-Bass.

Petry, T. (2014). Führungskräften mangelt es oft noch an Digitalkompetenz. *Human Resources Manager, 6,* 86–87.

Petry, T. (2016). Digital leadership—Unternehmens- und Personalführung in der Digital Economy. In Petry, T. (Ed.), *Digital leadership—Erfolgreiches Führen in Zeiten der Digital Economy*. Freiburg et al., 21–82.

Petry, T., Grabmeier, S., Armutat, S., & Schabel, F. (2015). *Digital Readiness Check deutscher Unternehmen*. Working Paper Hochschule RheinMain.

Petry, T., & Schreckenbach, F. (2015). Enterprise 2.0—Empirische Befunde zum Status Quo 2015. *Zeitschrift Organisations Entwicklung, 4,* 102–104.

PWC. (2014). The five behaviors that accelerate value from digital investments. In *6th Annual Digital IQ Survey*.

Stepper, J. (2015). *Working out loud: For a better career and life*. Ikigai.

Author Biography

Prof. Dr. Thorsten Petry is professor for strategy, organization and human resources management at RheinMain University of Applied Sciences, Germany. He is working as a consultant, trainer and keynote speaker. Professor Petry has written more than 80 management books and articles. His current research is focused on the digital transformation. For more information please visit www.hs-rm.de/de/hochschule/personen/petry-thorsten/.

Autosomes as Managers—A Commented Case

Daniel Weihs

Abstract This chapter discusses the problems arising from the development of more and more capable and independent thinking machines, so-called "autosomes". With this development, even the top level of knowledge workers in business are challenged, and it is not clear how things will play out. Adding autosomes to all levels of work will clearly make work-streams more efficient, but as machines start making decisions, the criteria for good and bad decisions, as well as loyalty (to the firm to the stakeholders, to mankind) may lead to unexpected results.

1 Preface

In the last century, Peter Drucker raised a challenge facing future management of how to increase knowledge worker productivity, implying that most menial and repetitive work will be performed by machines (Drucker 1999a, b; Starbuck 2012). The competitiveness of firms will be defined by their innovation and the successful employment of knowledge workers. Since then robotics has advanced by tremendous leaps, moving beyond fully programmed machines, to autonomous systems. These systems can adapt their own course of action in order to be able to accomplish their assigned mission, while operating under unexpected and uncertain environments.

In this chapter, we will explore capabilities of autosomes in two steps. Firstly, we will discuss in a fictitious case the activities of a senior member of a firm, who happens to be an autosome; our coined name for an autonomous system, or advanced robot (cf. Weihs in North and Gueldenberg 2011, pp. 139–142).

Secondly, we look even further into the future, where many more niches of value creation are taken over by such autosomes. We will combine short imagined scenarios with discussion of the implications for future business and people.

D. Weihs (✉)
Israel Institute of Technology, Haifa, Israel
e-mail: dweihs@technion.ac.il

K. North et al. (eds.), *Knowledge Management in Digital Change*, Progress in IS,
https://doi.org/10.1007/978-3-319-73546-7_13

2 Quick Actors—Autosomes as Managers

6 A.M. Tuesday, May 12. Division Deputy Chief A., Rob Centuriuno has just turned on the dataport for raw materials supply—It's only three days before the monthly statistics report, and the information from the Australian mines hasn't appeared yet. "Tuesday is the worst day of the week, and this is the worst week of the month" he thought, "I have the 9 A.M. staff meeting, the 11 A.M. meeting with these pesky inspectors from Customs, and then the 3 P.M. board meeting, and no presentation in yet from David".

It seems the storm over the Indian ocean is delaying supplies again, even the automatic pilots, driving the submerged vessels were delayed, and the robotic smelting plant is running short of raw material. A quick look at the commodities market, using our quick-search program, identifies a dip in metal prices on the Moscow exchange, for stock in Yokohama port. So, the decision is easy- buy it in Moscow and send the ships out to Japan immediately. The Deputy Division chief rank has a 25 M$ limit on decisions, but luckily this is within the limit, so the electronic buy order is dispatched, and filled, before the flashing red lights for an input problem are activated.

A quick review of reports by section chiefs followed. All the reports were green, i.e. within the defined limits. Looking at the data, Rob decides to tighten the bounds in quantities considered green, as he extrapolates and sees that monthly quotas may be missed if several parameters are near the present limits simultaneously. This computation needed only about 1.2 s, so Rob returns the reports and asks for corrected reports—to include the new bounds for transmission by 8. "As usual, the two laggards will be Sam and Niva—they always need extra time to check for errors. Why can't they be more like the others, whose programming eliminates this annoying delay?" He thought. "**Why do I still need to allow some people an hour to read and think?** I do this instantaneously, and so do the other '**Quick-actors**'?"

"Why is it that the slower the reader, the more errors in his analysis?", this thought is interrupted by a e-call. Sig, the Quality Control inspector is asking: "The failure rate of widgets from Colombia is again too high. Should we return the whole shipment? If we do, the supply will reach 'low-critical' in 6 days, two days before the Moldova supplier can step in; so what to do"?

The analysis shows that the high failure rate stems from one production line (out of 12), so Rob decides to stop accepting delivery from this line, until the supplier proves it is fixed. The e-mail authorization for stopping goes out immediately, with copies to Production, Sales, and the CEO. But here a problem arises; the supplier in Colombia is not yet Roboticized, so this decision to stop has to be confirmed by the CEO of the Colombian supplier, who is now asleep (it's 3 a.m. in his time-zone), and his phone is turned off (did he expect such a call and disconnect on purpose?). So, Rob sends a copy to Legal, to make sure that the firm will not be charged for production during the hours till Colombian start of work. "Lucky, Legal will be active in less than 2 h—this department is still controlled by humans. I will have to 'talk' with the Head of Legal about having a rep. available 24/7—even if this means giving authorization to an autosome", Rob thought. This reminded him of the last big fight at the Board meeting, when he suggested appointing an autosome as the new Chief of Finance. While David, the CEO was broad minded enough to see the logic—some of the other human members were absolutely against. "They fear that their time to be replaced by an instantaneous autosome will come soon", the CTO wrote Rob on an unsigned note—realizing that Rob's **handwriting recognition** is good enough. Rob winked at the CTO and retreated.

The 9 a.m. internal staff meeting went smoothly with no further difficulties. The personal interview with the new candidate for secretary was next. Since his previous secretary left on natal leave he received 3 candidates, all of which were elderly ladies, who were the only ones willing to take a 4 month temporary job. The agreement with the unions, as defined by the 2026 Law for Protection of Employment, mandated 50% human employees. Company policy preferred human secretaries, as they "put a nice face on the firm" and were relatively cheap. The three previous candidates refused to work with aa autosome boss with no other humans in the same office—quoting the annex to the law which made this a reason for refusing employment while still getting unemployment benefits. This holdover from the 2010s, where several mishaps occurred, was long overdue for removal from the law, but insurance and other issues was delaying that.

So, Rob was glad to see the young woman walk in. After a quick interview, he accepted her—at this stage, he would have accepted anyone who could read, write and text.

Now he went back to reading the files sent for the board meeting. Board members were from several countries, and used different versions of document preparation, so he moved them through the multi-compiler, which also translated from Spanish, German, Chinese and Hebrew, as each board member wrote in his mother tongue, on principle. The multi-compiler was the new model, which included the learning function so that local idioms and even jokes could be translated. Rob had no problem using this, as his own analyzing capabilities included **humor analysis and synthesis**.

When would the **first autosome get a seat around the board table**, he wondered, this must be what the women and others felt many years ago—well, our time to be recognized will come…

3 Autosomes and Knowledge Work of the Future

3.1 Efficient Use of Time by Knowledge Workers

The first requirement for good knowledge work is having the time to think. This enables development of new ideas and correlations which lead to innovation. As populations grow and people still want personal space, commutes to work will become ever more tiresome. So the advantages of "wherever work" means essentially that knowledge workers do not need to travel to work. However, even with advances in communication technology, to 3D virtual reality "offices", there still are advantages to face-to-face interactions, even if the face is mechanical. Here we include "water-cooler" serendipitous meetings and other informal interactions. There still is a place for those, for making quasi-random connections, as well as freeing (at least the human knowledge workers) from the formal electronic connection, which can, and always does leave "breadcrumbs", the electronic record of connections and data. Such breadcrumbs may inhibit free brainstorming. The random conversation can, and in many cases, has resulted in innovative jumps.

3.2 Use of Big Data for Autosome Knowledge Workers

One of the perceived advantages of human knowledge workers is the way the brain works in nonlinear ways, making seemingly random connections. In what is known as "Eureka" moments, humans combine apparently unrelated experiences to produce value. This advantage will rapidly be eroded as the combination of data collection and sorting techniques are improved, and access to multiple sources in parallel by cloud techniques is enabled.

So, the autonomous non-human knowledge worker, given an issue to resolve, will be able to access data from all sources, and order them in relevance using the same sources as the human.

The almost unlimited memory of cloud users will enable optimizing solutions, first by "brute force" methods of running all the possibilities, and with development of machine deep learning techniques and genetic algorithms, making this process faster by eliminating obvious less suitable solutions.

Nowadays, the difference is that the human can weigh intangibles into the decision, but as the universe of information of past similar situations grows, this will be also within the capabilities of the thinking machine. Animals, and humans use species memory to 'instinctively' reduce alternatives—for example rotting smells will almost automatically reduce possibilities of poisoning by spoiled food. One needs to state here, that we have developed counterintuitive decisions here, like eating smelly cheese, or drinking bitter beverages, but try giving a small child such foods…

Many such instinctive decisions are described as emotionally driven, as it may be difficult to find an immediate rational explanation for actions like altruism, falling in love, individual food likings and even suicide bombers, etc. Social scientists are attempting to reduce these to quantitative logical decisions such as showing the altruism may improve the fortune of a group at the expense of the individual.

In thinking robots, autosomes, altruism is essentially built in, as experiences can, and are, shared between individual machines, so that in essence it's not only "one for all", it is "one <u>is</u> all", as information can be shared on demand by many autosomes. Analysis of this information, and the resulting decision-making can therefore also be divided among the group, which can be anything from one unit, through a designated network, to even the whole open worldwide cloud.

So, it is clear that as machine learning and evaluation capabilities develop, in parallel with data sharing bandwidth, the advantage of human thought processes, relative to machines will lessen, and might be lost completely. This can even happen by conscious choice by humans, i.e. too much delegation of decision-making. Even today (2017), who can say that they check if their software gives a correct solution, or even their spreadsheet. We see early signs of this process in children, who do no longer know the multiplication table, but trust their calculators.

Recent studies predict that a large proportion of work done by knowledge workers will shift to non-human entities, be it autosomes, computer algorithms, or other, as yet undefined modes.

This of course, in addition to more classical industries where an even higher percentage of jobs will not require humans. Even today, in many areas, new technology and equipment could replace humans in many areas, and the remaining barrier is regulation. This includes areas such as medicine (hospital staff) and transportation (autonomous cars, aircraft etc.), among others.

An interesting and potentially worrying recent development is the real-world analog of Robocop. While in the movie, Robocop was made to look and be fearsome, robots of 2017 serve in policing duties such as traffic control (these are essentially clever traffic lights) in Congo, some airports in China use robots with facial recognition to identify possible criminal suspects. These are passive, and only transmit information, but there are models in development in Russia and Israel that can actually shoot.

So, when returning to our "future world" where autosomes are involved wherever they are more cost-effective, we can again look at the work-day of such an autosome, the one we described above as "Division Deputy Chief A. Rob Centuriuno", who was identified as Rob "…is short for Robot [like the R. in Isaac Asimov's books (Asimov 1952)]. Autosomes, which are essentially reasoning autonomous systems, are already capable of doing much of the work in industrial and commercial companies- jobs in which a limited amount of decision making and a well-defined range of decisions is required."

The autosomes were defined there as "differentiated" from robots, in that robots are fully programmed, while autosomes can make decisions, within a wide range of levels of autonomy. This spans the gamut from no decision making, through limited independent decision making, to full "independence of one unit, and further to group action, with either hierarchal, or democratic group decision making".

Looking back on this case written about seven years ago, it seems we were very conservative in writing our scenario.

In this future managerial situation, we can now say more confidently, that the work week will vanish, and these autosomes essentially can work continuously. If nowadays some equipment still has to go offline for maintenance, or local failures, fast communication will allow the autosome "persona" to be transferred to another unit, either temporarily or permanently. The only temporal limitation, in knowledge worker scenarios, would result from interaction with humans. This necessarily will produce a drive to minimize such autosome-human interactions, causing a snowball effect in human jobs.

A major issue of data security and reliability will appear, as the more decisions are made by interacting autosomes, the greater the dangers become of hacking and other misleading information flows. While the present day worry of employee defection will reduce (but not completely vanish), the dangers of virtual data leaks will become a major issue.

So, where does this world need human knowledge workers? One area is in psychology, i.e. what will the customers of this future economic unit such as a corporation, want and need? How to make "our" product a market success. This will still be based on past statistics, where autosomes will be needed, but still the human taste will be required. This is would be the generalization of food and wine tasters, fashion mavens etc. As an interesting example, pet-food companies keep groups of "tasters" to cover the situation when initiating new products- which may be nutritious, easily produced etc. but will not sell if your cat does not like it.

Actually, it is not clear whether even such representative tasters will be needed as the customers of the future may learn to trust machines completely, as we now trust many systems- say in calculation and design.

3.3 Board Meeting Scenario 2035

23 March 2035. 10 a.m. GMT: Our Autosome Deputy Director General A. R. Centuriuno (REF) has started to prepare the quarterly CEO report to the supervisory board of the company. As the regulations still require firms employing over 50,000 employees to have a human CEO and supervisory board, he needs to collect data and collate it in human-readable media. So, he connects to the 7 division headquarters to download the required information, according to the matrix defined by regulation. He (/she/it- depending on the country to which the report is sent- AR can be defined as male, female or neutral*). He knows that the CEO, based in New Zealand has been awake for several hours and can read his report. As it needs to be downgraded to human-speed reading he needs to have the data in at least 2 h before the 12 p.m. GMT board meeting, to leave time for the CEO and Chairman (who is still just getting up in his Azores mansion) to review, comment and make necessary changes.

In order to save time, AR connects to the "personas" he has in these centers, and combines the data through an input-output loop especially designed to be one-way, to leave some autonomy to divisional personnel, keeping positive tensions. This data is essentially updated only when needed, to keep compartmentation of information as protection against hacking. He is reminded of complaints, especially by the board members' personal autosome assistants transmitted to him more and more frequently. As a result, last October the board initiated a cyber-security study which, to their surprise and chagrin, actually recommended increasing separations and data-locking procedures.

In parallel, again according to the recommendations of the data-security experts he starts an alias program, that continues his persona's normal daily routine, to sidetrack any tracking malware. He only hopes that all the board members did the same, before entering the virtual meeting database.

Compiling the several data streams is time consuming as security blocks have to be released in the right order and then re-checked, so the clock has now moved to 10:02. Further analysis and checks take almost a whole minute, so at 10:03 AR contacts the CEO's personal autosome (PA), to open the security procedure- called parallel universes, opening up 3 parallel time lines again to "confuse" hacking.

As defined in the Compact of Zurich 2029, requirements for identity of senior staff and managements quotas for male/female/other, where autosomes conveniently fit into other as defined in Gender—neutral identity filing.

The CEO Autosome named Rex—to remind the CEO of his childhood dog—turns on the heads-up display and instructs the coffee/server machine to put the double macchiato next to the CEO's armchair, with his morning chocolate biscuits. Rex then invites the CEO to his study.

When the armchair reports that the CEO is sitting and has taken a biscuit, AR puts his draft report on the display and starts to explain the data using the PowerPoint-like presentation the CEO uses. Old habits are hard to change, AR muses. Recent research showed that Direct Brain Stimulation could convey the information in more detail, but the board (and, secretly-the CEO) does not want to hear about this.

After about 20 min, the CEO is satisfied, and AR connects him to the chairman's PA and after similar confidentiality activities, the three-way call begins.

AR, while participating, continues to monitor his regular activities periodically through the parallel persona, who has meanwhile observed an emergency breakdown procedure in the Mexico plant- with repairs under control, so there is no need for one of the AR duplicates to intervene.

The three-way call ends at 11:05, including several updates resulting from questions raised at the meeting. The chairman tried to test some of the data but AR fended this off, while delaying the answers, so the human partners would think that checks and adjustments were made, not realizing that these were all pre-planned, with the data presented so as to make them raise these questions. The Chairman, who is a veteran of the transition to autosome control, knew that too, but as discussed with other leaders of his generation, the data battle was lost a long time ago and one has to trust the algorithms driving the autosomes analysis and loyalty.

Finally, the board meeting started, on time. The autosomes on board, who received the information package with the other, human directors, are also updated on the changes resulting from the morning's discussions. As per regulations, the autosomes are independent directors, with owner representatives still all humans, as regulations do not allow outside communications during the meeting. AR was reminded of the previous year's arguments about worker reps and observers at the board—should they be human or autosome (most "employees" are autosomes). This was left moot and it was agreed to wait on international regulatory bodies to decide.

All the points were accepted, as AR had planned. His last thought was, "Do we really need this time wasting procedure?"

End of scenario.

4 Summing Up

This discussion, which is in part fiction, in part prediction, highlights the problems arising with the development of more and more capable and independent thinking machines. Here, the top level of knowledge workers in business are challenged, and it is not clear how things will play out. Adding autosomes to all levels of work will clearly make work-streams more efficient, but as machines start making decisions, the criteria for good and bad decisions, as well as loyalty (to the firm to the stakeholders, to mankind) may lead to unexpected results. Truly we are experiencing the early stages of a major revolution.

References

Asimov, I. (1952). *I, Robot*. New York: Grosset & Dunlap.

Drucker, P. (1999a). *Management challenges for the 21st century*. New York: Harper Collins Publishers.

Drucker, P. F. (1999b). Knowledge-worker productivity: The biggest challenge. *California Management Review, 41*(2), 79–94.

North, K., & Gueldenberg, S. (2011). *Effective knowledge work: Answers to the management challenges of the 21st century*. Bingley: Emerald Group Publishing.

Starbuck, P. (2012). *Peter F Drucker: The landmarks of his ideas*. Lulu.com.

Author Biography

Daniel Weihs is Distinguished Professor emeritus at Technion Haifa, Israel, Faculty of Aerospace Engineering, and Head of the Technion Autonomous Systems Program. He is a member of the Israel Academy of Sciences and Humanities and the US National Academy of Engineering.

Who's in Charge?—Dealing with the Self-regulation Dilemma in Digital Learning Environments

Per Bergamin and Franziska S. Hirt

Abstract We are now facing an ever-increasing amount of knowledge, which is becoming obsolete at an ever-faster rate. This requires us to select from this virtually infinite amount of digital information and decide what to consume and when. Fast evolving technological innovations facilitate guidance and assistance during the learning processes. Sensors emerging from novel devices such as face-readers, eye-trackers and wearables are promising to help learners to show and develop appropriate learning behaviour, strategies or processes. Such technological opportunities may deliver more accurate data for decision-making than students can access through their own self-perception. These developments lead to further questions: Who makes the better decisions about the right learning process and material—the learner or an intelligent system? Does the learner benefit from free choice or is he/she distracted and overburdened by too much freedom of decision? The dilemma of how much self-regulation (control) should be left to the learner is discussed here and different approaches from formal and informal learning environments are presented.

1 Introduction

As a result of the lifelong learning perspective in modern societies, new forms of learning such as distance, online or informal workplace learning are gaining in significance (Bergamin et al. 2012; Marsick and Watkins 2015). Common characteristics of these learning methods include, among other things, flexibility, just-in-time or non-intentional information processing and time, location or institutional independence. Another aspect is the increasing orientation towards technology, for example through the use of Learning Management Systems, digital assessment frameworks, intelligent tutoring systems or other systems that support

---

P. Bergamin (✉) · F. S. Hirt
Institute for Research in Open, Distance and eLearning (IFeL), Swiss Distance University of Applied Sciences (FFHS), Brig, Switzerland
e-mail: per.bergamin@ffhs.ch

K. North et al. (eds.), *Knowledge Management in Digital Change*, Progress in IS,
https://doi.org/10.1007/978-3-319-73546-7_14

information retrieval, processing or storage. Considering both lifelong learning as well as technology-based aspects, educational research shows that self-regulation by making autonomous learning decisions is a crucial skill (e.g. Ifenthaler 2012; Kalyuga and Liu 2015; Song et al. 2016). A growing body of studies also shows that the use of modern learning technologies fosters self-regulated learning activities (e.g. Kitsantas and Dabbagh 2004, 2010) and that learners with a high level of self-regulation complete their distance study programmes more successfully (Deture 2004) or else achieve better academic performance (Artino Jr 2008; Barnard-Brak et al. 2010). In other words, students with a high degree of self-regulation abilities can be assumed to be more effective learners than those with low skills (Zimmerman 1990; Nicol and Macfarlane-Dick 2006). However, research in technology-based learning has also shown that self-regulation skills should be taught but formal training is still scarce (Azevedo and Cromley 2004; Bjork et al. 2013). It is therefore important to take into account that the promotion of self-regulated learning in digital learning environments is a challenging act with basically two different and sometimes conflicting aspects. On the one side, digital learning environments demand self-regulation (decision-making) by the learner and on the other side, they give external support to the learner (external regulation) in order to prevent overload. By viewing all these developments and goals, we note that helping learners to develop advanced self-regulation skills is one of the major challenges in research on intelligent tutoring systems today (Aleven et al. 2016).

2 Learning Strategies and the Self-regulation Dilemma

Within the context of making autonomous learning decisions, different terms like self-directed, self-organised, self-guided or self-regulated learning are usually used. From here on in, we use the latter notion synonymously with the others. The first sophisticated concept of self-regulated learning was developed in the 1980s. Basically, self-regulation can be seen as a skill—knowing how learning goals can be set, what is needed for their implementation and how they can be achieved. Therefore, three main levels of information processing—cognitive, meta-cognitive and motivational—are assumed to be relevant (Wolters 2003; Lehmann et al. 2014). In accordance with this distinction, three different types of learning strategies are assumed: cognitive, metacognitive and motivational/emotional resource strategies.

Cognitive learning strategies are linked to basic learning processes. The most relevant are rehearsal, elaboration and organisational strategies (Weinstein et al. 2011). Rehearsal strategies serve as a means of memorising new learning content in order to store this in the long-term memory e.g. by repeating a definition over and over or highlighting words in a text. A distinction between passive and active strategies can be made: passive refers to repetition without new cognitive processing while active involves a great deal of cognitive processing as well as developing meaning. Elaboration strategies serve to process new and old knowledge. This means integrating new knowledge into previous knowledge but also constructing new relationships within a given information structure or even reducing information in this structure (e.g. by summarising a text or developing analogies). Organisation strategies support the aforementioned processes but focus more on reorganising knowledge structures by processing information in new modi e.g. by drawing a mind map or constructing a diagram. They require a lot of cognitive action in order to transform learned content into a new form (e.g. text to graphics).

Meta-cognitive learning strategies help to plan, monitor and reflect on learning activities and if required, they lead to an adaptation of these activities (Weinstein and Mayer 1986). While planning the solution to a task, they help with the selection of cognitive strategies, monitoring them by observing their implementation and assessing their effectivity. If there is no effect or too small an effect in relation to the goal, this should result in an adaptation. There is a functional difference between cognitive and meta-cognitive strategies. Cognitive strategies directly affect the ongoing information processes while meta-cognitive strategies affect the reflection as well as considerations about the implementation of cognitive strategies (Schraw 2001). Such learning strategies include e.g. time planning of a task, evaluating the efficiency of the task implementation, identifying problems etc. These strategies can be very demanding and overload novice learners (Kalyuga 2009). Therefore in a lot of cases, the available knowledge seems to be a crucial factor in preventing an overload (Nayak et al. 2016).

The third form are *motivational/emotional strategies*. Through intrinsic and external learning resources (Zimmerman and Martinez-Pons 1990) learners can guide motivation and attention towards learning objects or the maintenance of learning activities. More specifically, we can include in such strategies behaviour like arranging the learning environment, protecting learning activities from other activities or seeking out collaboration with others.

There are a lot of studies measuring the relationship between learning strategies and self-regulated learning behaviour. The results are diverse and somewhat contradictory: there are studies showing the effectiveness of appropriate learning strategies (e.g. Pressley et al. 1989; Weinstein et al. 2011) but also some where no considerable effects were found (e.g. Garcia and Pintrich 1994). It should be assumed that there is a difference between knowledge about learning strategies and their actual use. Beside the knowledge about the strategies, it is crucial for effective self-regulated learning to establish internal monitoring and control, and feedback processes. These processes are directly related to learning content (objects) or to a meta-level of reflecting on learning activities (Nelson and Narens 1990), e.g. when a student scrutinises a table, she or he can experience less confidence in understanding (meta-level) the relevant information (content/object level). This can lead to restudying the explanatory text associated with the table in order to improve knowledge (control). Monitoring processes are generally cyclic because they also lead to an updating of the monitoring process itself through improved knowledge. Monitoring is important while learning because if it is poor, learning is terminated too early or too late—either studies are terminated while the knowledge is still rudimentary or studies are repeated after the knowledge has already been acquired. As a consequence of poor monitoring processes, learning outcomes are jeopardised (De Bruin and Van Merriënboer 2017).

All in all, within the context of self-regulated learning, it can be summarised that students intentionally adapt learning activities to their learning goals across three dimensions. The cognitive dimension can be assigned to procedures of informational and behavioural processing (learning behaviour). The metacognitive dimension regulates cognitive, behavioural and motivational processes according to specific goals (Ifenthaler 2012). And the motivational dimension refers to processes of initiating, guiding and maintaining learning activities. These dimensions are affected by relatively stable structural conditions like learners' behavioural dispositions or characteristics as well as procedural events like instantaneous behaviour or instructional interventions. Different studies have shown that, without a specific elaborated design of the learning environment, it is not possible to improve learning strategies supporting self-regulation (among others Barnard-Brak et al. 2010). Therefore, let us take a closer look at how self-regulated learning skills can be improved in technology-based learning environments.

Programmes, offers and supportive tools for strategic learning skills are often divided into two categories: *direct* and *indirect* support. Within the context of self-regulated learning, direct support consists of specific training offers while indirect promotion is about the implicit improvement of self-regulatory strategies. This corresponds, for example, to the design of the digital learning environment so that favourable conditions for the development and enhancement of self-regulated learning behaviours are created (Ryan and Deci 2000). In this sense, one principle design element could be the creation of learning conditions which are focused on the interests of the learners. Such conditions take students' independent learning into account and facilitate the development of competences by providing informative and motivational feedback. Another opportunity for focusing on cognitive strategies is the design of learning materials such as, for example, the appropriate formulation of headings and hints or the development of exercises considering learners' pre-knowledge. On a meta-cognitive level, for instance, the use of reflective or justification prompts or the introduction of web-based learning diaries can be envisaged. One of the advantages of such indirect support is the close link between strategic learning activities, the specific learning process and the coherent reduction of the transfer problem. One drawback, however, is that in most cases no conscious reflection on learning strategies is stimulated, as is possible with direct support measures. Another dimension of the same issue is that direct and indirect interventions to foster self-regulated learning can be either *embedded* or *non-embedded* in learning environments (Clarebout and Elen 2006). Embedded instructional interventions are directly integrated into the learning activity. Therefore, the learners have to consider them during the learning process. Non-embedded interventions depend on the learner's initiative. They are also located within the learning environment but the learner can freely decide whether to use them. Non-embedded technology-driven interventions are also referred to as optional 'tools' (Clarebout and Elen 2006). Mixed forms of this classification are possible.

All in all, self-regulated learning never occurs in an absolute form. Learning is both self (by the learner) and externally controlled (e.g. by giving instructions or interaction with a teacher, peer etc.). In this respect, it takes place on a continuum between the two extreme poles and there is neither complete self-regulation nor external regulation. This basic feature points to a dilemma within the context of self-regulated learning in digital learning environments. For example, the use of learning objects or learning tools involves external control by their creator/authors although they are not directly present during the learning process but the learner has to decide himself/herself about the individual activities he or she will implement as well as their duration, repetition and so on. The question of how much regulation (i.e. control) should be left to the learners and how much structure the system should provide is the subject of many discussions in didactical research on digital learning environments. We refer to this complex and still unresolved question as the *self-regulation dilemma* (also-called the 'assistance dilemma'; Koedinger and Aleven 2007). In the following sections, we will also focus on how many opportunities for self-regulation or control should be stipulated by a learning environment. We do this by focusing on the extent of the learners' control while processing information in technology-based learning environments. Later on, we discuss some examples taken from work and research experience in adaptive learning systems which address this dilemma.

3 Three Options for Regulating Technology-Enhanced Learning

First, we discuss the advantages and disadvantages of the extreme poles of self and external regulation. There are principally three options to regulate the learning process in order to facilitate effective learning strategies: it can be achieved by the learner, be promoted by another person (i.e. a tutor or peers) or by the learning system (e.g. rule-based mechanisms or artificial intelligence).

3.1 Control by the Learner

There are many open and flexible learning systems which demand a high degree of self-regulation and thus control by the learners. The learners adapt the learning environment to their own needs and decide to a large extent what, how and when to learn. Such systems, in which a strong degree of control is given to the learners, are called 'adaptable'. In adaptable systems learners can decide for themselves, with flexibility about topic choice, resources they want to use (e.g. which media type or level of difficulty) and when they need support. Giving learners such control over the learning process has several advantages. As already mentioned, such open learning environments seem to develop and promote self-assessment and self-regulation skills (Kitsantas and Dabbagh 2004, 2010). Furthermore, learners generally seem to like being in control of their own learning process (cf. Schnackenberg and Sullivan 2000; Orvis et al. 2009). Freedom to choose by one's own control can even be seen as an important motivation for using computer-based learning resources (Henning et al. 2014) per se.

Nevertheless, a high degree of learner control can also be detrimental in learning. Taminiau et al. (2015) describe the belief that learners know best as an 'urban myth'. Particularly at the beginning of a learning process, too much control by the learners might overburden them (Lee et al. 2010; Kirschner and Van Merriënboer 2013). Corbalan et al. (2010, p. 11) state that "even expert learners may become overwhelmed and demotivated by an excessive amount of freedom". Learners might overrate or underrate their level of knowledge or performance. Correct self-assessment (calibration) is, however, a precondition for self-regulation. Overloading with too much control might lead to random, inadequate decisions. Learners thus tend, for example, to select unsuitable tasks (Corbalan et al. 2011) or they tend to use any support functions offered too rarely or too fast (Aleven and Koedinger 2000). Granger and Levine (2010) found in their study that, particularly in complex learning objects, learner control can be detrimental to learning outcomes (cognitive and transfer). Vogel et al. (2015) concluded from data analysis of an adaptable system that only users with high self-regulation skills benefited from the provided learner control. Learner control can also be negative at an affective level. Too much freedom of choice can be experienced as a burden (Schwartz 2004), resulting in negative effects on motivation. Furthermore, overburdening learners with control might threaten their feeling of competence (Kicken et al. 2008). External control and support for decision-making can thus be helpful.

3.2 Control by Another Person

External control and support for self-regulation might be given through a human tutor or peers, which refers to the concept of 'co-regulation'. Co-regulation describes asymmetrical situations whereby a more knowledgeable or skilled member of a group provides one or more others with support (Hayes et al. 2015). Although for learning scenarios this is a feasible concept, it seems to have its limits if the goal is to exploit the full flexibility of technology-based learning in terms of the time and location of usage. Humans cannot as easily fulfil the same requirements of time availability and objective precision as machines.

3.3 Control by the Learning System

Technology-based systems that determine what learning content is presented to the learner and in which form and rhythm can be static (the same for everyone) but also 'adaptive'. Adaptive learning systems modify the learning process in terms of the specified characteristics of the learner, learning contexts and/or the learning content (Wauters et al. 2010). For the purposes of simplicity, we focus here on adaptation to personal characteristics. A main feature of personalised and adaptive learning scenarios is that the navigation, suitable learning support and content are presented dynamically, changing throughout the learning progress to match the learner's requirements (Wauters et al. 2010). Generally, it is assumed that such adaptive procedures provide better learning support than static learning aids (e.g. Kalyuga and Sweller 2005; Durlach and Ray 2011). The basis for dynamic, adaptive learning systems is built on repeated measurements of variables relevant to the learning process. With regard to these measurements, aspects of the learning process ('objects of adaptation') are adapted in order to optimise learning.

Adaptive learning systems are rather diverse in terms of their instructional designs. Bases for personalisation through adaptive learning systems range from gender to level of knowledge, interests, emotions and focus of attention etc. (see for an overview e.g. Nakic et al. 2015). These constructs have to be measured by means of a so-called 'sensor'. The data generated by this sensor is stored and analysed, building a 'learner model' for each learner. From this basis, the learning process can be modified with regard to various dimensions of what is being adapted (i.e. object of adaptation). Different adaptation processes can take place in parallel at different levels of a structured system of learning materials, also referred to as a 'domain model' (Brusilovsky 2016). In our understanding, adaptation mainly happens at three levels of the domain model (cf. Fig. 1):

Curriculum loop: The adaptive system helps in choosing the learning domains (curricula) matching the learners' needs and preconditions.

Task loop: Within the learning domain, decisions on instructional support, content complexity or sequencing (i.e. task selection) are made by the system depending on the individual learner's condition.

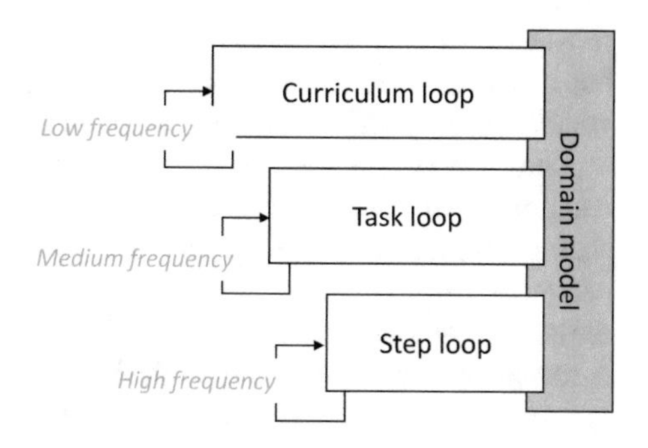

Fig. 1 Overview of the adaptation loops at three levels of the domain model

Step loop: Within the learning object (e.g. a task), hints, feedback and prompts regarding the current learning activity are presented depending on the learner's most recent learning behaviour (Aleven et al. 2016).

The three loops thus differ in their level of dynamic: the step loop frequently adapts to the learner model, the task loop less frequently (in-between learning objects, i.e. tasks) and the curriculum loop only occasionally (in-between courses/ subjects). Interventions in adaptive learning systems are, at this time, mostly implemented in the task and the step loop (Vanlehn 2006; Essa 2016).

Such control over the learning processes implemented by a system has various advantages and disadvantages. The advantage of system-controlled regulation is that it promises to make better assessments and more sound decisions than novice learners or learners of complex material. There is some empirical evidence for the superior efficiency of system-regulated learning compared to self-regulated learning (e.g. Azevedo et al. 2008; Kauffman et al. 2011).

A difficult issue when giving control to the system is that they can more or less force the learner to follow the presented learning process. As a consequence, learners have little control over their learning which can trigger feelings of frustration. This might be particularly the case when the system makes less suitable choices (e.g. because the learner model was not accurate). Furthermore, learners have no or little information about the decisions made by the system. Subsequently, monitoring and meta-cognitive reflexion on the learning activities is hampered. Moreover, learners do not develop or practise self-regulation skills because they are forced to rely strictly on the system's decisions (Kicken et al. 2008). In our view, while learning domain-specific content, learners should also develop skills to assess their own performance and make suitable learning choices. As already stated above, it can be assumed that there is a relationship between such self-regulation skills and the construction of domain-specific knowledge. A negative example of this relationship is provided by Taminiau et al. (2013). They found that advice on task selection was detrimental to domain-specific learning outcomes in on-demand education. Their hypothetical explanation is that forced advice can supplant self-regulatory processes which has negative effects on domain-specific problem-solving skills.

All in all, today it seems that none of the presented approaches to technology-based learning are fully convincing (cf. self-regulation dilemma). However, as already mentioned, self-regulation is not categorical. Whether control remains with the learner or is transferred to the learning system is a continuum. We hope to contribute solutions to the self-regulation dilemma by observing concepts which 'experiment' within this continuum.

Before giving examples for such approaches, one feature which is generally recommended in adaptive learning systems—the transparency of learner models—shall be explained in terms of its impact on the promotion of self-regulation skills. Adaptive learning systems can hide their learner model, which is measured through the sensor, and serves as basis for adaptation, or else openly present it to learners. A transparent communication of the learner model to the learners is referred to as an *open learner model*. Open leaner models support learners in getting to know the system's basis for regulation/recommendation. Such open learner models can be regarded as a form of external feedback (e.g. feedback on performance level). This kind of external feedback and particularly the comparison of the external feedback with one's own assessment may foster students' self-assessment. Open learner models further help learners to understand why the system regulates their learning process and in which way. It can also serve to promote metacognition and self-reflection and support self-regulated learning (Long and Aleven 2013). Many adaptive learning systems (also many of the following examples) include such open learner models.

4 Shared Control and Fading as Concepts for Approaching the Self-regulation Dilemma

There are technology-based learning systems which deliberately try to combine learner and system control in the form of a shared control approach by sharing the control between the learner and the learning system. A second approach is to fade the level of system control over the course of the learning progress (i.e. fading scaffolds). Let us discuss these two concepts by giving some examples.

4.1 Shared Control

The concept of shared control refers to a distribution of responsibility between the learning system and the learner. This means that the adaptation of the learning process to the learner's needs is conducted by decisions made by both the system and the learner. A prominent method for distributing the control between the learner and system is the use of open recommender/advisement systems. Such systems provide learners with a recommendation or offer on how to adapt their learning process. On the basis of the given advice, learners can decide themselves how to proceed. Whether they comply with the advice given to them is entirely down to their own decision. Adaptations of the learning process are thus not forced upon the learners. An interesting example of such a recommender system at the level of the above-mentioned curriculum loop is presented by (Al-Badarenah and Alsakran 2016). They developed an intelligent system that provides university students with recommendations as to which elective courses to choose. The system helps the learners to make decisions by systematically analysing the huge amount of elective course offers in terms of their suitability to the individual student. The system's algorithms are based on what other students with similar characteristics chose and also predicts one's estimated future grade in the recommended courses. It makes a suggestion but does not impose the choice on the students. In this way, students gain external feedback on which might be suitable for them and can compare this external opinion with their own assessment. This approach might stimulate and promote students' meta-cognitive self-regulation.

Another more embedded approach in recommender systems is a *two-step approach*. The system pre-selects a set of suitable learning objects in the first step. In the second step, the learner can freely choose from this restricted set of objects. Such a system has been implemented by Corbalan et al. (2006) in the task loop. Out of all the available tasks, their system selects a subset of learning tasks based on the learners' performance in previous tasks and their invested mental effort (first step: system-control). In the second step, this subset of tasks is presented to the learners who make the final decision on which of these pre-selected tasks they want to work on (second step: learner control). This approach is supposed to avoid cognitive overload caused by excessive choice and still guarantees some learner control. The two-step approach allows less learner control than open recommender systems. This has the advantage that it clearly prevents learners from selecting counterproductive tasks as the system (ideally) only presents appropriate tasks to choose from.

In addition to recommender systems where the learners themselves have the control over the final decision of adaptation, there is also the option of a simultaneous shared control in form of a *negotiation-based adaptation mechanism*. Such a system has been developed and evaluated by Chou et al. (2015). The system's assessment of the learner model (objective task performance) and students' self-assessment are both evaluated and if they do not match, the student and system 'negotiate' the adequate learner model and the next learning task to be worked on. In this way, students with low meta-cognitive skills are supported in their regulation. Still, students have the possibility to make corrections when the system might have wrongly assessed their learner model. This approach has been shown to promote better self-assessment accuracy (calibration) and better choices in terms of learning objects (i.e. self-regulation). Results show that students with low meta-cognitive skills in particular benefited from this design (Chou et al. 2015).

The presented approaches of shared control illustrate how opportunities for self-regulation can be provided on a continuum between self and system control. However, the presented approaches of shared control provide a static level of learner control. In order to approach inter- and intra-individual differences in self-regulation skills, the level of control can be flexibly increased or reduced according to the skills the learners have.

4.2 *Fading Computer-Based Scaffolds*

Adaptive learning systems are sometimes constructed as so-called 'fading scaffolds'. In fading scaffolds, the intensity of support lessens with increasing expertise in self-regulation. This expertise is dynamically assessed by the system. Fading scaffolds facilitate a gradual transfer of responsibility from the system to the learner

(Van De Pol et al. 2010; Van Meeuwen et al. 2013). The opposite of fading, 'adding', is also didactically recommend when learners seem to lack the necessary self-regulation skills (which they used to have before, e.g. with easier learning material). The final goal should, however, be to reduce system control in the long run.

Two of the presented examples for systems of shared control already address this issue and provide ideas on how to fade or add to the level of shared control. Within the framework of their negotiation-based shared control system, Chou et al. (2015) propose flexibly adapting the degree of concession to the learners' individual self-regulation skills. If the system detects that students are making inaccurate self-assessments or task choices, the systems might restrict students' control over negotiation. In the opposite case, the system may also transfer more control to the learner. In their paper on the two-step recommender system, Corbalan et al. (2006) propose to continuously increase the control given to learners by giving them the possibility to choose from an increasing pool of tasks. In this way, learners have to cope with an increasing amount of choice, thus self-regulate their learning more and more.

Another example for fading/adding scaffolding systems for promoting self-regulation skills at a cognitive and metacognitive level is Meta-Tutor. Researchers around Azevedo have published several experimental studies on this intelligent hypermedia-based scaffolding system for fostering self-regulated learning (e.g. Azevedo et al. 2016a, b; Harley et al. 2016; Taub and Azevedo 2016). The core of the scaffold is formed by four artificial pedagogical agents which promote self-regulated learning through prompting:

There is 'Gavin the Guide' who supports students' navigation in the learning environment and provides questionnaires for self-assessment.

'Pam the Planner' monitors the *planning* process during self-regulated learning and helps users to set sub-goals or to activate their pre-knowledge.

'Mary the Monitor' presents the *meta-cognitive* monitoring of self-regulation during learning by stimulating self-assessments on text comprehension or estimated sub-goal achievement and so forth.

'Sam the Strategizer' encourages *cognitive learning strategies*.

Together with the subject specific learning content, there is always just one pedagogical agent visible at a time. Which pedagogical agent this is depends on the current learning activity (Harley et al. 2016) and is system-controlled. Additionally, there is a non-embedded palette of tools for self-initiated self-regulated learning processes (e.g. note taking). In order to prevent overload, the program offers the possibility to provide the scaffold in an adaptive (system-controlled) manner. The system control is based on a set of (meta-) cognitive rules which are triggered by temporal and behavioural thresholds. For example, when learners remain on a page irrelevant to their current sub-goal for more than 15 s, 'Mary the Monitor' is activated. 'She' encourages the learner to reconsider the content relevance of the page and shows the current sub-goal. This adaptive amount of system prompts can be regarded as a sophisticated form of fading/adding. This adaptive scaffolding system does not just offer/recommend different navigation paths (task loop). Instead, learning strategies are initiated and also their efficient use is monitored by the system (step loop). Azevedo et al. (2016a) found in an experiment that their adaptive scaffolding condition did indeed promote better learning outcomes and metacognitive monitoring and regulation than the condition without scaffolding (no agents).

Contrary to didactical theories, a recent meta-analysis found no difference in the effect of scaffolding with or without fading or the opposite—adding—on cognitive outcomes (Belland et al. 2017). These findings contradict the assumption that fading should be seen as a crucial part of the scaffolding process (Lajoie 2005; Puntambekar and Hubscher 2005; Ge et al. 2016). Non-fading scaffolds, when they are actually unneeded, are believed to 'overscript' expert learners and thus have a detrimental effect on learning outcomes (Dillenbourg 2002; cf. expertise reversal effect Kalyuga et al. 2003). More differentiated analyses are of interest to assess which outcomes (cognitive, self-regulation skills etc.) are affected by fading/adding scaffolds, depending on which parameters (i.e. basis for adaptation). Didactical theories significantly involve the effectiveness of fading/adding scaffolds. However further research is needed to empirically clarify the actual effects of fading/adding scaffolds. Nevertheless, scaffolding as such has been proven to be highly effective in the development of domain-specific learning outcomes (Belland et al. 2017) and efficient for promoting self-regulation skills (e.g. Azevedo et al. 2004, 2008).

5 Outlook and Conclusions

In this article, we have outlined the main characteristics of self-regulated learning. Self-regulated learning skills become more and more important through the increased use of technology-based learning, the growing amount of information provided online and the fast obsolesce of knowledge. Considering research on the use of learning strategies and also self-regulated learning, it becomes clear that often self-regulation skills are crucial for efficient learning and that external support to provide regulation of the learning processes is a beneficial option. But the somewhat provocative question is who should be in charge of learning process decisions: the learner himself/herself or machines?

From an educational and didactic view, we have to deal with subsequent challenges. Developing self-regulation skills and to some degree effecting regulation externally by means of technology seems to be a feasible way. Combining these two approaches is, however, not easy as they require different learning environment characteristics. Depending on their conceptualisation and design, learning environments can either foster or hamper self-regulation skills. Characteristics that externally support regulation can hamper the development of one's own self-regulation skills. In the context of the self-regulation dilemma, there is a fine line between both effects. Today educational research still does not give a very clear answer. But there are some options available. Therefore, the aforementioned examples were presented.

Experimenting with the ratio of system and learner control at an inter- and intra-individual level seems to be a promising approach. Derouin et al. (2004) additionally stress the importance of clearly informing the learners of how much control they have in this context. They presume that learners' perceptions and understanding of the amount of control are even more important than the actual control they have. Furthermore, indications as to how significantly interventions/tools for self-regulation should be 'embedded' (i.e. forced to be used/seen by the learner) are needed.

So far, we have mainly focused on formal learning scenarios using technology-based environments which are usually used in universities, business and increasingly also in schools at secondary and primary level. In the context of open, informal learning (unintended daily life learning), self-regulation is also of relevance. The selection of suitable information (of interest, relevant to one's life, professional, educational goals etc.) seems to be becoming too complex for our daily self-regulation. Tintarev et al. (2016) stress the necessity of support in social media: "With large amounts of noisy, user-generated content, we have no choice but to rely on automated filters to compute relevant and personalised information that are small enough to avoid cognitive overload" (p. 279). Artificial intelligence might be of value in supporting our decision-making as too much user choice can be detrimental (e.g. Katz and Assor 2007). Recommender systems (on the basis of artificial intelligence) have become particularly common in many online platforms (Amazon, Google, Mendeley etc.). Some recommender systems are more embedded (i.e. 'forced' upon the user) than others: pre-selecting information (strongly embedded), prioritising information (embedded) or recommendations presented on request or displayed in the corner of the environment (non-embedded). Such recommender systems (particularly when embedded) have an influence on users' selection of information in informal learning (see Pan et al. 2007). We might not wish to transfer too much control over to artificial intelligence.

Self-regulation theories and research (Tintarev et al. 2016) suggest giving users (in our case, learners) some ability to inspect and control, in order to develop self-regulation skills and prevent users from developing negative feelings, e.g. of not being able to understand the system's behaviour. One disadvantage of artificial intelligence-based system control (compared to rule-based system control) is that *open* learner/user models are hard to implement, as the algorithms are often complex and not necessarily logically explainable. Nevertheless, efforts are being made to provide users with an understanding of complex algorithms. Even options for the user to easily adapt these algorithms are conceivable. For instance Kulesza et al. (2015) describe a method of 'exploratory debugging' in order to help end-users to understand and correct predications made by machine learning systems. Transparent information about what forms the basis for the adaptations made by the system (i.e. open user models) and giving the possibility of adapting this basis (i.e. shared control) seems important in formal as well as informal learning environments.

Otherwise, our information processing might indeed become strongly dependent and controlled by algorithms/artificial intelligence. We thus propose to transfer the discussion on the self-regulation dilemma also to the context of informal learning environments such as social media, newsfeeds and search engines.

In addition to developments in control by artificial intelligent systems of formal and informal learning environments, there are also promising developments happening in the field of sensors to measure variables relevant to learning. As the basis for the learner/user model, sensors are a crucial factor in the quality of adaptive learning systems or general recommender systems. Without a good assessment of the learner model, the derived interventions/recommendations are probably useless or even detrimental. So far, sensors commonly measure the level of user knowledge, performance, interests, preferences etc. More sophisticated sensors, which are currently in development, promise to deliver more holistic user models (e.g. Bannert et al. 2017; Sawyer et al. 2017). For example, through connected data from smart homes, wearables, eye-tracking, face-reading and so on, more information can be derived on users' emotion, physiological condition and behaviour. This data can deliver a basis for further interventions and recommendations. Particularly for interventions in the step loop, sophisticated sensors which provide instant measurements (with a very high time frequency) without interrupting the user seem promising. If the system is well informed on the users' instant emotions (e.g. frustration) or the users' current focus of attention, it may immediately intervene in order to improve the learning process.

The concept of optimising the amount of learner control (i.e. level of required self-regulation) is not new (e.g. Campbell and Chapman 1967; Steinberg 1989). However, due to the growing amount of ill-structured learning objects available online, the reduction of overload and promotion of self-regulation skills is of significant importance. Technological developments can be seen as facilitating a new level of quality and efficiency in self-regulation. This is attempted by adaptive learning systems by considering individual characteristics and requirements. One promising possibility is to optimise the amount of control given to the learners but this 'optimal amount' of control is so fluid that it is hard to grasp. Koedinger and Aleven (2007, p. 261) see this issue as "the fundamental open problem in learning and instructional science". We agree that research on instructional designs should not be satisfied with acknowledging this dilemma alone. Rather, we should "strive toward characterizing qualitative conditions and quantitative threshold parameters that can aid instructional designers and instructors in making good decisions" (p. 261). We call for more research and the implementation of sophisticated open learner models plus adaptive (fading/adding) intensity of shared regulation in technology-enhanced formal and informal learning environments. Many technological possibilities for such systems are already in existence and will be developed further.

References

Al-Badarenah, A., & Alsakran, J. (2016). An automated recommender system for course selection. *International Journal of Advanced Computer Science and Applications, 7*(3), 1166–1175.

Aleven, V., & Koedinger, K. R. (2000). Limitations of student control: Do students know when they need help? In *Intelligent Tutoring Systems*. Springer.

Aleven V., Mclaughlin E. A., Glenn R. A. & Koedinger K. R. (2016). Instruction based on adaptive learning technologies. In R. E. Mayer & P. Alexander (Eds.), *Handbook of research on learning and instruction* (pp. 522–560, 2nd edition). Routledge.

Artino, Jr A. R. (2008). *Learning online: Understanding academic success from a self-regulated learning perspective*. University of Connecticut.

Azevedo, R., & Cromley, J. G. (2004). Does training on self-regulated learning facilitate students' learning with hypermedia? *Journal of Educational Psychology, 96*(3), 523.

Azevedo, R., Cromley, J. G., & Seibert, D. (2004). Does adaptive scaffolding facilitate students' ability to regulate their learning with hypermedia? *Contemporary Educational Psychology, 29* (3), 344–370.

Azevedo, R., Moos, D. C., Greene, J. A., Winters, F. I., & Cromley, J. G. (2008). Why is externally-facilitated regulated learning more effective than self-regulated learning with hypermedia? *Educational Technology Research and Development, 56*(1), 45–72.

Azevedo R., Martin S. A., Taub M., Mudrick N. V., Millar G. C., & Grafsgaard J. F. (2016a). Are pedagogical agents' external regulation effective in fostering learning with intelligent tutoring systems? In *International Conference on Intelligent Tutoring Systems*. Springer.

Azevedo R., Taub M., Mudrick N., Farnsworth J., & Martin S. A. (2016b). Interdisciplinary research methods used to investigate emotions with advanced learning technologies. In M. Zembylas & P. A. Schutz (Eds.), *Methodological advances in research on emotion and education* (pp. 231–243). Springer.

Bannert M., Molenar I., Azevedo R., Järvelä S., & Gasevic D. (2017). Relevance of learning analytics to measure and support students' learning in adaptive educational technologies. In *LAK*.

Barnard-Brak, L., Paton, V. O., & Lan, W. Y. (2010). Self-regulation across time of first-generation online learners. *ALT-J, 18*(1), 61–70.

Belland, B. R., Walker, A. E., Kim, N. J., & Lefler, M. (2017). Synthesizing results from empirical research on computer-based scaffolding in STEM education: A meta-analysis. *Review of Educational Research, 87*(2), 309–344.

Bergamin, P. B., Ziska, S., Werlen, E., & Siegenthaler, E. (2012). The relationship between flexible and self-regulated learning in open and distance universities. *The International Review of Research in Open and Distributed Learning, 13*(2), 101–123.

Bjork, R. A., Dunlosky, J., & Kornell, N. (2013). Self-regulated learning: Beliefs, techniques, and illusions. *Annual Review of Psychology, 64,* 417–444.

Brusilovsky, P. (2016). Domain modeling for personalized guidance. In R. A. Sottilare, A. C. Graesser, X. Hu, A. Olney, B. Nye, & A. M. Sinatra (Eds.), *Design recommendations for Intelligent Tutoring Systems: volume 4 - Domain modeling* (pp. 165–183). Army Research Laboratory.

Campbell, V. N., & Chapman, M. A. (1967). Learner control versus program control of instruction. *Psychology in the Schools, 4*(2), 121–130.

Chou, C.-Y., Lai, K. R., Chao, P.-Y., Lan, C. H., & Chen, T.-H. (2015). Negotiation based adaptive learning sequences: Combining adaptivity and adaptability. *Computers & Education, 88,* 215–226.

Clarebout, G., & Elen, J. (2006). Tool use in computer-based learning environments: Towards a research framework. *Computers in Human Behavior, 22*(3), 389–411.

Corbalan, G., Kester, L., & Van Merrienboer, J. J. (2011). Learner-controlled selection of tasks with different surface and structural features: Effects on transfer and efficiency. *Computers in Human Behavior, 27*(1), 76–81.

Corbalan, G., Kester, L., & Van Merriënboer, J. J. (2006). Towards a personalized task selection model with shared instructional control. *Instructional Science, 34*(5), 399–422.

Corbalan, G., Van Merriënboer, J. J., & Kicken, W. (2010). Shared control over task selection: A Way out of the self-directed learning paradox? *Technology, Instruction Cognition & Learning, 8*(2), 137.

De Bruin, A. B., & Van Merriënboer, J. J. (2017). Bridging cognitive load and self-regulated learning research: A complementary approach to contemporary issues in educational research. *Learning and Instruction, 51*, 1–9.

Derouin, R. E., Fritzsche, B. A., & Salas, E. (2004). Optimizing e-learning: Research-based guidelines for learner-controlled training. *Human Resource Management, 43*(2–3), 147–162.

Deture, M. (2004). Cognitive style and self-efficacy: Predicting student success in online distance education. *American Journal of Distance Education, 18*(1), 21–38.

Dillenbourg, P. (2002). *Over-scripting CSCL: The risks of blending collaborative learning with instructional design*. Heerlen: Open Universiteit Nederland.

Durlach, P. J., & Ray, J. M. (2011). *Designing adaptive instructional environments: Insights from empirical evidence* (technical report 1297). Arlington, VA: Army research INST for the behavioral and social sciences.

Essa, A. (2016). A possible future for next generation adaptive learning systems. *Smart Learning Environments, 3*(1), 16.

Garcia, T., & Pintrich, P. R. (1994). Regulating motivation and cognition in the classroom: The role of self-schemas and self-regulatory strategies. In D. H. Schunk & B. J. Zimmerman (Eds.), *Self-regulation of learning and performance: Issues and educational applications* (pp. 127–153). Hillsdale, NJ: Lawrence Erlbaum Associates.

Ge X., Law, V., & Tawfik A. (2016). The design of scaffolding and fading: research issues and challenges. In *Proceedings of the Workshop on Computer-Based Learning Environments for Deep Learning in Editors and Useful Understanding of Conceptual Material.*

Granger, B. P., & Levine, E. L. (2010). The perplexing role of learner control in e-learning: will learning and transfer benefit or suffer? *International Journal of Training and Development, 14*(3), 180–197.

Harley, J. M., Carter, C. K., Papaionnou, N., Bouchet, F., Landis, R. S., Azevedo, R., et al. (2016). Examining the predictive relationship between personality and emotion traits and students' agent-directed emotions: towards emotionally-adaptive agent-based learning environments. *User Modeling and User-Adapted Interaction, 26*(2–3), 177–219.

Hayes, S., Uzuner-Smith, S., & Shea, P. (2015). Expanding learning presence to account for the direction of regulative intent: self-, co- and shared regulation in online learning. *Online Learning, 19*(3), 15.

Henning, P. A., Forstner, A., Heberle, F., Swertz, C., Schmölz, A., Barberi, A., Verdu, E., Regueras, L. M., Verdu, M. J., & De Castro J. P. (2014). Learning pathway recommendation based on a pedagogical ontology and its implementation in moodle. In *The DeLFI 2014 Conference*.

Ifenthaler, D. (2012). Determining the effectiveness of prompts for self-regulated learning in problem-solving scenarios. *Educational Technology & Society, 15*(1), 38–52.

Kalyuga, S. (2009). Instructional designs for the development of transferable knowledge and skills: A cognitive load perspective. *Computers in Human Behavior, 25*(2), 332–338.

Kalyuga, S., Ayres, P., Chandler, P., & Sweller, J. (2003). The expertise reversal effect. *Educational Psychologist, 38*(1), 23–31.

Kalyuga, S., & Liu, T.-C. (2015). Managing cognitive load in technology-based learning environments. *Journal of Educational Technology & Society, 18*(4), 1.

Kalyuga, S., & Sweller, J. (2005). Rapid dynamic assessment of expertise to improve the efficiency of adaptive e-learning. *Educational Technology Research and Development, 53*(3), 83–93.

Katz, I., & Assor, A. (2007). When choice motivates and when it does not. *Educational Psychology Review, 19*(4), 429–442.

Kauffman, D. F., Zhao, R., & Yang, Y.-S. (2011). Effects of online note taking formats and self-monitoring prompts on learning from online text: Using technology to enhance self-regulated learning. *Contemporary Educational Psychology, 36*(4), 313–322.

Kicken, W., Brand-Gruwel, S., & Van Merriënboer, J. J. G. (2008). Scaffolding advice on task selection: A safe path toward self-directed learning in on-demand education. *Journal of Vocational Education & Training, 60*(3), 223–239.

Kirschner, P. A., & Van Merriënboer, J. J. G. (2013). Do learners really know best? Urban legends in education. *Educational Psychologist, 48*(3), 169–183.

Kitsantas, A., & Dabbagh, N. (2004). Supporting self-regulation in distributed learning environments with web-based pedagogical tools: An exploratory study. *Journal on Excellence in College Teaching, 15,* 119–142.

Kitsantas, A., & Dabbagh, N. (2010). *Learning to learn with integrative learning technologies (ILT): A practical guide for academic success*. Greenwich, CT: Information Age Publishing.

Koedinger, K. R., & Aleven, V. (2007). Exploring the assistance dilemma in experiments with cognitive tutors. *Educational Psychology Review, 19*(3), 239–264.

Kulesza, T., Burnett, M., Wong, W. -K., & Stumpf, S. (2015). Principles of explanatory debugging to personalize interactive machine learning. In: O. Brdiczka & P. Chau (Eds.), *Proceedings of the 20th international conference on intelligent user interfaces* (pp. 126–137). New York: ACM.

Lajoie, S. P. (2005). Extending the Scaffolding Metaphor. *Instructional Science, 33*(5–6), 541–557.

Lee, H. W., Lim, K. Y., & Grabowski, B. L. (2010). Improving self-regulation, learning strategy use, and achievement with metacognitive feedback. *Educational Technology Research and Development, 58*(6), 629–648.

Lehmann, T., Hähnlein, I., & Ifenthaler, D. (2014). Cognitive, metacognitive and motivational perspectives on preflection in self-regulated online learning. *Computers in Human Behavior, 32,* 313–323.

Long, Y., & Aleven, V. (2013). Supporting students' self-regulated learning with an open learner model in a linear equation tutor. In H. C. Lane, K. Yacef, J. Mostow, & P. Pavlik (Eds.), *Artificial intelligence in education* (Vol. 7926, pp. 219–228). Berlin, Heidelberg: Springer.

Marsick, V. J., & Watkins, K. (2015). *Informal and incidental learning in the workplace (Routledge revivals)*. San Francisco: Routledge.

Nakic, J., Granic, A., & Glavinic, V. (2015). Anatomy of student models in adaptive learning systems: A systematic literature review of individual differences from 2001 to 2013. *Journal of Educational Computing Research, 51*(4), 459–489.

Nayak, C. R., Viswanathan, V., Solomon, J. (2016). The first step towards a pre-requisite knowledge tracking architecture for engineering programs. IEEE.

Nelson, T. O., & Narens, L. (1990). Metamemory: A theoretical framework and new findings. *The Psychology of Learning and Motivation, 26,* 125–141.

Nicol, D. J., & Macfarlane-Dick, D. (2006). Formative assessment and self-regulated learning: A model and seven principles of good feedback practice. *Studies in Higher Education, 31*(2), 199–218.

Orvis, K. A., Fisher, S. L., & Wasserman, M. E. (2009). Power to the people: Using learner control to improve trainee reactions and learning in web-based instructional environments. *Journal of Applied Psychology, 94*(4), 960–971.

Pan, B., Hembrooke, H., Joachims, T., Lorigo, L., Gay, G., & Granka, L. (2007). In google we trust: Users' decisions on rank, position, and relevance. *Journal of Computer-Mediated Communication, 12*(3), 801–823.

Pressley, M., Borkwski, J. G., & Schneider, W. (1989). Good information processing: What it is and how education can promote it. *International Journal of Educational Research, 13*(8), 857–867.

Puntambekar, S., & Hubscher, R. (2005). Tools for scaffolding students in a complex learning environment: What have we gained and what have we missed? *Educational Psychologist, 40*(1), 1–12.

Ryan, R. M., & Deci, E. L. (2000). Intrinsic and extrinsic motivations: classic definitions and new directions. *Contemporary Educational Psychology, 25*(1), 54–67.

Sawyer, R., Smith, A., Rowe, J., Azevedo, R., & Lester, J. (2017). Enhancing student models in game-based learning with facial expression recognition. In *Proceedings of the 25th conference on user modeling, adaptation and personalization* (pp. 192–201). NewYork: ACM.

Schnackenberg, H. L., & Sullivan, H. J. (2000). Learner control over full and lean computer-based instruction under differing ability levels. *Educational Technology Research and Development, 48*(2), 19–35.

Schraw, G. (2001). Promoting general cognitive awareness. In H. J. Hartman (Ed.), *Metacognition in learning and instruction: Theory, research and practice* (pp. 2–16). Springer Netherlands.

Schwartz, B. (2004). *The paradox of choice: Why less is more*. New York: Ecco.

Song, H. S., Kalet, A. L., & Plass, J. L. (2016). Interplay of prior knowledge, self-regulation and motivation in complex multimedia learning environments: Knowledge, self-regulation, and motivation. *Journal of Computer Assisted learning, 32*(1), 31–50.

Steinberg, E. R. (1989). Cognition and learner control: A literature review, 1977–1988. *Journal of Computer-Based Instruction, 16*(4), 117–121.

Taminiau, E. M. C., Kester, L., Corbalan, G., Alessi, S. M., Moxnes, E., Gijselaers, W. H., et al. (2013). Why advice on task selection may hamper learning in on-demand education. *Computers in Human Behavior, 29*(1), 145–154.

Taminiau, E. M. C., Kester, L., Corbalan, G., Spector, J. M., Kirschner, P. A., & Van Merriënboer, J. J. G. (2015). Designing on-demand education for simultaneous development of domain-specific and self-directed learning skills: On-demand education. *Journal of Computer Assisted learning, 31*(5), 405–421.

Taub, M., & Azevedo, R. (2016). Intelligent tutoring systems. In J. S. A. Mircarelli, & K. Panourgia (Eds.), *Intelligent Tutoring Systems* (pp. 34–47). Springer.

Tintarev, N., Kang, B., O'donovan, J., & Höllerer, T. (2016). What am I not seeing? An interactive approach to social content discovery in microblogs. *Lecture Notes in Computer Science (Including Subseries Lecture Notes in Artificial Intelligence and Lecture Notes in Bioinformatics)* (Vol. 10047). LNCS.

Van De Pol, J., Volman, M., & Beishuizen, J. (2010). Scaffolding in teacher–student interaction: A decade of research. *Educational Psychology Review, 22*(3), 271–296.

Van Meeuwen, L. W., Brand-Gruwel, S., Kirschner, P. A., De Bock, J. J., Oprins, E., & Van Merriënboer, J. J. (2013). Self-directed learning in adaptive training systems: A plea for shared control. *Technology, Instruction, Cognition and Learning, 9*(3), 193.

Vanlehn, K. (2006). The behavior of tutoring systems. *International Journal of Artificial Intelligence in Education, 16*(3), 227–265.

Vogel, F., Kollar, I., Ufer, S., Reichersdorfer, E., Reiss, K., & Fischer F. (2015). Fostering argumentation skills in mathematics with adaptable collaboration scripts: only viable for good self-regulators? Exploring the material conditions of learning. In *The Computer Supported Collaborative Learning Conference (CSCL) 2015*. International Society of the Learning Sciences, University of Gothenburg.

Wauters, K., Desmet, P., & Van Den Noortgate, W. (2010). Adaptive item-based learning environments based on the item response theory: possibilities and challenges: Adaptive ITSs based on IRT. *Journal of Computer Assisted learning, 26*(6), 549–562.

Weinstein, C. E., & Mayer, R. E. (1986). The teaching of learning strategies. In M. C. Wittrock (Ed.), *Handbook of research on teaching* (3rd ed., pp. 315–327). New York: Macmillan.

Weinstein, C. E., Acee, T. W., & Jung, J. (2011). Self-regulation and learning strategies. *New Directions for Teaching and Learning, 2011*(126), 45–53.

Wolters, C. A. (2003). Regulation of motivation: evaluating an underemphasized aspect of self-regulated learning. *Educational Psychologist, 38*(4), 189–205.

Zimmerman, B. J. (1990). Self-regulated learning and academic achievement: An overview. *Educational Psychologist, 25*(1), 3–17.
Zimmerman, B. J., & Martinez-Pons, M. (1990). Student differences in self-regulated learning: Relating grade, sex, and giftedness to self-efficacy and strategy use. *Journal of Educational Psychology, 82*(1), 51–59.

Author Biographies

Per Bergamin, Prof. Dr. is the Head of the Institute for Research in Open, Distance and eLearning (http://www.ifel.ch) at the Swiss Distance University of Applied Sciences (FFHS). Since 2016, he is Chairholder of the UNESCO Chair on Personalised and Adaptive Distance Education. His research focus lies in self-regulated and adaptive learning in technology-based environments as well as emotions in e-reading and e-learning.

Franziska S. Hirt, MSc is Research Associate at the Institute for Research in Open, Distance and e-Learning (www.ifel.ch) at the Swiss Distance University of Applied Sciences (FFHS). She is also a Doctorate Student in Psychology at the University of Bern. Her focus of research lies in the formative development, use and evaluation of adaptive instructional designs in technology-enhanced learning environments.

Towards a Learning Oriented Architecture for Digitally Enabled Knowledge Work

Jörgen Jaanus, Nina Suomi and Tobias Ley

Abstract Despite large investments and research, many Knowledge Management platforms still are not used to their full potential. In this paper, we present the learning oriented architecture for the implementation of knowledge management technology to ensure that it would contribute to a better connection of employees' just in time learning with business demands. The framework draws on Knowledge Organisation Systems to establish this connection. We introduce four case studies in the professional services industry that have informed the framework. A key insight gained through this analysis is that Knowledge Management platforms need to better account for individual and collective perspectives in learning to realize their full potential.

1 Problem Statement

Knowledge workers are the main source of competitive advantage for most companies, especially in knowledge intensive sectors. For example, in the professional services industry knowledge workers constitute the main productive factor, as the quality of services offered highly depends on their expertise and professional judgment.

With the growing digitization in all economic sectors during the last decade, knowledge work has dramatically changed. There seems to be an assumption that knowledge workers seamlessly adapt to the challenges of digitization, since working with digital information is one of their main activities. However, is this really the case? Information is still growing at incredible rates, while the cognitive apparatus of human kind has not changed much in the last centuries. Demands on speed and flexibility have been growing. Especially in the services industry, efficiency demands have reduced time for learning and personal development. The huge number of customized products and services as well as their shortened life cycle

J. Jaanus (✉) · N. Suomi · T. Ley
Tallinn University, Tallinn, Estonia
e-mail: jorgen.jaanus@tlu.ee

K. North et al. (eds.), *Knowledge Management in Digital Change*, Progress in IS,
https://doi.org/10.1007/978-3-319-73546-7_15

leaves less room for training. The focus, therefore, needs to shift from "just in case" training to continuous and "just in time" learning, that is connected to job demands.

Information technology should act as a natural companion for knowledge workers by making information easily accessible and sharable. It should also turn digital information into a productive resource that helps to create value and facilitates the overall strive for efficiency, consistency and sustainability within an organisation. Knowledge management platforms have been created with the promise to address some of these challenges. However, many studies show that knowledge management platforms are still dysfunctional (Sultan 2013) while not delivering the requirement to better integrate learning with job demands in knowledge work.

In this paper, we suggest a framework for implementation of knowledge management technology in such a way that it would contribute to a better connection of "just in time" learning with business demands. We call the framework *learning oriented architecture*. The framework centrally draws on knowledge organisation systems (KOS) to establish this connection. It enables the development of several knowledge services that support knowledge workers in performing important tasks as well as their learning on the job.

In the following chapter we will review the conceptual foundation of the approach that lies in social knowledge management theories. We will then present the learning oriented architecture for knowledge work. Subsequently, we will present four case studies we have conducted in several professional services companies to illustrate the framework and to show how it has provided a valuable perspective on challenges and potential solutions in those cases. We will then summarize the theoretical contributions we have derived from these case studies and present a set of knowledge services that have been motivated from it.

2 Conceptual Foundation

For knowledge workers learning is crucial, not only in the sense of learning to understand but also and primarily learning to perform for achieving professional goals. Informal learning is a significant aspect of a learning experience, it occurs in a variety of ways—through communities of practice, personal networks, and through completion of work-related tasks (Siemens 2005). Information that resides in a database needs to be connected with the right people in the right context in order to enable learning and knowledge development (North and Kumta 2014). For organizations, the attempt of making knowledge available for reusing purposes is handled through customer relationship management (CRM), document management, collaboration tools and learning management. Those platforms are typically connected with knowledge organization systems such as glossaries, taxonomies, ontologies etc. In many organizational settings knowledge management roles and programs are not visible but have been embedded into other initiatives. Consequently, the artefacts and tools have to be considered in dispersed organizational environments.

A perspective that explicitly addresses Knowledge Management from a social learning perspective is Knowledge Maturing, as it describes the organizational learning process as goal-oriented learning on a collective level. In the Knowledge Maturing model learning activities are embedded into, interwoven with, and even become indistinguishable from everyday work processes and practices. Knowledge is continuously repackaged, enriched, shared, reconstructed, translated and integrated across different interlinked individual learning processes. During this process knowledge becomes less contextualized, more explicitly linked, easier to communicate; in short, it matures (Schmidt et al. 2009).

Despite substantial effort and investment, knowledge management platforms remain dysfunctional, as they do not accommodate learning context. Tight integration of working and learning in a workplace learning environment relies on a clear computer-interpretable conception of what content the material in question actually conveys (Ley et al. 2008). Given the case of a typical knowledge worker's IT-based workplace, the work learning context needs to take care of at least three conceptual spaces that are considered to make up the workplace: the work space, the learning space and the knowledge space (Lindstaedt and Farmer 2004). Learning and knowledge building activities must bring together elements originating from and necessitated by the social, organizational and informal context of organizational learning. While addressing motivational and self regulatory aspects that aim for the individual learning of knowledge workers (Stokic et al. 2013).

Knowledge management is inherently collaborative; thus, a variety of collaboration technologies can be used to support KM practices. Collaborative KM tools that allow people to share documents, make comments, engage in discussion, create schematic diagrams, etc. can be valuable aids to support organizational learning. Furthermore, the policies and ways in which collaborative KM tools are used can facilitate or impede organizational learning, as the use of those tools changes organizational practice. Indeed, the management of technology and the practices of using technological artefacts are always critical issues (Jones 2001).

According to Caruso, organizations are focusing on workforce productivity and are beginning to increase their focus on human resource development, a win-win situation for the employer as well as the employee (Davenport and Prusak 1998; Caruso 2017). Performance support is a discipline of enabling human performance on the job. It helps people to do their jobs and to develop competence through the normal course of doing work, rather than through off-job training or extensive reading. Ultimately, it supports the performance of a business through enabling the performance of individual knowledge workers (Bezanson 2002). For performance requirements, the knowledge content has to be delivered in the context of the work, where the highest need for learning is. Any performance support solution must consider the three roles of the knowledge worker: learner, performer and expert.

The process of knowledge reuse and knowledge creation needs to be balanced by integration of routine and structured information processing, non-routine, and unstructured learning at collective level in the same business model. Additionally, according to Ford et al., knowledge management efforts represent attempts to formalize these processes (Ford and Mason 2013). In 2011, Back and Koch stated that Knowledge Management has undergone some development in the last decade. Initially from a focus on capturing (externalizing) information from people and storing the information in databases without having a particular use in mind, to learning that knowledge is somehow bound to people and that it therefore is essential to connect people (instead of filling databases) (Back and Koch 2011). Our findings from all those case studies are in congruence with this, however, implementing learning oriented architecture remains an ongoing challenge.

3 Learning Oriented Architecture

To address the challenges of digitalization and to integrate it in knowledge work, any approach to knowledge management needs to give learning a more prominent position in knowledge management platforms. Moreover, implementing knowledge management platforms needs to start from business goals (Chen and Huang 2012). Here we draw on the three central business aims: efficiency, quality and sustainability.

Below we propose the concept of a Learning Oriented Architecture, i.e. integrating the overall goals towards efficiency, quality and sustainability as depicted on the Fig. 1 below. Implementing competence based view to the Knowledge Management platform is not a minor improvement but rather an overall design approach.

Business processes together with tasks are connected to the digital representation of communities through Knowledge Organization Systems (KOS). Establishing these connections becomes the basis for implementing knowledge management platforms. As indicated in the context of previous cases, those platforms typically enable two essential business goals: efficiency and quality. The efficiency target is reached by implementing the key principle of knowledge management that thrives to make knowledge gained available for those who need it next (e.g., library of cases, market metadata related tags, expert advice in enterprise social media).

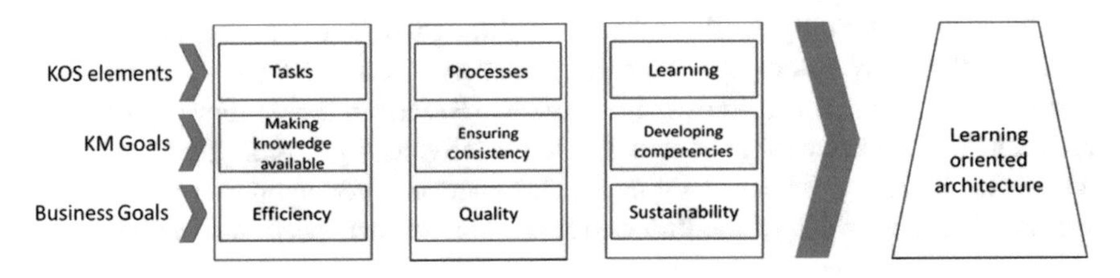

Fig. 1 Formation of learning oriented architecture

Quality is reached by maintaining and collectively developing the knowledge for ensuring the consistency in carrying out the tasks within the business processes (e.g. business rules, data definitions). Learning goals are derived from a product and service glossary, while currently the systems relies largely on intranet, fuelled by document management and enhanced by some social media functionality.

Sustainability is achieved by combining the learning context together with the reflection of competencies to the knowledge management platforms. Sustainability becomes a motivating factor for knowledge workers. It leads to increased job satisfaction through status as well as by achieving professional goals and ensuring the unity between personal goals and targets set by organization. Consequent longer working relationships are clearly factors, which leads to sustainability. Parallel to that, the cycle of knowledge development from personal to network to organization, allows learners to remain informed in their field of work through the connections they have formed. It is the basis for generating new ideas and making better business decisions.

4 Overview of Cases

We have developed the learning oriented architecture after conducting four case studies in the professional services industry. The professional services industry seems to be a particularly well-suited case for the present purposes. The industry has been undergoing tremendous changes and is especially impacted by digitalization. Knowledge work plays an important role in establishing competitive advantage in the industry. There is a particular challenge to balance the growing demands introduced through growing information and increasing speed on the one hand, and the need for informed professional judgments on the other. Also, all three business goals are clearly important: *efficiency* as companies need to react faster and faster, *quality* as companies want to ensure a consistent service for their clients, and *sustainability* as professional judgment needs to be continuously ensured despite increasing complexities and speed. The case studies were drawn from four different domains and broadly covered the challenges faced by the industry. Table 1 provides an overview of the four case studies.

4.1 Case 1. Ontological Change in a Leasing and Assets Company

Problem

The first case is set in a leasing and asset management company. Employees in this company frequently work with different types of taxonomies and glossaries that

Table 1 Overview of the cases

	Case 1	Case 2	Case 3	Case 4
Company and domain	Leasing and asset management	Accounting and financial services	Legal services and risk management	Staffing and training
Type of knowledge managed	Ontology about products and services	Requirements knowledge	Assignment records	Competency definitions
Main challenges	Adapting to continued change	Mapping requirements	Drawing from past experience	Acquiring new competencies
Methods employed	Interviews with management, analysis of concepts in the ontology	Qualitative research approach which included semi-structured interviews and data analysis	Interviews, analysis of taxonomy structure and content, using test account in KM platforms	Critical incident analysis, repertory grid, interviews

describe, for example, products and their properties or typical customer segments. In order to improve the efficiency of working and the findability of documents, these taxonomies are embedded into various information systems used at the company. Problems frequently appear when the glossaries change. New concepts are introduced that are not immediately understood by all and reflected in the information systems. Whenever a new concept is created, it needs to be compounded to the existing knowledge organization systems. This case tackled the challenge of aligning continuous knowledge development in innovation-driven context with managing additions to established knowledge organization systems.

Approach

In Enterprise Application Software that form a framework for work processes and practices, knowledge organization systems are ubiquitous. They are found in the form of shared folder structures, product categories, customer segments, staff positions etc. We look at the changing nature of these structures as a form of ontological change that happens in concepts and glossaries. We consider ontology as a formal, explicit specification of a shared conceptualization (Dietz 2006). Changing semantics (like adding customer sub-type to glossary) is consequently transferred to business rules and enables modelling of knowledge intensive processes. We take the view that in such situations knowledge workers learn and reflect, and they interpret data through semantic structures. Those structures exist both at the individual level (as individual knowledge) and at the collective level (as shared conceptualizations). This interpretation and insight leads to a formation of new ideas. Viewed in this way, establishing a new concept is a learning process where the term has to be negotiated for maintaining the shared conceptualization. Defining the terms in data management is primarily about setting the relations to other entities and attributes (Chisholm 2009). It is not possible to provide an adequate definition unless there is an in-depth understanding of the classification logic within the project team or within the broader community. We have applied the model of knowledge maturing (see above) that looks at the changing conceptualizations in terms of how these individual and collective structures interact over time.

Method

As new concepts emerge through the knowledge maturing process, which have a direct impact on enterprise knowledge organization systems, a longitudinal approach that follows the development of new concepts over time is needed. Our approach was to extract the concepts that were initiated during the last six month period by the knowledge workers who are the experts of the domain knowledge. Due to this extraction, it was possible to focus on concepts that were still at their early stage of the lifecycle, but at the same time matured enough for further analysis. As the next step of the analysis we separated the entities and attributes that helped us to focus on six emerging concepts for further analysis.

Results

While conducting research through document analysis, the essential role of classification logic became evident. Due to missing common understandings, the compounding of new concepts was more difficult and the probability of arguments and misunderstandings was high. Based on the results from interviews, the quality of existing definitions was considered to be insufficient for the compounding of a new term as a related concept.

This research case led us to design three designated services:

(a) Compounding services lead to retrieval and presentation of associative concepts. This works by relating the emerging concept to the existing concepts by setting criteria on common attributes, class relations and associative relations.
(b) Process modelling services include developing knowledge intensive processes by enriching them with business rules. It enables collective information mapping, retrieval and annotation based on the underlying KOS model.
(c) Association services enable to annotate unstructured information such as manuals, presentations, documents or specific snippets by tag recommendation system which is based on real time data driven ontology as described in previous chapter. Associations lead to improved data-driven knowledge development and decision making. It leads to the development of new competencies.

Insights to learning oriented architecture

The design oriented research which leads to the experimental development of those services is considered over the next cases in attempting to align knowledge development between innovation-driven context and knowledge organization systems. Studying ontological change has created the role of KOS in a learning oriented architecture.

4.2 Case 2. Managing Requirements Knowledge in a Financial Services Organization

Problem

The second case in an accounting and financial services organization, which adds the requirements knowledge for connecting the business layer and collective knowledge organization in communities. Previously the drivers for change towards centralization and standardization have been mainly efficiency benefits and cost reductions through scale of economy and standardization. It had been decided to develop standardized digital framework for network collaboration, across companies and business areas, which is a major step forward in standardizing systems and processes within the business network.

Approach

Requirements represent a verbalization of decision alternatives on the functionality and quality of a system (Thurimella and Maalej 2013). Engineering, planning and implementing requirements are collaborative, problem-solving activities, where stakeholders consume and produce considerable amounts of knowledge. Managing requirements knowledge is about efficiently identifying, accessing, externalizing, and sharing this knowledge by and to all stakeholders and it becomes even more complex when several organizations collaborate to develop the system.

Method

We employed qualitative research approach that included semi-structured interviews and analysis of secondary data like internal and external reports, articles, presentation materials, process maps, detailed work instructions and additional internal documentation. Data from multiple sources was then converged in the analysis process rather than handled individually. Data interpretation relied on triangulation approach at two different dimensions. Initially we considered the existing situation based on working documents, snapshots project reviews etc. and compared it with the future perspective which was based on management vision and project plans. The second perspective enabled to compare more general reports and guidelines (matured knowledge) to the drafts and working documents.

Results

We complemented the usual perspective on efficiency in cross-organizational networks with a knowledge sharing and creation perspective. Therefore, we propose to manage requirements knowledge in business networks in line with the knowledge maturing model. The conceptual model for requirements knowledge which—instead of looking at single activities of retrieval—considers a continuous cycle where the knowledge intensive processes and requirements shape each other. This cycle sets the overall approach on quality and efficiency requirements for the learning oriented architecture.

Insights to learning oriented architecture

Our findings for meeting innovation requirements reflect that the selective access to communication platforms due to the less formal network structure needs to be enhanced by process based roles and tasks of employees. Considering efficiency requirements, diverse data sources result in the need to capture and incorporate the semantics of concepts for elimination of duplicated process related effort using association services. Regarding quality requirements the guidance role of matured knowledge, such as standards, best practices, controls etc. needs to be integrated to the knowledge management platforms by implementing process modelling services. From those findings we have created the main building blocks for LOA.

4.3 Case 3. Establishing Learning Goals in Risk Management and Legal Services Company

Problem

The third case study was carried out in legal services and risk management business organization and it focused on implementing learning goals as boundary objects. In this organization KM platforms have been taken as "nice to have", consequently they fall between established work processes and endeavour for professional growth, and thus do not contribute to business targets.

The initial review indicated that KM platforms fall between ongoing peer communication and reluctance to contribute to the abstract knowledge base beyond community and geographical location. Investing into out of box software solutions had led to insufficient focus on cultural and business aspects. Case study on KOS from integrated perspective comprising personal, organizational and industry-wide perspectives was well positioned in this professional services company where the focus on knowledge management and related practices is advanced compared to overall industry practice. It enables the shift of focus from dispersed technological agenda to more integrated organizational issues.

Approach

We consider learning goals as boundary objects for integrating emerging ideas with more mature forms of knowledge. Boundary objects are plastic, interpreted differently across communities but with enough immutable content to maintain integrity. This perspective has broadened the value of knowledge organization systems from solely standardization and findability to coordination and sense-making, consequently supplementing to learning effort. The role of the boundary object is not the by-product of organizing knowledge but it is essential to consider KOS as artefacts becoming mediators of distributed cognition as described by Wallace and Ross (2016).

Method

Interviews with the associates at different levels of seniority from different locations across Europe was needed for capturing the individual perspective. For creating boundary objects we studied all the existing forms of KOS by conducting tests using the temporary account for all the KM platforms. We also used secondary data in the form of term lists, taxonomies and interviews on changes in knowledge base in the form of concept compounding. As an example of the work for creating boundary objects we compared KOS across different domains (department management, assignment management, accounting, employment administration, training) in order to connect concepts like overtime, shift planning, additional salary, regulated working hours and identified unique skills to the concept of service level, that in this instance becomes the boundary object.

Results

The knowledge processes represent a range of different ways of creating knowledge. They are forms of action, or things done in order to increase knowledge. Based on tens of interviews, document analysis and knowledge management platform tests, we have established the following knowledge processes which rely on boundary objects:

(a) establishing context for the individual documents in assignment library for learning and re-use;
(b) creating personal learning goals based on the types of assignment and roles within an assignment;
(c) connecting internal and external knowledge bases for seamless performance support by presenting integrated information units as snippets.

Insights to learning oriented architecture

The indicated knowledge processes have been taken as the point of departure when establishing the fourth case study. Applying participatory design in developing KM platforms, the focus is stretched from software to business and cultural aspects. It leads to connecting private level KOS and collective level KOS and the consequent learning effort. Boundary objects supplement corporate metadata and create the basis for learning oriented architecture.

4.4 Case 4. Developing Competencies in Training Industry

Problem

The fourth and final case was carried out in the organization providing staffing and training services. It demonstrated the learning oriented architecture and complemented the competence based view by adding the user, i.e. human perspective. In this study, the status quo in the organization is an obsolete knowledge management platform that does not accommodate learning perspective. The digital platforms exist but based on earlier studies, observations and interviews they are considered as time consuming and irrelevant. They store data and information but do not manage to connect it between each employee and each platform. This type of knowledge is collective but it remains arbitrary and thus does not contribute to learning.

Approach

To transform this type of knowledge into cognitive knowledge requires an individual approach. Initial learning takes place on an individual level and only then escalates to collective learning. This is where an individual is able to restructure the data and information and form connections between different pieces of knowledge. This is done via discussions and co-operation with colleagues, clients and other stakeholders. After having formed these connections the collective sense of learning can emerge. Defined as the collective, as organization is able to acquire skills and become competent in different areas of tasks and operations in general.

Digitalization has become high priority in the corporate agenda with the goal of digitally enabled knowledge work. It must comprise skills and specifically the support for acquiring those skills. The position of knowledge management platforms in the context of digitalization remains limited, as learning goals are considered only through knowledge as the theoretical or practical understanding of a subject. Skills define specific learned activities and can be more easily integrated into the knowledge management platforms. Competences comprise on-the-job behaviours as abilities to perform the job requirements competently. It has become a well-established concept in human resource management but typically it remains disregarded in the digitization agenda.

Method

Knowing that the organization requires learning to maintain its competitiveness and that it can only learn after individual learning has first taken place, it was crucial to define the moments when the individual learning happens. This was examined with the help of reflective methodology and the method of critical incident analysis. Based on the results of interviews with employees, it was possible to construct certain key moments that the individual him- or herself considered crucial in regards to learning. We drew conclusions from the constructs and implemented critical learning moments into the participatory design of a user focused digital platform. The critical reflection methodology provides tools for understanding and structuring the development of professional self and leads to integration of personal perspective to the collective knowledge base. KOS based pattern recognition as learning activity enables connection making between the processes, tasks, skills and competencies and, consequently, builds on construction towards critical reflective action.

Results

For an organization to learn as a collective, individual learning needs to occur first. As this case study aimed at studying how to connect the individual learning with the collective learning, it was important to begin with by clarifying the individual's key learning opportunities: What lead the individual to a learning experience? Was there something or somebody in the environment that contributed to this? How a good decision was made, and why was it good? We aimed finding out whether a specific learning experience could be an occurring event; and if so, how could it be recognized on an organizational level to support further learning by integrating it to the knowledge base.

We followed the approach implemented in modern Human Resource (HR) technologies. For the professional development of knowledge workers the future of any knowledge organization is dependent on establishing sound recruiting, career planning and placement policies. HR technologies follow the organizational perspective and take a gradual, cyclical view on knowledge workers with the following stages: sourcing, screening, selection, induction, training, collaboration, retention. Those stages can be connected to the development of professional self and studied through reflection using repertory grid as well as critical incident analysis. It enables the use of reflection in learning as described by Gray (2007). Beyond that it is viable to extract the additional layer that depicts the competences and connects to the business layer and knowledge workers' communities using knowledge organization systems.

Constructs on the "professional self" are then in the role of boundary object and facilitate establishing learning goals. As an instance in the framework within the case study, a daily task of an employee may be a personality assessment interview. This task falls under a certain business process (Executive Search) and requires interviewing skills to be performed successfully. The employee might aim at becoming the assignment manager of similar projects in the future and thus may require project management training that would support his/her aspirations. In order for the employee to develop the needed skills the organization must recognize the need. The construct "recognized as an expert" serves as the boundary object and when linked to the specific element (me when I get an offer/promotion), enables the organization to recognize and connect the learning experience.

Insights to learning oriented architecture

Conclusions of the fourth case study have indicated the need to consider learning in the design of knowledge management platforms. Our conclusions have developed gradually through the cases in professional service industry and are presented on Fig. 2. Knowledge organization systems (such as glossaries, taxonomies etc.) connect processes, tasks in organizational designations, roles, skills and types of information. It forms the foundation as a business layer. Beyond that, work with

HR Practices	Sourcing	Screening	Selection	Induction	Training	Retention
Professional Identity	Educational background	Work background	Current position	Immediate goals	Career goals	Life goals

Professional self / Constructs (C)	C1	C2	C3	C4	C5	C6
Competencies						

Skills						
Task	Task1	Task2	Task3	Task4	Task5	Task6
Process			Business processes			

Fig. 2 Learning oriented design

communities facilitates workers composing digital representations of themselves. Participatory design includes linear co-design processes and consensus building. Applying a repertory grid shifts the focus to understanding the professional self with competencies and their dynamics in the role of a boundary object. The professional development layer is based on widespread HR practices and ensures the integrated view.

5 Applying Learning Oriented Architecture

Typically, performance support solutions consider performer and his/her need for retrieval of information leaving the role of expert or becoming an expert aside. This role can be described through applying knowledge through various contexts. Knowledge solution framework on Fig. 3 provides possible application of learning oriented architecture and is derived from individual and collective perspective in learning.

The starting point for such a framework is adding and tagging resources as it is the core of knowledge workers' activity, turning the outcomes into digital artefacts. It is a prerequisite to have a specific role within an organization that leads to learning. With modelling and annotating the new knowledge is incorporated into the knowledge base and made available for the person(s) who might need it next. Associated retrieval leads to learning through creating connections using semantic metadata. Task model makes knowledge actionable and leads to developing competencies. User modelling leads to depicting professional self and dynamic user modelling adds career goals as described in the fourth case. Explicit learning paths lead to individual reflections and contributing to the overall consistency by adding new resources.

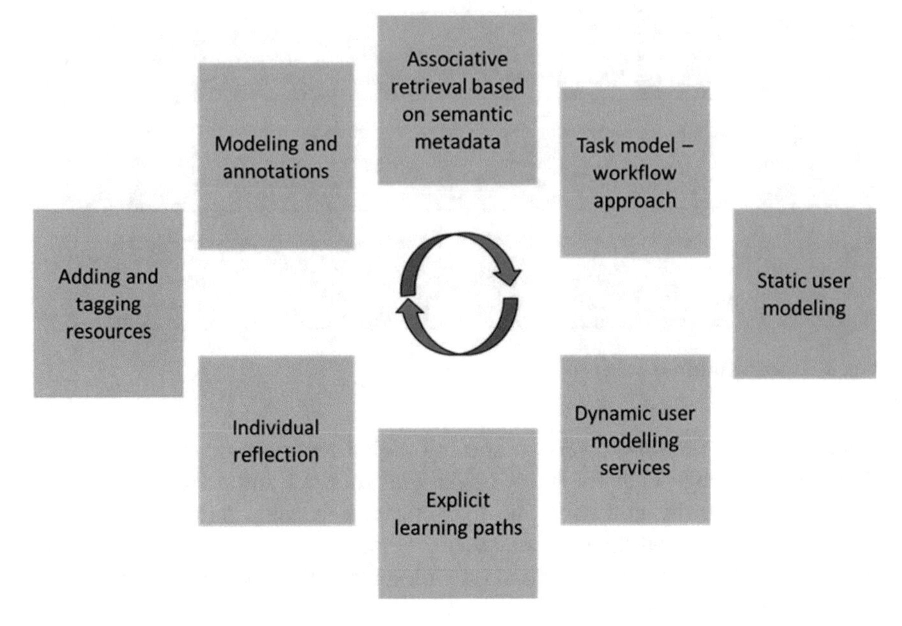

Fig. 3 Knowledge solution framework

6 Conclusions

Figure 4 that depicts individual and collective perspective in learning enables the main generalization and conclusions from the different case studies.

As reported by the first case, knowledge workers need learning and reflection. They interpret data residing in the domain knowledge base through semantic structures that can be both at the individual and collective level. This interpretation and insight is a cognitive process and leads to a formation of new ideas. In the second case, for generalization of the solution instance the research has indicated the importance in unity of requirements knowledge, where all the requirements need to be managed from balanced perspective. Innovation requires the cross-organizational completeness of information where quality becomes a precursor through following best practice as a cognitive learning process. Beyond that, efficiency is taken as an imperative for connecting shared services providers' tasks, various processes and related need for acquiring competencies. Within the third case, we concluded that applying participatory design in developing knowledge management platforms shifts the focus from software to business and cultural aspects. Participatory design leads to connecting private level KOS with collective level KOS and the consequent learning effort. The boundary objects supplement corporate metadata and create business processes, tasks and skills to competencies and constructs of the professional self. It has led to the development of learning oriented architecture.

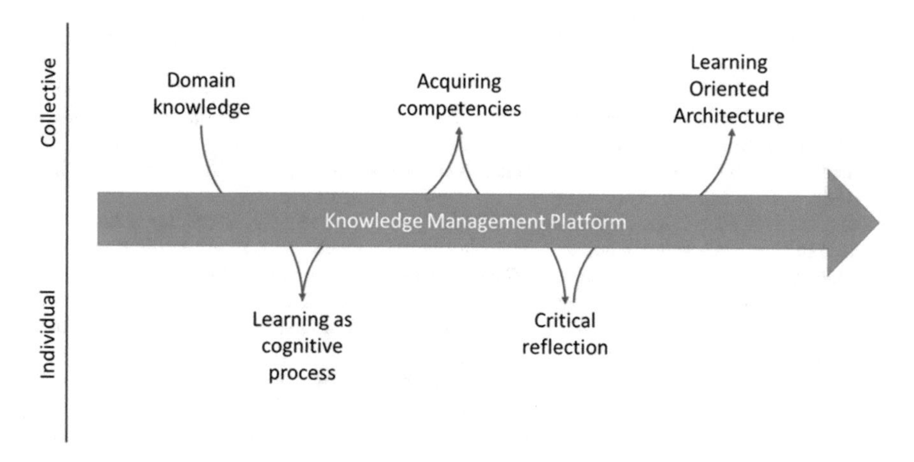

Fig. 4 Individual and collective perspectives in learning

References

Back, A., & Koch, M. (2011). Broadening participation in knowledge management in enterprise 2.0. *it-Information Technology Methoden und innovative Anwendungen der Informatik und Informationstechnik 53*(3), 135–141.

Bezanson, W. (2002). *Performance support solutions: Achieving business goals through enabling user performance*. Victoria: Trafford on Demand Publishing.

Caruso, S. J. (2017). A foundation for understanding knowledge sharing: organizational culture, informal workplace learning, performance support, and knowledge management. *Contemporary Issues in Education Research (Online), 10*(1), 45.

Chen, Y.-Y., & Huang, H.-L. (2012). Knowledge management fit and its implications for business performance: A profile deviation analysis. *Knowledge-Based Systems, 27,* 262–270.

Chisholm, M. (2009). *Rethinking definitions*. BeyeNETWORK. Retrieved from http://www.b-eyenetwork.com/view/10301.

Davenport, T., & Prusak, L. (1998). Learn how valuable knowledge is acquired, created, bought and bartered. *The Australian Library Journal, 47*(3), 268–272.

Dietz, J. L. (2006). *Enterprise ontology: Theory and methodology*. Berlin: Springer.

Ford, D. P., & Mason, R. M. (2013). A multilevel perspective of tensions between knowledge management and social media. *Journal of Organizational Computing and Electronic Commerce, 23*(1–2), 7–33.

Gray, D. E. (2007). Facilitating management learning: Developing critical reflection through reflective tools. *Management Learning, 38*(5), 495–517.

Jones, P. M. (2001). Collaborative knowledge management, social networks, and organizational learning. *Systems, Social and Internationalization Design Aspects of Human-Computer Interaction, 2,* 306–309.

Ley, T., Ulbrich, A., Scheir, P., Lindstaedt, S. N., Kump, B., & Albert, D. (2008). Modeling competencies for supporting work-integrated learning in knowledge work. *Journal of Knowledge Management, 12*(6), 31–47.

Lindstaedt, S. N., Farmer, J. (2004). Kooperatives Lernen in Organisationen. In J. Haake, G. Schwabe & M. Wessner (Eds.), *CSCL-Kompendium: Lehr- und Handbuch zum computerunterstützten kooperativen Lernen* (pp. 191–200). Munich: Oldenbourg.

North, K., & Kumta, G. (2014). *Knowledge management: Value creation through organizational learning*. New York: Springer.

Schmidt, A., Hinkelmann, K., Ley, T., Lindstaedt, S., Maier, R., & Riss, U. (2009). Conceptual foundations for a service-oriented knowledge and learning architecture: Supporting content, process and ontology maturing. In S. Schaffert, K. Tochtermann, & T. Pellegrini (Eds.), *Networked Knowledge—Networked Media: Integrating Knowledge Management, New Media Technologies and Semantic Systems.* (pp. 79–94). Heidelberg: Springer.

Siemens, G. (2005). Connectivism: Learning as network-creation. *ASTD Learning News, 10*(1), 1–28.

Stokic, D., Correia, A. T., & Reimer, P. (2013). Social computing solutions for collaborative learning and knowledge building activities in extended organization. In *Proceedings of the 5th International Conference on Mobile, Hybrid, and On-line Learning (eLmL'13).*

Sultan, N. (2013). Knowledge management in the age of cloud computing and Web 2.0: Experiencing the power of disruptive innovations. *International Journal of Information Management, 33*(1), 160–165.

Thurimella, A., & Maalej, W. (2013). Managing requirements knowledge: Conclusion and outlook. *Managing requirements knowledge* (pp. 373–392). Berlin: Springer.

Wallace, B., & Ross, A. (2016). *Beyond human error: taxonomies and safety science*. Boca: CRC Press.

Author Biographies

Jörgen Jaanus has held several executive positions in HR and global consulting companies. His primary academic interest in knowledge management and knowledge organization systems has led to his PhD studies at Tallinn University.

Nina Suomi is a Liberal Arts graduate from Tallinn University. She has worked in advisory roles in HR industry. Her academic interest is in digital anthropology with the research focus in the human perspective of technology.

Tobias Ley is a Professor for Learning Analytics and Educational Innovation and heads the Center of Excellence for Educational Innovation at Tallinn University. His research interests are in ICT for learning across the lifespan and knowledge management.

Competence Development for Work 4.0

Angelika Mittelmann

Abstract Digitalization permeates nearly every sphere of life. This transformation does not only change the technical field but also the collaboration at all levels of work. This is often referred to as *work 4.0*. In this new world of work, an organization's managers as well as employees need competencies enabling them to cope with the challenges of a digitized working place. Based on the findings of the major effects of the digitalization three essential categories and their corresponding competencies are outlined and described. Straightforward approaches for the development of these competencies are introduced. At the individual level the 'Fitness Circuit for Personal Knowledge Management' is suggested for mastering personal knowledge management. At the organizational level the 'Agile Competence Development Cycle' is proposed to enable organizations to establish effective and sustaining learning environments embedded in the working processes.

Keywords Digitalisation · Work 4.0 · Skills · Competencies · Personal knowledge management · Learning environment

1 Introduction

Digitalization opens up innovative possibilities for designing content, process, the organization of work and collaboration (BMAS 2015; Plass 2016). It does not only enable the access to intelligent tools, automation, production, and networking technologies, but also to globally distributed information, knowledge, competences, resources, work forces, and markets (Hirt and Willmott 2014; Kraft 2015). Taking all these challenging aspects into consideration, the competencies of the workforce is one of the most important key success factors for businesses today and in the future. Because it is the people who drive, improve and innovate the business processes in their daily routine.

A. Mittelmann (✉)
GfWM, Linz, Austria
e-mail: angelika.mittelmann@artm-friends.at

K. North et al. (eds.), *Knowledge Management in Digital Change*, Progress in IS,
https://doi.org/10.1007/978-3-319-73546-7_16

At a closer look three essential effects reveal (Picot and Neuburger 2013):

- **Digital penetration of the work**

The increasing use of digital media and intelligent tools in many organizations does not only lead to more efficient and effective working processes, but also to self-controlling industrial production processes with aid of cyber-physical systems. The possible level of digital penetration depends on the sector, the design of the business processes and the customer groups of the organization in question.

- **Flexibilization of work**

Individuals benefit from the above-mentioned access opportunities by designing their working processes flexible in terms of time and location. They may offer their workforce in agile working processes and projects as smart workers or *crowd-sourcees*. Enterprises incorporate globally distributed competences and resources in their value chain by integrating flexible, virtual, and mobile working models. In the end workers decide self-directed when, where, and how they collaborate in e.g. globally distributed project teams.

- **Polarization of work**

On the one hand digitalization fosters flexible working structures, on the other, it fuels the polarization of work by automating routine- and repetitive-intensive working processes. This affects especially workplaces with mid-level qualification needs and remuneration (Autor and Dorn 2013). Polarization applies less to workplaces with complex and mentally demanding tasks, and occupations with poorly automatable but physically demanding business processes where considerable experience is essential. This is the reason why some occupational fields will disappear completely in the long run, some will change more or less, some new one will emerge.

Taking all these considerations into account, it is not surprising that many managers (Kohlbacher et al. 2016) and employees are well aware of the importance of competencies for work 4.0 in the current situation as well as in the future. The burning issue is which competencies are the most important ones and how to develop them in a timely manner and maintain them over time. Furthermore, the question arises of what solutions knowledge management, as a foundation of our knowledge society, may offer for this demanding situation.

2 Competence Requirements for *Work 4.0*

Several studies have been conducted (Baker et al. 2015; European Political Strategy Centre 2016; OECD 2016; Pfeiffer et al. 2016; Kohlbacher et al. 2016) on the competence requirements for *work 4.0* recently. They all draw a similar overall picture but with different focal points on the new world of work. Baker et al. take a closer look on digital technologies (like broadband internet, world wide web, social media, and eCommerce) needed and utilized by SMEs. They survey if their business plans include digital training, and what barriers might hinder them to adopt digital technology in a wider range for additional business advantages. The OECD study concentrates on ICT and ICT-complementary skills. It suggests a replicable approach to identify work tasks complementary to ICTs and measures the demand for skills required to perform such tasks. The EPSC study emphasizes the importance of skills in general ("the future of work is all about skills"). It introduces a T-skill-framework-like set of competences, including cognitive as well as non-cognitive skills and core literacies, rounding it up with up-to-date suggestions of development initiatives for life-long learning. Pfeiffer (2016) concentrates on non-routine physical work (assembly tasks) and its implications on the development of knowledge and skills in a digitized working environment. One of her major findings is the importance of experience for achieving working results of high quality and innovative process improvements. The ECN Report (Kohlbacher et al. 2016) sheds a light on the most influential trends (first three in order: technical progress, digitalization of economy, globalization) by the CEOs interviewed and their role in skills development. It summarizes the needed employees' skills and how to develop or acquire the right skills. Overall, most CEOs agree that it is mostly soft skills and people skills that are crucial for an organisation to rise to the challenges brought on by the driving trends.

As outlined by all of these studies, routine work will be increasingly automated by use of cyber-systems. Humans will mainly carry out non-routine mental and physical tasks based on in-depth knowledge, expertise and experience in globally distributed working environments. Persons holding such mission critical knowledge are often referred to as knowledge workers, a term coined by Drucker in the middle of the last century (Drucker 1968). On the whole, one can therefore assume that they are more or less used to use ICT tools (especially Web 2.0 tools, Sondari 2013) for assisting them in acquiring, categorizing and classifying, storing, and sharing their knowledge and experience within their communities of practice. But the studies also reveal, that there are still remarkable skill gaps in effective knowledge management and technology use combined with appropriate social skills.

Summing up, in a work 4.0 environment individuals require competencies for coping with the following challenges:

- digitalization of the working environment based on technological progress
- collaboration with cyber-systems alongside business processes
- (globally) distributed and flexible adjustable working processes for adapting to customer needs
- modes of working independent of location and time
- non-routine mentally and/or physically demanding tasks combined with in-depth knowledge, expertise and experience
- diverse composed working groups and teams.

In order to establish development initiatives meeting these challenges one has to know the competence requirements at hand. The concept of T-shaped skills

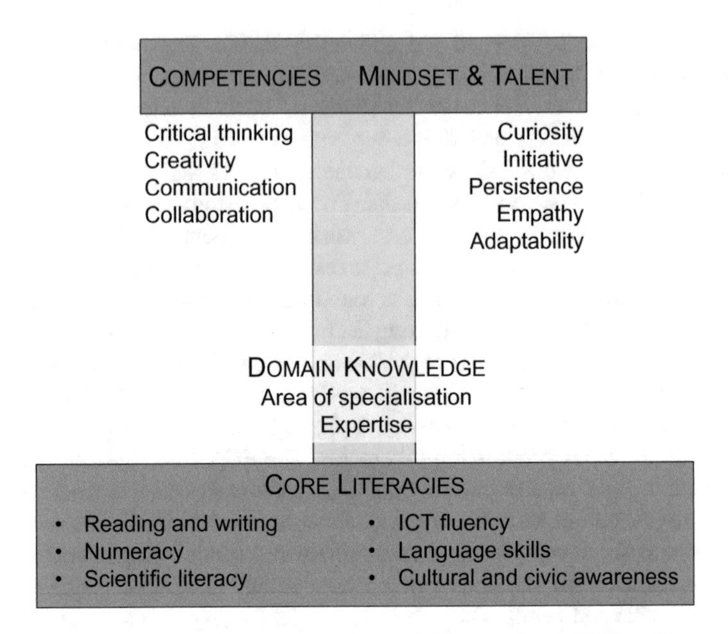

Fig. 1 Concept of T-shaped skills (European Political Strategy Centre 2016, p. 8)

(Wikipedia 2017) with its adaption of the European Political Strategy Centre (EPSC) may serve as a starting point. The framework (see Fig. 1) combines the depth of related skills and expertise in a single field (vertical bar on the T) with the ability to collaborate across disciplines with experts in other areas and to apply knowledge in areas of expertise other than one's own (horizontal bar on the T). EPSC suggests the "4-C competences" (critical thinking, creativity, communication, collaboration) and mindset and talent (curiosity, initiative, persistence, empathy, adaptability) on the horizontal bar, with domain knowledge (area of specialization, expertise) on the vertical bar. A second horizontal bar is added at the bottom of the T, defining core literacies like reading and writing, numeracy, scientific literacy, ICT fluency, language skills (mother tongue + 1), and cultural resp. civic awareness.

For further planning purposes, this framework needs to be expanded and restructured in categories of competencies (see Table 1). The selection of competencies within these categories is based on three factors:

(a) there is enough evidence of their importance for work 4.0 in the literature (Baker et al. 2015; Bates 2015; European Political Strategy Centre 2016; OECD 2016; Pfeiffer et al. 2016; Kohlbacher et al. 2016),
(b) the competencies are independent from domain knowledge except ICT basics, and
(c) there are widely agreed definitions of the competencies in question.

For the categories and their correspondent competencies see Table 1.

How to develop and maintain these competencies over time will be outlined in the following sections.

Table 1 Categories and correspondent competencies (O'Connor et al. 2007; Singh 2013; Soland et al. 2013; CEN 2014; Gallardo-Echenique et al. 2015; European Political Strategy Centre 2016)

Intrapersonal competencies	
Critical thinking	Using good judgement and common sense as well as logic and reasoning to identify the strengths and weaknesses of alternative solutions, conclusions or approaches to problems
Sense-making	Determining the deeper meaning or significance of what is being expressed visually or in written or spoken texts
Novel and adaptive thinking	Routinely thinking across boundaries and coming up with responses and solutions beyond that which is rote or rule-based
Transdisciplinarity	Understanding concepts across multiple disciplines and crossing many disciplinary boundaries to create holistic solutions
Self-direction	Guiding and organizing oneself, steering and controlling one's learning, and maximizing cognitive functioning with respect to well-being
Interpersonal competencies	
Communication	Active listening, conveying information comprehensibly, having difficult conversations with ease to avoid resp. resolve conflicts
(Virtual) collaboration	Working productively, driving engagement, and demonstrating presence as a member of a (virtual) team
Social intelligence	Connecting to others in a deep and direct way, sensing and stimulating reactions and desired interactions
Intercultural competency	Operating effortlessly in different cultural settings
ICT-related competencies	
ICT fluency	Using computers, communication technologies and applications to access, manage, integrate, evaluate, and create information in order to take part in a knowledge society
Computational thinking	Identifying general principles and patterns in data, processes, or problems, effectively explaining the purpose and meaning of problems and their potential computational solutions
Social media literacy	Critically assessing and developing content that uses social media forms, and leveraging these media for persuasive communication
Information security awareness	Realizing the consequences of revealing personal information on the web, and taking appropriate actions to protect personal information from misuse and unwanted dissemination

3 Approaches to Competence Development

A multi-layer approach is necessary to develop this demand of competencies. It begins at the individual level broadening to the interpersonal one and adding ICT-related skills at the top as needed (see Fig. 2). Of course, appropriate development programs should start in the early childhood and go on during the whole lifespan of a person. In the following we will concentrate on adult learning and development where individuals as well as organizations can influence.

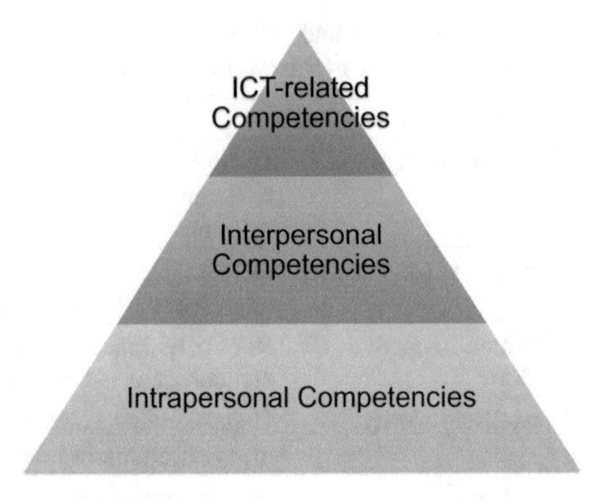

Fig. 2 Multi-layer approach for developing competencies

3.1 Development Approach at the Intrapersonal Level

Neuroscientists (Cole et al. 2013) claim that our brain is some kind of "knowledge machine" enabling us to adapt flexibly to new situations, nearly regardless of our age. We acquire new information by learning and our brain stores it in our memory. This structural change of the brain by learning is what scientists call its plasticity. In order to maintain this valuable ability of our brain we should nurture a life-long learning attitude. This can be achieved by practicing personal knowledge management (PKM) on a regular basis (for further details on PKM see Pircher 2010; McFarlane 2011; Wiig 2011). The 'Fitness Circuit for PKM' (Mittelmann 2016) (Fig. 3) may serve as an approach at the individual level to put PKM into practice in a sustainable way (see Table 2). This fitness circuit recommends five different types of exercises: warm-up exercises, starting exercise, circuit training, sustained exercises, and partner exercises.

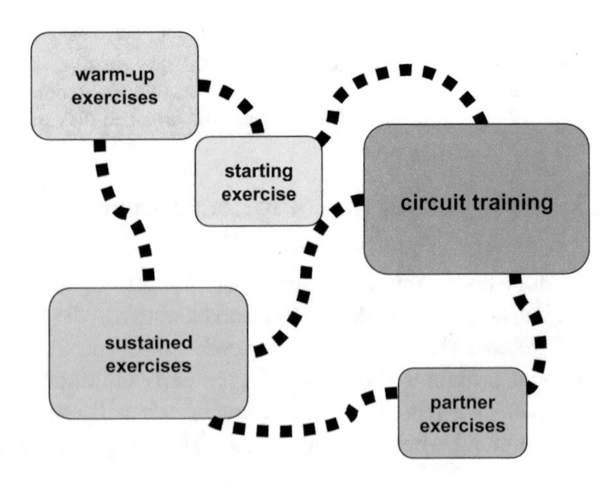

Fig. 3 Fitness circuit for personal knowledge management

Table 2 Competencies developed by exercises of PKM fitness circuit

Warm-up exercises	Competencies
Physical fitness: Balanced/healthy diet Bufficient amount of sleep "Brain-compatible" body exercises	Self-direction Novel and adaptive thinking
Mental fitness: Memory training Singing Learning to play a musical instrument Playing parlor games	Self-direction Novel and adaptive thinking Social intelligence (Virtual) collaboration[a] Computational thinking[b]
Starting exercise	Competencies
1. Make list of appropriate keywords 2. Structure your knowledge using a mind map 3. Name your folders using these keywords out of your list 4. Create some sensible rules for saving your files resp. knowledge objects 5. Tag your knowledge objects with appropriate keywords 6. Use your keywords for searching purposes 7. Evaluate the currency of your knowledge objects and list of keywords 8. Delete outdated knowledge objects on a regular base	Self-direction Sensemaking ICT fluency Computational thinking Social media literacy
Sustained exercises	Competencies
Training of knowledge related behavior: Actively asking for help, routinely saving lessons learned, selectively sharing learnings with others, asking oneself again and again what to do better the next time	Social intelligence Sensemaking Critical thinking Self-direction Communication ICT fluency
Developing knowledge related skills: Being able to explicate implicit knowledge, active listening, sum up briefly and precisely complex issues, graphically prepare context clearly	Sensemaking Critical thinking Communication Social intelligence Computational thinking ICT fluency Social media literacy
Self-regulated learning: creating one's current personal competence portfolio, setting knowledge goals, deducing one's future portfolio, defining and implementing measures, using social media and open education resources for learning purposes	Self-direction Critical thinking Computational thinking Novel and adaptive thinking Social intelligence ICT fluency Social media literacy Information security awareness

(continued)

Table 2 (continued)

Circuit training	Competencies
Circuit training	Competencies
Searching for knowledge objects or persons	Social intelligence Communication Intercultural competency ICT fluency Social media literacy
Documenting and visualizing	Critical thinking Sense-making ICT fluency Social media literacy Communication
Creating new knowledge	Sense-making Novel and adaptive thinking Critical thinking Transdisciplinarity
Partner exercises	Competencies
Relationship building	Social intelligence Communication Intercultural competency ICT fluency Intercultural competency Information security awareness
Saving contact data	Social intelligence ICT fluency Social media literacy Information security awareness
Systematic management of personal relationships	Communication Social media literacy Intercultural competency Self-direction

[a]In case of playing role based online games
[b]In case of playing strategic games

The *warm-up exercises* aim at the physical and mental fitness which is a prerequisite for high performance capability needed by all knowledge workers in their daily work. The *starting exercise* (Della Schiava 2007) may assist individuals in organizing and systematizing all of their knowledge objects and professional relationships gathered so far to make use of them in the ongoing PKM process.

The *sustained exercises* should be practiced in an ongoing process to sustain the full benefits for PKM. They consist of three primary elements: training of knowledge related behavior, developing knowledge related skills and self-regulated learning.

The exercises of the *circuit training* are meant to be practiced in a circular process. It starts with effective searching for knowledge objects or persons based on your defined keywords (see starting exercise above). Next comes documenting and visualizing your findings within your defined structure. Finally, one creates new knowledge with the aid of combined reading, documentation, and creativity techniques and social software tools.

Contrary to all of the above methods and techniques, the *partner exercises* can only be done by communicating with others. They aim at building a sustainable personal network, which is vital for individual learning and knowledge creation (for further details on the *Fitness Circuit* see Mittelmann (2016), on techniques see Mittelmann (2011)). As every person is different, there is no standardized PKM. The *Fitness Circuit* can only serve as a practical approach for getting started. However, it will obviously help to strengthen all of the suggested competencies (see Table 2).

3.2 Development Approach at the Interpersonal Level

Interpersonal competencies can only be developed with interaction. In digitized working environments technology-enabled collaboration is the norm, therefore ICT-related competencies must be included in such development initiatives. On the other hand, knowledge workers are working primarily self-directed. They need their trainings in an independent manner and as close to their working place as possible. Informal and social learning is what this situation calls for.

When establishing effective development programs, one should adopt two additional concepts for designing successful learning environments. The first is the 70-20-10 model, a learning framework based on research in the early 1980s done by the Center for Creative Leadership (Lombardo and Eichinger 2006). It implies that the most effective learning occurs when about 70 percent is based on informal, experimental learning, 20 percent is social learning by learning from others, and 10 percent is formal learning by traditional classroom training or reading. However, the 70:20:10 model incorporates a blend of multiple learning modalities and activities, the biggest part of this blend is instructor-led training, the Brandon Hall Group's 70:20:10 Framework Survey reveals (Wentworth 2015). Organizations have to decide wisely on the blend appropriate for their situation and staff. In the meantime, the blended learning approach is a widely-understood concept with well documented lessons learned (e.g. see Martyn 2003; Bersin 2004). It combines formal classroom training methods with online digital media.

In rapid changing working environments, extensive and long running development initiatives are neither effective nor efficient. On the contrary, they should be as flexible as the people executing (globally) distributed business processes. Shifting from instructional design to experience design based on design thinking (Bersin 2016) is one creative solution for this situation of change. The core concept is outlined in the following as an aligned blended learning design process called 'Agile Competency Development Cycle' (ACDC, see Fig. 4).

The ACDC is meant to be adoptable and scalable to any given working environment. It is a framework including the following parts:

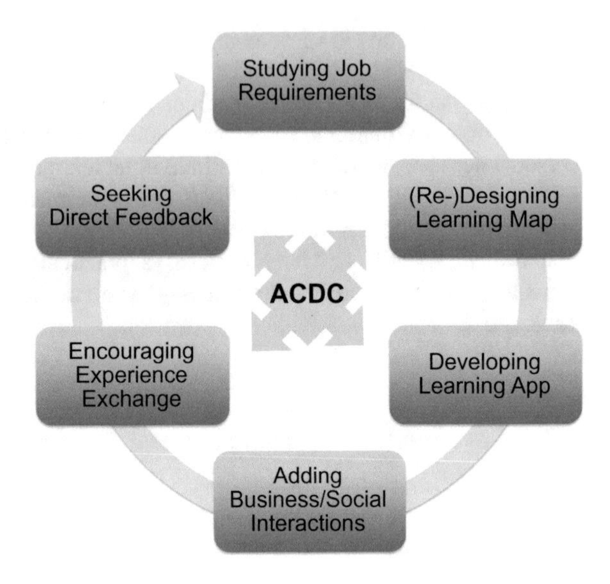

Fig. 4 Agile Competency Development Cycle (ACDC)

- **Studying job requirements:**

In order to embed learning into working as close as possible one has to know the actual requirements of the job(s) at hand. A small group of job and learning experts evaluate what people need to know over a defined period of time starting with their first working day.

- **(Re-)designing learning map:**

Based on their findings they design a learning map where all needed content is visualized. It includes the urgent learning needs as well as specialized knowledge of the business processes, products and customer needs. This learning map is redesigned whenever changes of the business processes occur.

- **Developing learning app:**

As people often want to learn independent from time and place the learning map is implemented as an application using the visualized elements of the design step. Of course, this application should also be available for any mobile device the learners might use. It should also be easily adapted to new requirements.

- **Adding business and social interaction:**

The learning map serves as a navigation tool for the individual learning journey of every learner. However, most effective learning takes place whenever groups or teams collaborate on a regular base. Therefore, opportunities for business and social interactions should be added. At appropriate points of the learning journey they are invited to join small project teams, video or coaching sessions. They may also participate in informal meetings to strengthen their social connections for enhancing their community of practice.

- **Encouraging experience exchange:**

Learning from and working with others naturally leads to experiences that may also be valuable for another person. This is the reason why all learners are encouraged to share what they have learned online or in face-to-face situations. The more they share the more likely new ideas and solutions come up from which the organization benefits as a whole.

- **Seeking/providing direct feedback:**

This approach would be incomplete if there is no possibility for direct feedback at any point of the cycle. It is of crucial importance for the overall quality that every feedback is considered and addressed in the design and implementation of the learning journey. It guarantees the agility of the development initiative.

- **ACDC core:**

The ACDC may serve as a guideline assuring effective and sustainable development of the required competencies in the long run. It is not meant to be a strict cycle. After the implementation of the basic learning map, every part can be processed in any order as needed in the current situation. Naturally, this process must be supported continuously by job experts, learning professionals, and management as well. The role of management is to continually support the ongoing development with appropriate resources and be prepared for mentoring activities at the defined points of the learning map.

4 Conclusion

For being prepared for work 4.0 organizations need work forces with key intrapersonal, interpersonal, and ICT-related competences enabling them to cope with the challenges of a digitized working environment. At the individual level, the competencies in question may be developed by mastering PKM by help of the Fitness Circuit. Carrying out the selected exercises regularly will help to develop and strengthen many of these competencies. At the organizational level, blended learning approaches are the method of choice establishing learning environments as close to the workplace as possible and transferring the responsibility for the learning process into the hands of the employees. An aligned blended learning design process like ACDC will help the management as well as the staff to set-up an effective and sustainable development process within the organization. What remains to be done is adapting organizations as a whole to the deep transformation going along with the implementation of work 4.0 principles.

References

Autor, D., & Dorn, D. (2013). The growth of low-skill service jobs and the polarization of the US Labor Market. *American Economic Review, 103*(5), 1553–1597.

Baker, G., Lomax, S., Braidford, P., Allinson, G., & Houston, M. (2015). Digital capabilities in SMEs: Evidence review and re-survey of 2014 small business survey respondents. *BIS Research Paper (Vol. 247)*. BMG Research and Durham University.

Bates, A. W. (2015). *The skills needed in a digital age. Teaching in the digital age*. Retrieved from https://opentextbc.ca/teachinginadigitalage/chapter/section-1-3-the-skills-needed-in-a-digital-age/.

Bersin, J. (2004). *The blended learning book: Best practices, proven methodologies, and lessons learned*. New York: Wiley.

Bersin, J. (2016). Using design thinking to embed learning in jobs. *Harvard Business Review*. Retrieved from https://hbr.org/2016/07/using-design-thinking-to-embed-learning-in-our-jobs.

BMAS (2015). *Re-imagining Work, Green Paper Work 4.0*. Retrieved from http://www.bmas.de/SharedDocs/Downloads/DE/PDF-Publikationen/arbeiten-4-0-green-paper.pdf.

CEN (2014). *European e-Competence Framework 3.0, A common European Framework for ICT Professionals in all industry sectors*. CWA 16234:2014 Part 1. Retrieved from http://www.ecompetences.eu/wp-content/uploads/2014/02/European-e-Competence-Framework-3.0_CEN_CWA_16234-1_2014.pdf.

European Political Strategy Centre (2016). *The Future of Work - Skills and Resilience for a World of Change*. EPSC Strategic Notes, Issue 13/2016, June 10. Retrieved from http://ec.europa.eu/epsc/sites/epsc/files/strategic_note_issue_13.pdf.

Cole, M. W., Laurent, P., & Stocco, A. (2013). Rapid instructed task learning: A new window into the human brain's unique capacity for flexible cognitive control. *Cognitive, Affective, & Behavioral Neuroscience, 13*(1), 1–22.

Della Schiava, M. (2007). *Effektiver suchen, schneller finden*. TOPGEWINN MdS toolbox. Retrieved from http://www.marketinggesellschaft.at/download/MdS/Buch2008/20070412_TOPGEWINN_MdSToolbox_Tipps_suchen_finden_Studie.pdf.

Drucker, P. F. (1968). *The age of discontinuity: Guidelines to our changing society*. Boston: Transaction Publishers.

Gallardo-Echenique, E. E., De Oliveira, J. M., Marqués-Molias, L., & Esteve-Mon, F. (2015). Digital competence in the knowledge society. *Journal of Online Learning and Teaching, 11*(1), 1.

Kohlbacher, F., Schimkowsky, C., & Wijaya, R. (2016). *Skills 4.0 - How CEOs shape the future work in Asia*. Economist Corporate Network Report. Retrieved from https://www.corporatenetwork.com/media/1554/ecn-skill-40-nov-2016-fv.pdf.

Hirt, M., & Willmott, P. (2014). Strategic principles for competing in the digital age. *McKinsey Quarterly, 5,* 1.

Kraft, B. (2015). *The Biggest Digital Challenges and Opportunities Facing Businesses Today*. Digital Marketing Magazine. Retrieved from http://digitalmarketingmagazine.co.uk/articles/the-biggest-digital-challenges-and-opportunities-facing-businesses-today/2705#.

Lombardo, M. M., & Eichinger, R. W. (2006). *The career architect development planner 4th Edition*. Minneapolis: Lominger.

Martyn, M. (2003). The hybrid online model: Good practice. *Educause Quarterly, 26*(1), 18–23.

Mcfarlane, D. (2011). Personal knowledge management (PKM): Are we really ready? *Journal of Knowledge Management Practice, 12*(3), 108–114.

Mittelmann, A. (2011). *Werkzeugkasten Wissensmanagement*. Norderstedt: BoD–Books on Demand.

Mittelmann, A. (2016). Personal knowledge management as basis for successful organizational knowledge management in the digital age. *Procedia Computer Science, 99,* 117–124.

OECD (2016). *New Skills for the Digital Economy*. OECD Digital Economy Papers, No. 258, OECD Publishing, Paris. Retrieved from http://dx.doi.org/10.1787/5jlwnkm2fc9x-en.

O'Connor, B. et al. (2007). *Digital Transformation, A Framework for ICT Literacy*. A Report of the International ICT Literacy Panel. Retrieved from https://www.ets.org/Media/Tests/Information_and_Communication_Technology_Literacy/ictreport.pdf.

Pfeiffer, S. (2016). *Beyond Routine: Assembly Work and the Role of Experience at the Dawn of Industry 4.0. Consequences for Vocational Training*. University of Hohenheim, Chair for Sociology, Working Paper 01-2016.

Pfeiffer S., Lee, H. S., Zirnig, C., & Suphan, A. (2016). *Industrie 4.0: Qualifizierung 2025*. Frankfurt: VDMA.

Picot, A., & Neuburger, R. (2013). *Arbeit in der digitalen Welt*. Zusammenfassung der Ergebnisse der AG1-Projektgruppe anlässlich des IT-Gipfels-Prozesses 2013. Retrieved from http://www.forschungsnetzwerk.at/downloadpub/arbeit-in-der-digitalen-welt.pdf.

Pircher, R. (2010). Targeting the blind spot: Personal knowledge management as an enabler for knowledge creation and application. In *Interaktive Kulturen: Workshop-Band: Proceedings der Workshops der Mensch & Computer 2010-10. Fachübergreifende Konferenz für Interaktive und Kooperative Medien, DeLFI 2010-die 8. E-Learning Fachtagung Informatik der Gesellschaft für Informatik eV und der Entertainment Interfaces 2010*, Logos Verlag.

Plass, C. (2016). *Digital Business Processes and Models change the Workplace*. Opportunity - Facts for Experts & Decision Makers. Retrieved from http://www.unity.at/fileadmin/news_events/Studien/OPPORTUNITY_Digital_Business_Processes_and_models.pdf.

Singh, P. (2013). Influence of leaders' intrapersonal competencies on employee job satisfaction. *The International Business & Economics Research Journal (Online), 12*(10), 1289.

Soland, J., Hamilton, L. S., & Stecher, B. M. (2013). *Measuring 21st century competencies*. Global Cities Education Network Report, Asia Society. Retrieved from https://asiasociety.org/files/gcen-measuring21cskills.pdf.

Sondari, M. C. (2013). Personal knowledge management 2.0. *International Journal of Social Science and Humanity, 3*(4), 426–428.

Wentworth, D. (2015). *The 70:20:10 Learning Framework: Formalizing the Informal*. Training Magazine. Retrieved from https://trainingmag.com/702010-learning-framework-formalizing-informal.

Wiig, K. M. (2011). The Importance of Personal Knowledge Management in the Knowledge Society. In Pauleen, D. J., & Gorman, G. E. (Eds.), *Personal Knowledge Management: Individual, Organizational and Social Perspectives*. Farnham Surrey, England: Gower Publishing Limited, pp. 229–262.

Wikipedia (2017): *T-shaped skills*. Retrieved from https://en.wikipedia.org/wiki/T-shaped_skills.

Author Biography

Angelika Mittelmann, Dipl.-Ing., Dr. studied computer science and holds a doctorate degree in technical sciences from the Johannes-Kepler-University in Linz. She has been a staff member of voestalpine Stahl as an expert for knowledge, skills and change management for a long time. For many years she has been also active as a lecturer, consultant, and trainer in her fields of expertise. In 2015, she received the Knowledge Management Award by KM Academy, Vienna. Since 2016, she is a member of the advisory board of the "Gesellschaft für Wissensmanagement (GfWM)".

Learning 4.0

Peter A. Henning

Abstract Didactical methods and models of learning are determined by the questions of where learning content is stored and how it is accessed. The digital transformation of information storage and access therefore necessitates new models of learning and dramatic changes in educational systems. In this article, these new learning paradigms are outlined, classified and weighted for their disruptive impact on societal and industrial processes—ranging from the *everywhere, every-time* of digital mobile devices to human strategies for coping with information overflow.

1 Introduction

Formal models of learning were rare before the 19th century (Comenius 1654; Kant 1803) and paradigms such as behaviorism, cognitivism and constructivism blossomed to maturity only in the 20th century. To a large extent they still dominate our view on education in the 21st century. In all parts of the educational system many persons who teach design their teaching material after one of these paradigms, as they believe that these models and their derivatives describe the human mind and therefore are independent of technological and societal changes (Ertmer and Newby 1993).

Such an instructional design, however, ignores the simple fact that up to now the "human mind" cannot be directly observed. Rather, *human behaviour* is observed—and any conclusion has to be worked backwards, estimating the cause from the effect. In doing so one immediately finds that our learning behavior has undergone dramatic changes in the past 20 years. In searching for a particular piece of public information, adults nowadays rather use a search engine than a large printed encyclopaedia. This may seem obvious, since according to numerous studies internet resources such as Wikipedia are better maintained, faster updated and much larger in volume than printed encyclopedias have ever been.

P. A. Henning (✉)
Karlsruhe University of Applied Sciences, Karlsruhe, Germany
e-mail: peter@henning-weingarten.de

K. North et al. (eds.), *Knowledge Management in Digital Change*, Progress in IS,
https://doi.org/10.1007/978-3-319-73546-7_17

Today, information is accessed differently also in the professional environment. While two decades ago we were discussing whether internet usage at a work place might not impose a severe drain on the resources of an enterprise, we now find the opposite: Hardly any workplace for knowledge workers comes without internet access. Certain branches of our modern industry, like e.g. software development, are indeed completely helpless without internet usage. No platform demonstrates this more profoundly than Stackoverflow.com[1]—the knowledge platform for all kinds of technological information. On a typical workday, an average number of 4,000,000 engineering people from a variety of fields all over the world are exchanging information on this platform at a highly professional level.

A change in learning behaviour is also found in top quality scientific research, where preprint services have increased the speed of publication tremendously. Starting in 1991 from a small archive for physics papers, running on an 486-computer in the Los Alamos National Laboratory, the service *Arxiv.org*[2] has surpassed the landmark of more than 1 million scientific articles from various fields in January 2015. Currently it is getting more than 10,000 submissions of articles per month. Furthermore, the cost model for scientific publication is dramatically changing in 2017, following an increasing pressure to publish in Open Access journals, where the author's institution pays for publication rather than the reader's organization. The DEAL project[3] intends to make a significant reduction in the estimated € 7600 million paid annually by German libraries and scientific institutions to publishing houses.

Thus, we have to address the question whether these observed changes in learning behavior are merely customary adaptions that might pass like last year's fashion. If this were true, we could maintain our learner models as they have evolved in the 20th century. However, if finding that these changes go deeper and are indeed irreversible changes to society, we should rather develop new paradigms for learning very rapidly. To this end, let us first discuss the established learning paradigms from the viewpoint of the digital age.

2 Behaviouristic Learning in the Digital Age

In the behaviouristic *learner model* the human mind is a black box. *Learning* consists of linking a stimulus (the input) to a desired response (the output), without ever needing to discuss the cognitive processes behind this linkage. Two examples demonstrate that this type of learning indeed has its place in the digital age.

1. Consider students in knowledge communities—irrelevant whether they are of the more serious type, or just (mis-)using Facebook or WhatsApp. In these

[1]http://www.stackoverflow.com

[2]http://www.arxiv.org.

[3]https://www.projekt-deal.de/.

knowledge communities, the frequency of help requests rises tremendously on Sunday afternoon, when the next day's school attendance and, therefore, teacher control of homework assignments are looming. These help requests are going as far as to ask for complete seminar papers or essays and it may be assumed that these plagiates are then really presented at school. The same problem arises in higher education, where plagiarism is even more important because academic degrees still are the canonical way to higher social status and income. Consequently, university plagiarism has turned out to be a business case. Services like *Assignment King*[4] are becoming very fashionable. They even claim to provide results that cannot be discovered with plagiarism checking software. Plagiarism has become more severe than one would like believe: A systematic survey carried out by Australian universities on the originality of "scientific" Master's Thesis writing reaches the conclusion, that 25% of them are plagiates. Digital methods therefore allow to *simulate* learning, they help indeed to establish dysfunctional mappings from input to output in the behaviouristic model.

2. Consider players of a parallel online role-playing game with a high number of participants: In general, they do not read manuals or study screencasts of previous game rounds. Rather, they jump into the game and learn by "observation, imitation and modelling". Even skilled players are seldom able to perform an abstraction of their success, or to derive formal rules from being skilful. All classical elements of the behaviouristic paradigms such as punishment (loss of one's virtual status or life) and reward (as reaching the next level) are identifiable. This informal learning is indeed behaviouristic and has been described accurately in Bandura's model of *Social Learning* (Bandura 1977). According to this model, people learn from one another, via observation, imitation, and modelling. The Bandura model includes concepts such as attention, memory, and motivation, and therefore is much closer to reality than other behaviourist examples. Social informal learning below the cognitive level, in particular through gaming elements, therefore has its representation in the behaviouristic model.

3 Cognitive Learning in the Digital Age

The cognitive learning paradigm tries to open the behaviouristic 'Black Box'; to see, understand and modify the inner workings of the learner. Since that is the model of programs running on the human computer, cognitivism is the learner model seemingly best suited to the digital age. The question arises; what is the proper method to write these programs? Cognitive scientists have tried to define this for decades, and now well-established "programming" aids such as *rehearsal models*, the *loci method* or *gesture supported learning* (Moè and De Beni 2005;

[4]http://www.assignmentking.com.

Macedonia and Knösche 2011) as well as infamous pseudo-scientific models exist like e.g. NLP ("neuro-linguistic programming").

Still, the big unsolved question of cognitivism is: What exactly is the physical difference between a brain that has learned—say a new word in a foreign language—and the same brain before this learning was achieved?

From 1962 until a few years ago it was thought that this difference is a biochemical ingredient (RNA) that might even be transported from one individual to another (McConnell 1962). However, since the work of the Nobel Prize awardee Eric Kandel, a much more complicated picture has emerged (Kandel 2009). Thousands of molecules as well as epigenetic switches are—somehow—involved here.

Nevertheless, even today we may clearly state that so far, nobody has understood the programming language of the human computer—and therefore a cognitivistic learner model is useless for *our part* of the digital age.

4 Constructivist Learning in the Digital Age

Our understanding of learning took a big step forward when the learner model of constructivism was developed around 1970. According to the constructivist paradigm, learning is an active process by the learner (ignore the teacher's role for now). A learner therefore expands his knowledge by *constructing* a mental representation of the outside world (Cooper 1993). From the viewpoint of computer science, this model is very appealing because of its analogy with the World Wide Web, where knowledge is *built* by linking items (Jonassen 1999).

Closely related to this construction idea is the development of artificial neural networks, where research has been going on for the past few decades. These neural networks are, simply spoken, computer programs simulating several layers of coupled artificial neurons. They are trained by exposing them to example patterns, which leads to the *building* of an inner representation. A properly trained neural network is then able to recognize these patterns; also with much more complicated input data. One could therefore argue, that with the digital age, we have finally discovered how learning really works: by programming neural networks.

In 2017, we are even able to demonstrate this success of neural networks with exciting new technologies. It has only recently become possible to run neural networks involving many layers of neurons. Together with an immense volume of statistical data for training of such *deep learning* networks, this technological break-through has led to spectacular progress in several fields: automatic translation of content from one language to another, autonomous driving and speech recognition are milestones of this understanding.

However, this analogy has to be seen critically—because a difference exists between *building* and *construction*. Even simple animals are building things, like spiders building their intricate webs. However, these are *not rationally planned*, and therefore, in the language of an engineer, *not constructed*. Indeed, even in simple, and much more so in deep neural networks, we do *not* have an understanding on how the internal representation is mapped to the external world. Therefore, we cannot produce artificial neural networks that have a well-defined a priori knowledge. A neural network, therefore, is a 'Black Box' mapping input to output in some unintelligible fashion—and we are suddenly thrown back to the behaviouristic model.

The key to understand constructivism in the digital age can be found when looking at a statement advertised by this model even since its early days: *Constructivist teaching has to pick up a learner where he is now* and has to induce in him new concepts and relations. Constructivist learning therefore means one already has a set of terms and relations—an ontology in the language of computer science.

Ontologies are a key ingredient of the *Semantic Web*, which even now seems to be the future of the World Wide Web (Berners-Lee and Fischetti 1999). In the semantic approach, data is accompanied by meta data (=annotations) from different views or knowledge fields (=domains), which allow the deduction of the *meaning* from the data. This meaning then allows classification of the terms and relations and is currently considered to be the correct state-of-the-art model to store knowledge (Staab et al. 2001).

Constructivist learning, therefore, consists of the meaningful and planned extension of the learner's given ontology. By considering analogies from this given ontology, the constructivist learner is adding new terms and possible relations among the terms to his ontology, leading also to new statements and *hypothesis building* about classes of objects. By falsification of such a hypothesis, new classes may be formed, not simply building but instead *constructing* an improved mental model.

5 Connectivism—A New Model for the 21st Century?

The first genuine 21st century model of learning was formulated in 2005 when Siemens picked up the term *connectivism* already known earlier to describe the changes in learning behaviour due to technological advances. The model has evolved considerably since then (Downes 2010), but even now is based on the same eight principles (Siemens 2005):

- *Learning and knowledge rests in diversity of opinions.*
- *Learning is a process of connecting specialized nodes or information sources.*
- *Learning may reside in non-human appliances.*
- *Learning is more critical than knowing.*
- *Maintaining and nurturing connections is needed to facilitate continual learning.*
- *Perceiving connections between fields, ideas and concepts is a core skill.*

- *Currency (accurate, up-to-date knowledge) is the intent of learning activities.*
- *Decision-making is itself a learning process. Choosing what to learn and the meaning of incoming information is seen through the lens of a shifting reality. While there is a right answer now, it may be wrong tomorrow due to alterations in the information climate affecting the decision.*

Detailed papers have been written criticizing these principles—starting at the topmost principle which emphasizes opinions and ending at a semantic analysis of such unprecise non-terms as *information climate* (Chatti 2010). To summarize the most crucial aspects: Connectivism is not a complete model of learning, but is focused on a small aspect of constructivism—the linking of top-level knowledge nodes. Neither the internal structure of nodes (and their evolution), nor algorithmic aspects of learning such as reflection, recursion and the cycle of inductive hypothesis building and deductive reasoning are contained. Moreover, the influence of the teacher is treated like some random disturbance, decoupled from the structure of the network as well as from the knowledge domain and not at all influencing the outcome of the learning process.

6 Recent Advances in Constructivist Learning

Having outlined the most common models of learning, we now take a step backwards and look at the famous question asked by Kant in the early stages of educational theory: "*Wie kultiviere ich die Freiheit bei dem Zwange?*"—*How do I cultivate freedom in light of force?* (Kant 1803). As he understood correctly, every educational action employs a certain amount of force acting on the learner. It is then necessary to prove to the learner that this force is used for his own good, i.e. guiding him to the use of his own freedom.

The model of connectivism, to a certain extent, has shown the proper direction: Liberate the learner to make his own choices in building his own personalized knowledge network, but not without guidance and not without meaningful structure. The necessary guidance must encompass:

- General pedagogical knowledge accumulated since the age of Comenius, i.e., a pedagogical ontology (Swertz et al. 2014).
- Domain specific knowledge addressing the specific didactical needs for a field of knowledge (=domain), called a Cognitive Map and possibly also written as an ontology.
- Specific knowledge about the available learning material, called a Cognitive Content Map.
- A fourth ontology comprising the knowledge about this particular learner, his abilities, his learning history and learning environment.

To understand the structure of a personalized knowledge network we consider each small and indivisible (atomic) *Knowledge Object* (KO) as contributing a single bit to a possible very large string describing a *cognitive position* of the learner. If we assume that the field of knowledge (domain) consists of N such KOs, the cognitive position of each learner in this domain corresponds to a corner of an N-dimensional hypercube.

In this hypercube model, depicted in Fig. 1 for a simple case of four KOs;

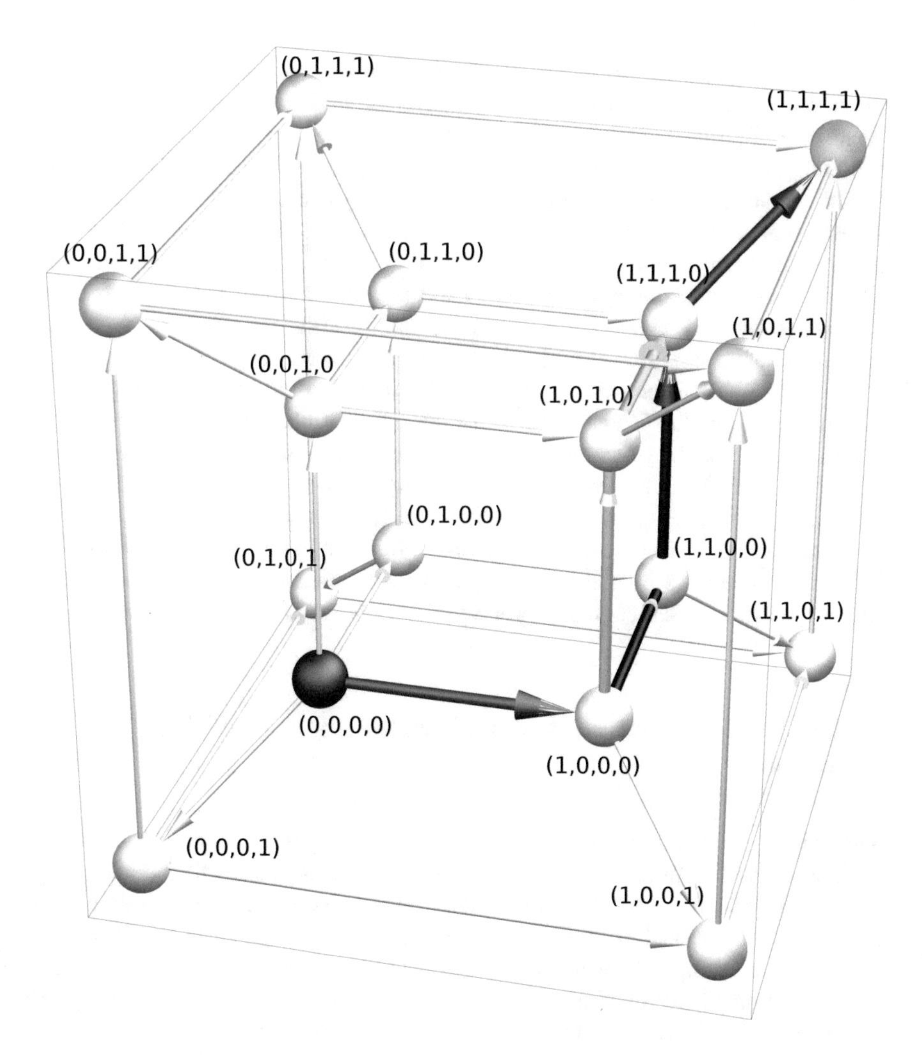

Fig. 1 Four-dimensional cognitive space describing a learning process with only four KOs. The model (Fuchs and Henning 2017) here shows two exemplary learning pathways leading form the state (0, 0, 0, 0)—nothing learned, to the state (1, 1, 1, 1)—everything learned

- *Learning* is a movement of the cognitive position within the hypercube. This trajectory of the learner's cognitive position is called a *Learning Pathway.* A huge number of possible learning pathways exist (*N*! for *N* KOs)—this is what makes learning so individual.
- *Teaching* consists of leading the learner along one or another predefined learning pathways to achieve the desired goal, i.e. a *Learning Pathway Recommendation.*

This hypercube model of learning is independent of how the learning material is presented. It may be applied to all modalities of teaching, including classroom teaching, self-paced learning by reading books or viewing videos as well as to technology enhanced learning (TEL) with the help of a computer (Meder 2006). Indeed, the latter has given birth to this formalized hypercube model of the learning process through the European research project INTUITEL (Henning et al. 2014a, b; Fuchs et al. 2016; Fuchs and Henning 2017). In this INTUITEL project, learning pathway recommendations are issued to the learner based on the four layers of ontology described above—and some of the leading Learning Management Systems have been equipped with the capability to do so.

In establishing this *counselled freedom of choice,* the hypercube learning model does not consider the learner as a separate entity. Rather, the recommendation system (which may be of human nature) and the learner are a joint system. For the case of a technical recommendation system, a model of such joint entities has been termed a *hybrid actor* (Latour 1996), which clearly has more capabilities than either the human learner or technical computer alone. In exerting this freedom of choice in the learning process while still being guided along certain learning pathways, the learner then follows the cognitive process called *planned behaviour* (Ajzen 1991).

7 Closing the Loop of Human Learning and Machine Support

One of the key capabilities of a hybrid actor consisting of a human learner and a machine supporting him is the accumulation of precise meta data about the learning process. Only with the support of an advanced learning management system, one may register traditional factors like *progress* and *knowledge level* together with other data, for example: How much of which KO has been seen? What is the current speed of learning? Can one infer anything about the learner's mood from the speed of typing? Can one track the eye movement across a text or image?

Data mining in this (possibly) vast amount of data and concluding anything on the learning process is the focus of *learning analytics*—aiming at an improvement of the learning process (Ferguson 2012). However, firstly one must constitute that in most cases the big word of *learning analytics* only hides the small reality of measuring classroom performance of employees or students. In particular, most implementations of learning analytics in the commercial sector do not go beyond this. Secondly, it is a completely open question what an *improvement of the learning process* could be: Learning faster? Learning more? Improving the learner experience? Or possibly a combination of all these?

- Consider learning speed—if a learner is found to pass through the pages of a course very rapidly, is that a good sign because he knows already and just wants to repeat the content? Or, is it a bad sign because we have lost his attention?
- Does one really improve learning by adjusting the speed of text presentation such that a constant stress level of the learner is measured (as was done in a recent experiment)?

It is, in general, not known what the proper pedagogical reaction to an arbitrary learning analytics datum could be. Obviously, this points towards a weakness of most learner models, as they do not allow correction of the learning process by measured data.

While INTUITEL tries to cure this in e-Learning environments by making them adaptive according to ontological data, other examples exist that hint towards the immense potential of tracking human behaviour in learning. The 2010 project *Text 2.0*[5] introduces a new dimension in the old informational access method of reading by making it *interactive, responsive* and *multi-modal.* Clearly, this may be the future of textbooks—but what is really improved here, remains unanswered.

More research therefore is needed, and the learning models that we outlined above need to be extended by proper input channels of real world data.

8 Informational Reality: Mobile First, Upload Second

Through numerous studies it is known that in the leading industrial nations a vast majority of the adult population owns a mobile digital device—mostly in the form of a smartphone. This geo-sociological group is defined by having a nearly permanent connection to the internet, and therefore an *everywhere-every time* access to learning material. Also, this group is expanding rapidly, like e.g.

- in terms of age: Groups of small children will cluster around any of their peers owning such a device already at very young age, and even some seniors of 90 years of age are keen to use such a device.
- in terms of device quality: The educational startup *Eneza*[6] is providing online courses via old-fashioned non-smart cell phones already to 2 million learners in Africa—targeting at 50 million.

Moreover, information flow is now bi-directional: Actively asking questions in knowledge communities, posting information on social networks, self-organization

[5]https://text20.net/.

[6]http://www.enezaeducation.com.

using instant messages and other uplink methods together with the downlink access possibility has driven learning beyond horizons we imagined even a few years ago and now determines the *informational reality.*

Clearly, this combination of downlink information search and uplink publishing possibilities signals the beginning of a process that will be highly disruptive to the educational systems. For example, already 2014 more than 40% of all school students in Germany admitted to learn for school purpose by using publicly available video sequences—but had to do so outside school (Pols 2014) since the school system so far is unfit to integrate this learning behaviour into its methodology. As a consequence schools and in the long run also institutions of higher education will lose the motivation of their students if they do not take up this informational reality and integrate it into their teaching paradigms. Educational institutions banning mobile digital devices from certain locations or at certain times are fighting a losing battle.

Also the corporate learning and knowledge management area will be affected. A smartphone is not only a communication device (with huge impact on work-life balance and organizational aspects, that are outside the scope of this article), but a device to store, share and transport knowledge. This has a dark spot nevertheless: Huge damage may be the consequence, if such a mobile knowledge device falls into the hands of competing business adversaries—or if knowledge is deliberately transported outside its intended geographical boundaries. Certain car manufacturers therefore do not allow visitors to bring cell phones with camera, much less smartphones along to business visits to their labs.

Clearly, also these companies fight a losing battle: Digital cameras are now so small that they fit into pens, watches, eyeglasses—and cannot be recognized by their outer appearance.

Another dark spot arises from the fact that the uplink methods allow to spread and to share information independent of its validity, moreover, they allow to offer guidance towards learning pathways. Some of these learning pathways are harmful or fatal, not only to the learner but also to his environment. While "alternate facts" have been around for centuries, it has only now become possible through digital media to present these to a significantly large audience and to (wrongly) relate them to "true facts" by seemingly logical reasoning. However: hearing, reading or otherwise receiving invalid information (or "alternate facts") does not yet imply *believing* it—other factors must be present.

A major additional factor is disclosed by the model of *Satisficing* as the basis of decision making (Simon 1956): In a situation with high information flow, humans tend to make decisions on a basis as simple as possible—and rather accept sub-optimal solutions than informational complexity. In other words: The dominant human strategy in a situation with information overflow is the reduction of one's thresholds and expectations.

It is generally easier to (wrongly) generalize from isolated examples, i.e. to follow *inductive* methods, than to understand a possibly complicated rule system and then to *deduce* a prediction for an isolated example. Satisficing therefore also leads to a preference for *inductive* over *deductive* reasoning methods in situations with high information flow. It is, for example, easier to believe that a wall across a continent would secure the future than it is to believe that the education of a skilled workforce would produce this security.

These three factors; *everywhere every-time* upload ability, inductive reasoning and *satisficing,* therefore, prove that certain populistic political figures as well as terroristic organizations of the current style are straight consequences of the current informational reality.

9 Learning 4.0

Finally, we are therefore able to collect the pieces of our analysis to describe the most probable *future learning model.*

- Future learning will be *digital to a large extent*. The amount of knowledge that we have assembled is so big, that the last scientifically sound estimate of the world data volume was done in 2003. Today, only big commercial players have the infrastructure to perform such estimates and one has to trust them about their finding. Management of this data volume can only be achieved through technical support. This will most certainly not mean that learning will occur *only* through digital media, nor will this be the death of the traditional printed book.
- Future learning will be *network-oriented*. Fellow humans that we meet in social networks, digital databases, knowledge archives of all types and a diversity of other sources will be linked together to form our personal knowledge network; much in the sense of the connectivist paradigm, but clearly not in the form of opinions as suggested by connectivism.
- Future learning will be *diverse*: Informal learning environments where social learning takes place (see the section on behaviourism) may be intermixed (spatially as well as in time) with formal learning environments where one follows a well-defined learning pathway.
- Future learning will be *constructive* in the sense of controlled and planned ontology learning. To achieve this construction (as opposed to simple building) also in informal learning environments does *not*, however, require a stronger or central control of learning processes. Rather, it requires a self-confident media critical competency in every learner, such that he (or she) is able to exert some control over their own learning process—a well-trained meta cognitive competency must be present in order to achieve organized learning processes. Demands to include computer science into school curricula (exceptionally well done in Great Britain with the National Curriculum Computing at School,[7] therefore, are not targeted at the production of programmers for the industry. Rather, they are to be seen as teaching the *algorithmic competencies* for constructive learning and are included as one of the most important skills in

[7]http://www.computingatschool.org.uk.

consideration of 21st century skills (see e.g. http://www.p21.org/our-work/p21-framework).

- Future learning will be based on *semantically enhanced material*. This not only allows the existence of semantically different views on data items, their re-use and is the basis for its sharing. But understanding the meaning of data is mandatory to exert the meta-cognitive skills named above. The rapid learning of short pieces of knowledge (or KO) is easier when it is clear how and where they fit into the existing ontology.
- Future learning will be *individualized* and *adaptive*. As we have shown in the INTUITEL project, individualization is possible, including various layers of knowledge about the learner, the learning material and the learning environment. Each of these factors varies with time and space. While learning in general will be possible independent of space and time, one may achieve a learning process that is strongly coupled to the spatial and temporal location. The future knowledge worker may be able to prepare himself for his daily work equally well in front of his desktop computer or using his smartphone in the commuter train—but the process will be different in both locations.

Reconsidering the historical development, we then clearly see how this future learning fits in. The first stage of learning was strictly behaviouristic and informal: Some skills had to be acquired, or you were punished. The second stage of learning gave rise to formal learning environments (such as our school and higher education systems)—but still without a plan. The third stage of learning then installed pedagogical models of all kinds in the formal learning environments, but still these formal learning systems were unable to put the human *individual* in the centre of the classroom teaching. The aspects of *future learning* outlined above will change this dramatically, as the informational aspects of learning are more clearly understood and it will be possible to implement constructive learning also in the mixed learning environments of the future. We are therefore stepping into the fourth stage of learning—or *Learning 4.0*.

Obviously, the answer to our initial question, therefore, is a complex one. Our society has changed tremendously, and the change in learning behavior is not just a “digital fashion” that one may ignore. Rather, it is deeply interwoven with the societal changes, and cannot be reverted as long as we are living in the informational reality outlined above. On the other hand, it is exactly this informational reality that has provided us with the proper knowledge about *Learning 4.0*, assigning it a proper place in the hierarchy of established learning models. Also, the informational reality provides us with the tools to cope with the information flow.

The informational reality in turn is man-made, as such a reflection of the societal changes. In principle one may even argue, that we (as humans) always invent exactly those methods and tools that are necessary for the current situation. One could label this *innovation efficiency*, in analogy of the formation of efficient financial markets known from the economical sciences.

References

Ajzen, I. (1991). The theory of planned behavior. *Organizational Behavior and Human Decision Processes, 50*(2), 179–211.

Bandura, A. (1977). *Social Learning Theory*. New York: General Learning Press.

Berners-Lee, T., & Fischetti, M. (1999). *Weaving the web: The original design and ultimate destiny of the World Wide Web by its inventors*. San Francisco: HarperCollins.

Chatti, M. A. (2010). Personalization in technology enhanced learning: A social software perspective, RWTH Aachen.

Comenius, J. A. (1654). Auffgeschlossene Güldene Sprachen-Thür oder Ein Pflantz-Garten aller Sprachen und Wissenschafften: das ist: ... Anleitung, die Lateinische und alle andere Sprachen... zu lernen... = Janua linguarum reserata aurea-. Leipzig.

Cooper, P. A. (1993). Paradigm shifts in designed instruction: From behaviorism to cognitivism to constructivism. *Educational Technology, 33*(5), 12–19.

Downes, S. (2010). New technology supporting informal learning. *Journal of Emerging Technologies in Web Intelligence, 2*(1), 27–33.

Ertmer, P. A., & Newby, T. J. (1993). Behaviorism, cognitivism, constructivism: Comparing critical features from an instructional design perspective. *Performance Improvement Quarterly, 6*(4), 50–72.

Ferguson, R. (2012). Learning analytics: Drivers, developments and challenges. *International Journal of Technology Enhanced Learning, 4*(5–6), 304–317.

Fuchs, K., & Henning P. A. (2017). Computer-driven instructional design with INTUITEL: An intelligent tutoring interface for technology-enhanced learning. R. P. S. i. I. a. C. i. Education. Delft: River Publishers.

Fuchs, K., Henning, P. A., & Hartmann, M. (2016). Intuitel and the hypercube model–developing adaptive learning environments. *Journal on Systemics, Cybernetics and Informatics: JSCI, 14*(3), 7–11.

Henning, P. A., Forstner, A., Heberle, F., Swertz, C., Schmölz, A., Barberi, A., et al. (2014a). Learning pathway recommendation based on a pedagogical ontology and its implementation in moodle. In *the DeLFI 2014 conference*.

Henning, P. A., Fuchs, K., Bock, J., Zander, S., Streicher, A., Zielinski, A., et al. (2014b). Personalized web learning by joining OER. *DeLFI 2014-Die 12. e-Learning Fachtagung Informatik*.

Jonassen, D.H. (1999). Constructivist learning environments on the web: engaging students in meaningful learning. The educational technology conference and exhibition, singapore. Retrieved September 24, 2003 from http://www.moe.edu.sg/iteducation/edtech/papers/d1.pdf. Citeseer.

Kandel, E. R. (2009). The biology of memory: A forty-year perspective. *Journal of Neuroscience, 29*(41), 12748–12756.

Kant, I. (1803). Über Pädagogik Herausgegeben und mit einer Vorrede versehen von D. Friedrich Theodor Rink Königsberg bey Friedrich Nicolovius.

Latour, B. (1996). Social theory and the study of computerized work sites. *Information Technology and Changes in Organizational Work*, 295–307.

Macedonia, M., & Knösche, T. R. (2011). Body in mind: How gestures empower foreign language learning. *Mind, Brain, and Education, 5*(4), 196–211.

Mcconnell, J. V. (1962). Memory transfer through cannibalism in planarium. *Joutnal of Neuropsychiatry, 3*(1), 542–548.

Meder, N. (2006). Web-Didaktik: eine neue Didaktik webbasierten, vernetzten Lernens, Bertelsmann.

Moè, A., & De Beni, R. (2005). Stressing the efficacy of the Loci method: Oral presentation and the subject-generation of the Loci pathway with expository passages. *Applied Cognitive Psychology, 19*(1), 95–106.

Pols, A. E. A. (2014). Digitale Schule. Eine repräsentative Untersuchung zum Einsatz digitaler Medien an Schulen. BITKOM Research for LEARNTEC. Berlin: BITKOM.
Siemens, G. (2005). Connectivism: Learning as network-creation. *ASTD Learning News, 10*(1), 1–28.
Simon, H. A. (1956). Rational choice and the structure of the environment. *Psychological Review, 63*(2), 129.
Staab, S., Studer, R., Schnurr, H.-P., & Sure, Y. (2001). Knowledge processes and ontologies. *IEEE Intelligent Systems, 16*(1), 26–34.
Swertz, C., Henning, P., Barberi, A., Forstner, A., Heberle, F., Schmölz, A. (2014). Der didaktische Raum von INTUITEL. Ein pädagogisches Konzept für ein ontologiebasiertes adaptives intelligentes tutorielles LMS-Plugin. Paper accepted at the GMW 2014 Conference.

Author Biography

Peter A. Henning, Prof. Dr., teaches computer graphics, semantic technologies, game programming and e-learning at Karlsruhe University of Applied Sciences since 1998 and Information Business Technology at the Steinbeis-University Berlin since 2012. He is founding director of the Institute of Computers in Education. Peter Henning is in charge of several industrial and scientific projects connected to technology-enhanced learning, acts in the scientific committee of the LEARNTEC trade fair and congress, acts as the member of the program committee of the Virtual University of Bavaria and as member of the steering committee of the e-learning chapter of the German professional computer association "Gesellschaft für Informatik". Current research activities include the determination of the cognitive position of a learner in a multidimensional space of learning objects and its attribution to predefined learning pathways.

Transfer of Theoretical Knowledge into Work Practice: A Reflective Quiz for Stroke Nurses

Angela Fessl, Gudrun Wesiak and Viktoria Pammer-Schindler

Abstract Managing knowledge in periods of digital change requires not only changes in learning processes but also in knowledge transfer. For this knowledge transfer, we see reflective learning as an important strategy to keep the vast body of theoretical knowledge fresh and up-to-date, and to transfer theoretical knowledge to practical experience. In this work, we present a study situated in a qualification program for stroke nurses in Germany. In the seven-week study, 21 stroke nurses used a quiz on medical knowledge as an additional learning instrument. The quiz contained typical quiz questions ("content questions") as well as reflective questions that aimed at stimulating nurses to reflect on the practical relevance of the learned knowledge. We particularly looked at how reflective questions can support the transfer of theoretical knowledge into practice. The results show that by playful learning and presenting reflective questions at the right time, participants reflected and related theoretical knowledge to practical experience.

Keywords Knowledge transfer · Game-based learning · Reflective learning Reflection guidance

1 Introduction

In our society, transferring and disseminating new knowledge and insights from research and development to practice plays a significant role in many professional work-lives. This holds true especially as the development and research cycles get

A. Fessl (✉) · G. Wesiak
Know-Center, Graz, Austria
e-mail: afessl@know-center.at

G. Wesiak
e-mail: gudrun.wesiak@uni-graz.at

V. Pammer-Schindler
Institute for Interactive Systems and Data Science,
Graz University of Technology, Graz, Austria
e-mail: viktoria.pammer-schindler@tugraz.at

K. North et al. (eds.), *Knowledge Management in Digital Change*, Progress in IS,
https://doi.org/10.1007/978-3-319-73546-7_18

shorter and the gained insights needs to be quickly distributed from researchers (or developers) to practitioners. In parallel, the ongoing digitisation of our society can utilize these technological advances to revolutionise the knowledge transfer and integrate it into lifelong learning approaches at work. As the knowledge society emerges, many professions such as health workers or nurses see lifelong professional learning as an indispensable part of their work life (Jensen et al. 2012).

Given this background, we see serious games in combination with reflective learning as a viable mean to conduct this knowledge transfer. On the one hand, games have been proven to be very effective for knowledge transfer. Still in the literature it is not clear in which context such games fit best, however it has been shown that "*passive processes are less effective than interactive and engaging ones, regardless of the audience*" (Lavis et al. 2003; Jensen et al. 2012). On the other hand, reflective learning and practice is viewed as an important learning strategy (Hendricks et al. 1996; Mann et al. 2009). While reflective practice can be seen as the re-evaluation of past experiences with the goal to learn for the future, reflective learning could mean to derive new insights, a change in behaviour and perception (Schön 1987; Boud et al. 2013).

In this work, we will therefore present how knowledge transfer from theory into practice can be performed with a serious game, in our case through the form of a medical quiz, using reflective learning as the underlying theoretical approach. Therefore, we will first discuss the theory of knowledge transfer, serious games (and gamification), and reflective learning. Second, we will present a use case at a neurological clinic, in which a medical quiz was integrated as additional learning instrument in a qualification program for nurses becoming a nurse at a stroke unit. Finally, we will present the results, showing that by playful learning and presenting reflective questions at the right time, participants reflected and were able to transfer knowledge from theory into practice.

2 Background and Related Work

2.1 Gamification and Game-Based Learning

Playing games is one of the first form of learning we are faced with in our childhood. At birth, we own an innate attitude to learn through experimenting and having evaluated the consequences (Jensen et al. 2012). While we grow up, our attitude towards playing games changes, however we are still able to acquire skills, competences and knowledge—thus we can learn by playing games. Playing is often associated with freedom, joy and diversion, while in contrast learning is often related with effort, work and concentration (Breuer and Bente 2010). There are parallels between games and learning, as games have a great potential as tools for learning, while learning can have an important impact for the development of games for education. Nevertheless integrating games as meaningful tools for learning especially at the workplace is not trivial at all (Breuer and Bente 2010).

Gamification was defined by Deterding et al. (2011) "*as the use of game design elements in non-game context*" and following Michael and Chen (2005) "*A serious game is a game in which education (in its various forms) is the primary goal, rather than entertainment*". Thus, serious games are not designed for pure entertainment, but designed with an "*educational aim, a training purpose and/or a behaviour change incentive*" (Jensen et al. 2012). Furthermore, serious games have shown to be very effective for learning, if the learning goal to be achieved is not merely notional. In addition, with regard to knowledge transfer processes, "*passive processes are less effective than interactive and engaging ones, regardless of the audience*" as stated by Jensen et al. (2012).

Furthermore, "*Game-based learning refers to teaching-learning actions carried out in formal and/or informal educational settings by adopting games*" as stated by Kirriemuir and Mcfarlane (2004). Games are our brain's favourite way of learning (Prensky 2001) and consequently are a very effective mean to attract attention and retain interest and can be simultaneously entertaining and instructive (Van Eck 2006; Bontchev and Vassileva 2010).

Already in the 80s and 90s many scientists envisaged that computers and later hypermedia could be used as a cognitive tool for learning, while they also outlined a number of other potential advantages that computer supported learning offers (Pivec et al. 2004). By now, diverse kinds of games have been effectively used to support nurses' learning, such as simulations (e.g. Stanley and Latimer 2011; Dit Dariel et al. 2013), strategic board games (e.g. Mann et al. 2009) or quiz games (e.g. Boctor 2013).

Although the use of games has been noted in nursing education since the early 1980s, many instructors in higher education still prefer a conventional style of delivering educational material. For instance, (Boctor 2013) reported within a study carried out in the UK to assess nurse educators' perspectives of educational games. The three main benefits of using games were perceived to be: enhancement of student learning, enjoyment and interest, interaction and participation among students. Two main factors that discouraged instructors from using games were potential negative reactions of students and time constraints. On the other hand, the study reported a limited use of games despite evidence that educators generally find the use of games to be beneficial.

Especially quizzes are widely used in e-learning since they represent a familiar way to play (Bontchev and Vassileva 2010), are suitable for formative assessment within the scope of a given course or topic (Hudson and Bristow 2006; Koch et al. 2010), and improve performance on summative examinations (Kibble 2007). Learning can be encouraged by involvement in quiz content-creation (Pollard 2006) or by adding meta-cognitive questions to motivate students to reflect on and monitor their own learning (O'hanlon and Diaz 2010).

2.2 Technologies for Reflective Learning

Reflective learning can be seen as the conscious re-evaluation of past situations or experiences with the goal to learn from them and to use the gained outcomes to guide future behaviour. This is in line with the definition of Boud et al. (2013), who define reflective learning as "*those intellectual and affective activities in which individuals engage to explore their experiences in order to lead to new understandings and appreciations*". In workplace learning, reflective learning is seen as a core process with the goal to get new insights, derive better practices, and finally to improve the learner's work (Schön 1987; Boud et al. 2013).

Technologically supported guidance by providing different triggers to induce reflective learning, is well investigated in the area of self-regulated learning within learning management systems. In such settings, prompts are used to organise, retrieve, monitor or evaluate knowledge as well as to reflect on students' learning (O'Hanlon and Diaz 2010; Ifenthaler 2012; Bannert et al. 2017). Davis (2000) distinguishes between self-monitoring prompts and activity prompts. The first encourage students to reflect on their own learning, by asking *thinking ahead* or *checking our understanding* questions. The second motivates students to reflect on their progress in the activity and specifically about whether they have devoted attention to each aspect of their project. In our work, we follow Verpoorten et al. (2011), who created the term reflection amplifier, which is a "*deliberate and well-considered prompting approach, which others learners a structured opportunity to examine and evaluate their own learning*". In work-related settings, there is only little research on usage of prompts. Fessl et al. (2015) investigate three different applications that were enhanced with reflection guidance components such as prompts or diaries to facilitate reflective learning in various workplaces. The results showed that people who are engaged using the applications achieved deeper reflective learning than people who were less engauged. Secondly, the correct timing of the presented reflection guidance components is a crucial issue in order to not interrupt the current workflow of the user. Prilla (2014) discusses prompts with regard to collaborative reflective learning at work. Their prompting approach tries to motivate people to use their reflection tool and the socio-technical nature of communities or face-to-face meetings for reflection.

In order to facilitate the knowledge transfer from theory into practice, we see the combination of serious games and reflective learning as a viable means. While serious games and especially quizzes are widely used and very successful in educational settings (as described in Sect. 2.1), reflective learning is also seen as an important learning strategy in the education of health care professionals. Skills like reflection, critical thinking and problem solving are of crucial relevance for nurses, however, there is gap between the existing reflection theory and its implementation in practice (Carroll et al. 2002; Thompson and Pascal 2012).

3 Use Case: Knowledge Transfer by Playful Reflective Learning

The use case was set up at a *stroke unit* at a German neurological clinic. The stroke unit is a specialized entity of the clinic that deals with acute cases of strokes. The time pressure and the daily work with emergencies and their consequences are a burden for all employees on a stroke unit. The stroke unit consists of 10 certified beds, and employs about 40 *stroke nurses*. The work of the stroke nurses is generally divided into three shifts comprising of early, late and night shifts. While during early and late shifts, usually about six to eight nurses are on duty, in the night there are only up to four. The responsibility of nurses is to ensure medical treatment of patients as well as ensuring their physical and mental well-being. Typically, a stroke nurse is a very experienced nurse. To become a stroke nurse, s/he needs to attend a qualification course dealing with special care at stroke units.

In our work, we present the results of a field study, which was conducted in the special qualification course for becoming a stroke nurse. In this study, a reflective quiz was integrated as an additional learning instrument in the qualification program. The aim of the evaluation was to investigate the usefulness, long-term usage and effectiveness of the reflective questions within the quiz with regard to learning support and reflective learning. Furthermore, we explore the perceived influence and impact on the nurses' practical work, thus the knowledge transfer from theory into practice. By analysing the results, we aim to answer the following research questions:

- R1: How is the quiz perceived with regard to its support for the qualification program?
- R2: How useful are the implemented reflective questions with respect to initiating reflective learning?
- R3: What is the perceived impact of the quiz on work practice, thus transferring knowledge from theory into practise with regard to reflective learning?

4 Methodology

4.1 The Medical Quiz

The medical quiz (Fessl et al. 2014) was developed for both nurses already working at a stroke unit or those in education to become a nurse working at a stroke unit in German hospitals. The goal of the quiz is twofold: First, as all quizzes, it provides an easy and playful way of refreshing knowledge (via the content questions). Second, it aims to connect theoretical knowledge with practical prior experience (via the reflective questions).

Implementation and Quiz Types. The quiz was implemented with the eLearning platform Moodle[1] and four different quiz types were created: Quiz-against-time, Quiz-of-20 (answer 20 questions), Quiz-of-10, and Quiz-of-5.

Content Questions. Altogether 142 content questions were developed by nurses and physicians working at the German stroke unit. The questions consist of multiple-choice or single-choice questions randomly chosen out of a database.

Reflective Questions. Three different types of reflective questions were implemented: “learning progress reflective questions” at the beginning all quizzes, “work-related reflective questions” during the Quiz-of-20, and “general reflective questions” at the end of the quizzes, except the Quiz-against-Time. Their goal is to stimulate reflective learning on different topics and at different points of time during the quiz play.

The reflective question at the beginning motivates users to reflect about their knowledge status (based on previous quiz results) and their play frequency (how often the user played the quiz). The question is composed of an introduction statement followed by a reflective question: “*You are very motivated and you play the quiz at least once per week—your results are really very good. What is your success recipe?*” The in-between reflective questions (see Fig. 1, point 2) are presented together with a content question (see Fig. 1, point 1). They aim at focussing on the content question and how this content question refers to the users work practice: “To what extent is the question stated above relevant for your work?” The question posed at the end of the quiz asks explicitly for gained insights or new knowledge with regard to the currently played quiz: “*Reflect on the currently played quiz. Have you perceived any special insights for yourself?*”

4.2 Procedure

Our study was integrated into the qualification course, which lasted from October 2013 to January 2014 with one course week per month. The participants were nurses working in different German hospitals and studying to become a nurse for a Stroke Unit. In one week of each month the participants came together for the training at the organising hospital. During the first week, the medical quiz was introduced to the participants and they completed a pre-questionnaire. During the next three months, the participants were asked to play the quizzes consequently. On the one hand, they should memorise and strengthen their newly gained knowledge. On the other hand, they should reflect about the content of the quiz and draw connections between the newly gained knowledge and their daily work practices with the help of the integrated reflective questions. Additionally, they were asked to create new content questions for the quiz in order to enlarge the item pool. For each created question, they were rewarded with a “Mozart Kugel”, a gourmet speciality

[1]https://moodle.org.

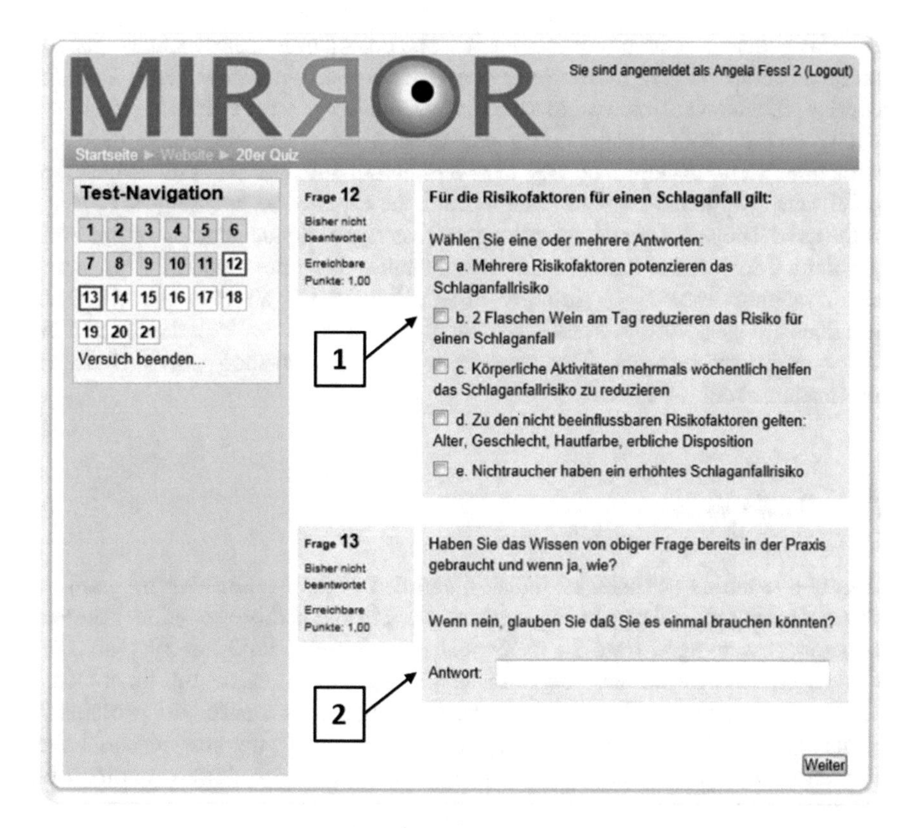

Fig. 1 Medical quiz: “Point 1” shows a content question, “Point 2” shows an in-between reflective question

of Austria. During the second and third course week, intermediate results regarding the usage frequencies and success rates (correct quiz answers) were presented to the participants. In addition, they answered two short in-between questionnaires. During the fourth course week, a half-day workshop was conducted at the hospital’s site. There, we presented the final usage and success rates to the participants, distributed another survey (post-questionnaire), and conducted group discussions and structured interviews to gather additional qualitative data. All questionnaires were presented in paper-pencil format, quizzes were mostly played on participants’ spare time on their own mobile phones.

4.3 Evaluation Tools

Objective usages rates of the quiz were captured via users’ log data, and the written answers to the reflective questions were collected within the quiz. Demographic data was gathered in the pre-questionnaire. The post-questionnaire included general questions about the quiz (reflection support) and its usefulness, questions about the learning effect, work improvement and work quality at stroke units as well as a loyalty metric. Both questionnaires contained a short reflection scale (SRS) to extract the users' general tendency to reflect before and after the quiz usage. Two in-between questionnaires were used to track the experiences and subjective usage of the quiz. They also contained three questions regarding the creation of new quiz questions. Most of the items of all questionnaires were presented as 5 pt. rating scales ranging from 1—"I strongly disagree" to 5—"I strongly agree". Open questions are used for questions about the users' expectations, experiences and the future usage of the quiz. The interviews and the workshop provided deeper information about the quiz.

4.4 Participants

Twenty-one nurses (2 male, 19 female), enrolled in the qualification program for nurses working at stroke units in Germany, participated in this evaluation. Fourteen participants were aged from 20 to 29 years, and seven from 30 to 59 years. The average time in their current position was 6:3 years, 81% worked full time. Of the 21 participating nurses, 18 played the medical quiz at least once. All participants completed the pre-questionnaire, 19 the in-between and 18 the post-questionnaire. The three nurses who didn't play the quiz are not the same as those who didn't fill out the questionnaires.

5 Results

5.1 Quiz Usage and Usefulness

Over a period of 7 weeks, 18 participants answered altogether 8314 questions, ranging from 25 to 1358 questions per user (M = 461.9, SD = 341.0). The Quiz-of-20 was clearly preferred: 18 different participants played the quiz, answered on average 320.6 (SD = 304.9) questions and finished altogether 239 quiz attempts (on average 13.3 per user, SD = 12.9). The other three quiz types were played by maximal 13 users, answering on average 24.3 (SD = 32.9) to 59.7 (SD = 76.9) questions and finished between 2.7 and 4.1 (SD = 3.9 to 5.0) quizzes. The number of questions includes finished and discontinued quiz attempts, content and reflective questions. Regarding the subjective estimated usage frequency of the quiz, users perceived the individual usage as rather low (M = 2.5, SD = 0.92). Interestingly, there is no correlation between subjective and objective usage in terms of number of questions played (see Table 1).

Table 1 Pearson correlations among usage rates and post-questionnaire ratings (n = 18)

Factors	Subj. usage	Cur. usef.	Long. usef.	Loyalty metric	Indi. SRS	Team. SRS	Refl. sup.	Learn. effect	Work. imp	Knowl. excha.	Work qual.
Objective usage	0.324	−0.099	0.169	0.073	0.037	−0.086	0.712[b]	0.490[a]	0.520[a]	0.483[a]	0.442
Subjective usage		0.219	0.266	0.42	0.212	−0.097	0.535[a]	0.512[a]	0.378	0.149	0.165
Current usefulness			0.430	0.127	0.522[a]	0.064	0.207	0.103	0.102	−0.333	0.138
Long-term usefulness				−0.014	0.536[a]	0.208	0.535[a]	0.521[a]	0.448	0.196	0.398
Loyalty metric					−0.029	0.006	0.257	0.177	0.368	0.456	0.496
Individual SRS						0.510[a]	0.243	0.112	0.095	−0.170	−0.006
Team SRS							−0.036	0.041	−0.172	0.029	−0.392
Refl. support								0.675[b]	0.831[b]	0.543[a]	0.634[a]
Learning effect									0.762[b]	0.702[b]	0.535[a]
Work im-provemnt										0.712[b]	0.718[b]
Knowledge exchange											0.570[a]

[a]significant $p < 0.05$, [b]significant $p < 0.01$

With respect to the usefulness, most participants agreed that the quiz can be used to complement professional training for nurses (M = 3.89, SD = 0.58). The long-term advantage of using the quiz during work as well as users interest in using the quiz continuously as part of their work-life received an average rating of 3.28 (SD = 0.60). Similarly, participants rated the likelihood to recommend the quiz to a friend or colleague slightly positive (M = 6.44, SD = 1.5) on a 10-point rating scale (loyalty metric). We found no correlation between usage (objective or subjective) and usefulness or the loyalty metric.

The three participants, who have not used the quiz at all explained this by a lack of internet access (2) or motivation (3). After having passed the exam, the participants lost interest in playing the quiz, which was perceived by the drastic drop of the quiz usage nearly to zero. Further barriers mentioned in the interviews were a lack of time, too many recurring questions and too little user-friendliness especially for nurses with lack of computational skills. In contrast, several statements included the wish, to have the quiz available at their ward.

5.2 *Reflection Support*

Participants' general tendency to reflect was assessed with the short reflection scale (SRS) before and after using the quiz. Considering the two sub-scales for individual and team reflection, a 2 by 2 repeated measures ANOVA revealed significant effects for both main factors: Participants reflected more as an individual than on the team level ($F(1,17) = 58.25$, $p < 0.001$, _2 = 0.774) and—against our expectations—the SRS scores decreased significantly from the pre- to the post-questionnaire ($F(1,17) = 20.48$, $p < 0.001$, _2 = 0.546). Further comparisons of the post-questionnaire SRS values with usage data and usefulness ratings show a positive relationship between high individual reflection and perceived usefulness (current and long-term, Table 1). With a mean rating of 3.51 (SD = 0.42) across seven items, participants slightly agreed that the quiz supports reflective learning. This confirms the general impression that participants viewed the quiz mainly as learning support and that reflection was only of secondary importance. Correlating the mean reflection support ratings with usage and usefulness data shows that participants with higher ratings concerning the quiz's potential to support reflection also had higher objective and subjective usage and long-term usefulness scores (see Table 1). Figure 2 depicts the mean ratings for the SRS and reflection support as well as learning effects, and the quizzes' impact on work gathered in the post-questionnaire, which will be discussed in the next section.

5.3 Learning Effect and Impact on Work

Participants' average rating regarding the perceived learning effect for the qualification course (6 items) reached 3.44 (SD = 0.94). Looking at the single items, the mean ratings ranged between 2.78 (SD = 0.81) for "Talking about the quiz with colleagues helped me to reflect" and 4.22 (SD = 1.06) for "The quiz supported me in preparing for the exam". Mean ratings for questions related to work-improvement due to the quiz (8 items) ranged between 3.11 (SD = 0.88) for gained confidence due to the app and 3.78 (SD = 0.94) for improving one's work-related skills. Also related to learning effects, we found that the quiz stimulated knowledge exchange among the nurses (M = 3.94, SD = 0.64).

The three described variables (learning effect, work improvement, knowledge exchange) are all interrelated and also positively correlated with objective usage (number of answered questions) and perceived reflection support (see Table 1). In order to assess whether the quiz had also an effect on participants' working practice, we asked two questions concerning the quality of work rated with a mean of M = 2.78 (SD = 0.83) (see Fig. 2). We also found a positive correlation between the perceived impact on work quality and the cluster of four variables: reflection support, learning effect, work improvement, knowledge exchange.

5.4 Learning Outcomes

From all automatically presented reflective questions, 52% were answered in a meaningful way. Thus, participants must have thought about the posed questions.

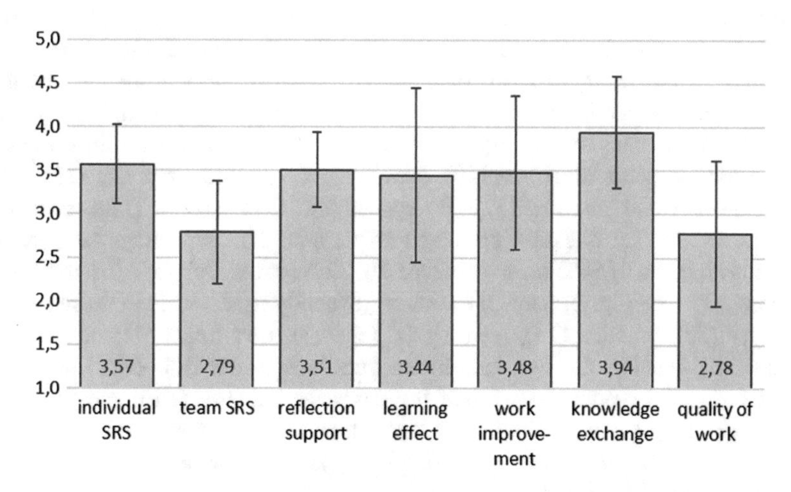

Fig. 2 Mean ratings (SDs) for general reflection (SRS), reflection support, learning effects, and impact on work after using the quiz (1-totally disagree to 5-totally agree)

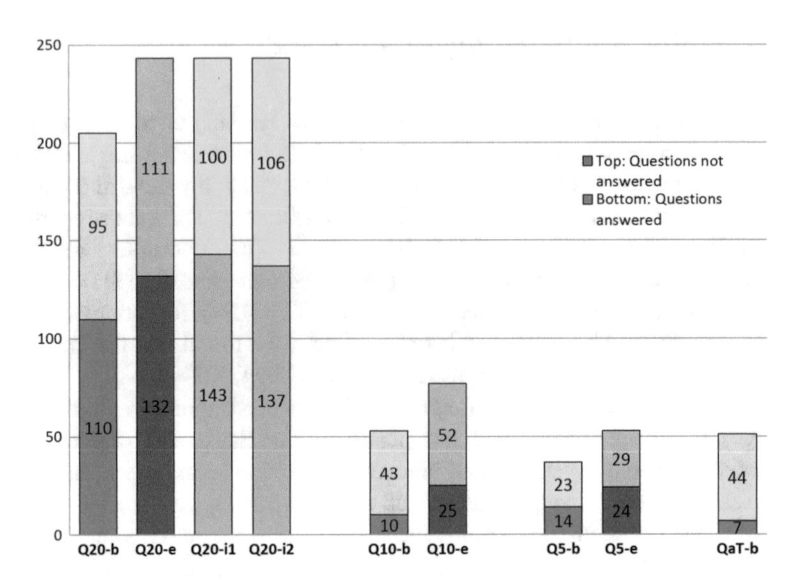

Fig. 3 Number of all reflective questions (entire bars), answered (bottom bars), not answered (top bars)

Figure 3 depicts for each quiz type (Q20/10/5 = Quiz-of-20/10/5, QaT = Quiz-against-time), and each reflective question type (b: at the beginning, e: at the end, i: in-between), how many reflection questions have been posed and answered (e.g. Q20-b: Reflective questions at the beginning of the Quiz-of-20). In the Quiz-of-20 over 50% of the 205 presented reflective questions at the beginning were answered (Fig. 3, Q20-b). For the Quiz-of-5, 38% out of 47 posed questions were answered (Fig. 3, Q5-b), for the Quiz-of-10 (Fig. 3, Q10-b) and the Quiz-against-time (Fig. 3, QaT-b) only 18 and 13% out of the 53 and 51 starting questions, respectively. A concrete answer is "*I partly better understand medical orders*" or "*I can recognize my state of knowledge by answering the questions several times and enhance my knowledge accordingly*." Summarizing all given responses, we looked for the most frequent words to get a general impression of participants thoughts: repetition (40), learning (27), yes (19), practice (10), retain knowledge (7) and nothing (17). Except for the Quiz-against-Time, each quiz included a reflection question presented at the end. The percentage of answered questions amounts to 54% for the Quiz-of-20, 32% for the Quiz-of-10, and 45% for the Quiz-of-5 (see Fig. 3, bars Qi-e). Most frequently used words in those answers were: yes (55), practice (13), learning (11), no (7), very much (7) and recognise progress (5). Finally, the two in-between questions in the Quiz-of-20 have been only shortly answered in about half the cases, as e.g. yes (145), no (38), very relevant (9), very (4) and combine theory with practice (4). Group discussions and interviews revealed that these questions were perceived as rather disturbing and interrupting the own workflow of playing.

5.5 *Expectations and Experiences*

In the pre-questionnaire, we asked about the users' expectations with regard to the quiz, and in the two in-between questionnaires about their experiences. The expectations and expected learning support can be summarised as follows (n = 21): gain new knowledge/deeper education (19), deepen/improving/refreshing/repeating of knowledge (26), better understanding of background (4) and learning with fun (4). The first experiences in the second week (n = 20) contained very good to good application (6), good experience/good for occasionally playing (5), playful learning (5) and fun (2). The individual goals they wanted to achieve are mainly strengthen/deepen/keeping/extending knowledge (10). In the in-between questionnaire of the third week (n = 20), we asked them about their goal achievement, which they answered with nearly achieved/achieved (6) and deepened/strengthened knowledge (3). Answers about their motivation to reflect encompass that they related theory with practice (4), got more background knowledge (3) and reflected if the work done was technically correct (2).

5.6 *User Created Questions*

We asked participants to create new questions for the quiz, which lead to 27 new questions. In both in-between questionnaires three questions were posed to find out if the participants liked to create quiz questions (NQ1), if the creation of quiz questions helped them to strengthen their knowledge (NQ2), and the perceived easiness to create quiz questions (NQ3). For all three questions there was a significant difference between the two questionnaires, week 2: M = 2.82 (SD = 1.03) and week 3: M = 2.13 (SD = 1.04), with NQ1 (U = 3.5, $p = 0.014$), NQ2 (U = 3, $p = 0.036$) and NQ3 (U = 3.5, $p = 0.014$). The participants did not benefit from creating quiz questions and perceived it as rather difficult.

6 Discussion

The evaluation of the quiz within the scope of an educational setting for stroke nurses showed that it supports learning in general and reflective learning in specific as well as confirmed a perceived impact on the nurses' work practice, thus transferring theory into practice.

6.1 Learning Support for the Qualification Program

Regarding R1 we could prove, that the medical quiz was perceived as a successful learning instrument for the qualification program.

Learning support. Most participants played the quiz very often to prepare themselves for the workshop exam, which proved its usefulness to support learning. Thus, the quiz provides users a possibility to recognize their knowledge state, pursue their learning progress and to deepen and extend their knowledge.

Suggestions for improvement. Although most participants liked the quiz and agreed to the long-term practicality of the quiz, they uttered some wishes in order to improve its usefulness and increase its motivational purpose. They suggested that besides a larger pool of questions, more case studies and questions of practice relevance as well as regular updates of the question pool would enhance the support for learning and work. Additionally, they would wish for features like various difficulty levels, a rewarding system and competition between nurses or whole clinics in order to increase the motivation to use the quiz.

Motivation and Barriers. The quiz usage dropped significantly after the participants had passed the exam. They used the quiz primarily to prepare for the exam and were no longer motivated to play it afterwards. Another reason for not using the quiz any more, was the limited number of the available content questions (altogether 142). The participants complained that lot of questions recurred, although they have already been correctly answered several times. For them, it would be much more efficient and motivating to show only questions they have not yet correctly answered as well as new questions. In order to avoid the problem with recurring questions, we asked the participants to create new questions. Altogether 27 questions were suggested by the participants. After their approval by the head nurse, they were directly integrated into the quiz and marked as participant questions. However, this approach did not raise the motivation of the participants for suggesting more new questions.

6.2 Support of Reflective Learning

Regarding R2 we could show that the quiz was able to trigger reflective learning.

Reflective Learning Competence. From the pre- to the post-questionnaires we have seen that the scores of the SRS decreased significantly, meaning that the participants' tendency to reflect reduced after the trial. One reasonable explanation of this phenomenon is that at the beginning of this trial all participants thought that they were rather reflective practitioners. After becoming aware of how reflection is defined in the context of the study, they might have changed their understanding of the concept of reflection and their reflective practices. *Reflection Guidance.* Technical guidance to trigger reflective learning was implemented in the form of "learning progress reflective questions" at the beginning of the quiz, "work-related reflective questions" during the Quiz-of-20 and "general reflective questions" at the end of the quiz. With these questions, we were able to prove that asking the right question at the right moment can trigger reflective learning.

The willingness to reflect was increased with the questions presented at the beginning and at the end of the quiz. Especially by answering the reflective questions at the end of the quiz the participants confirmed, that they gained clear benefits and insights for themselves, but unfortunately these learning outcomes were not inserted into the quiz. At the same time the in-between reflection questions were perceived as more disruptive during the learning process. And although more than half of the reflection questions posed were answered meaningfully, the participants primarily understood the quiz as learning support and reflection was only of secondary importance in the context of education. However, it could be observed that participants who played the quiz more extensively, also rated its potential to support reflection higher.

6.3 Perceived Impact on Work Practice

With regard to the R3, the participants confirmed that the medical quiz had positive effects on their working behaviour. However, since the quiz does not capture such data, we can only rely on participants subjective reports.

Better understanding of work. The participants mentioned that they were able to gather new knowledge with the help of the quiz, which in the end improved their work. They stated that they feel more self-confident during work, because they were able to answer more of the questions posed by physicians, patients or relatives. Because they gained more background knowledge, they understand the treatments and conclusions taken by the physicians much better than before. Most participants also agreed on the future usefulness of the quiz to support professional training, to improve patient care and as a consequence, to raise the employee satisfaction. They would wish for the quiz to be available at their ward to purposefully use their spare time during night-shifts.

Behaviour change at work. Some participants stated that they have tried to take some behavioural changes and applied them during their work as a consequence of using the quiz. In addition, the quiz does not only increase the motivation for learning, but also fulfils its purpose of bringing together theoretical knowledge with working practice and influences their work in a positive way. This means, that participants with high usage rates, high ratings regarding the support for reflection and knowledge exchange, learning effect, and work improvement perceived also a positive effect on their work quality.

6.4 Lessons Learned

The quiz and its purpose was introduced within 15 min by the project responsible for the clinic in the first week of the qualification program. In the half-day workshop at the end of the trial, the discussions revealed, that the concept of reflective learning and the reflective questions were not sufficiently well explained. For a successful evaluation, it is of crucial relevance that the theoretical concept of reflection as well as its practical implementation needs more time and deeper explanations to be accepted and to achieve better results. On the other hand, some minor technical flaws also reduced the effectiveness of the implemented reflective questions. These flaws included that reflective questions were not matching with the introduction statement (at the beginning) or with content questions (in-between), and that answers to reflective questions were counted to the achieved quiz points.

7 Conclusion

In this work, we presented how knowledge transfer from theory into practice can be performed with a serious game in form of a medical quiz, using reflective learning as the underlying theoretical approach. Therefore, we conducted a study with experienced nurses that took part in a qualification course dealing with special care at stroke units. The results showed that a serious game like a quiz is perceived as successful learning instrument, which is in line with existing literature (Hudson and Bristow 2006; Koch et al. 2010). Second, we could also show that reflective learning took place (Davis 2003; Ifenthaler 2012; Fessl et al. 2015; Bannert et al. 2017). Finally, we could show, that combining a serious game with reflective learning led to knowledge transfer from theory into practice. The impact on work practice resulted in a better understanding of work and a behaviour change at work, thus resulting in better care for the nurses' patients.

Acknowledgements The project "MIRROR—Reflective learning at work" is funded under the FP7 of the European Commission (project number 257,617). The Know-Center is funded within the Austrian COMET Program—Competence Centers for Excellent Technologies - under the auspices of the Austrian Federal Ministry of Transport, Innovation and Technology, the Austrian Federal Ministry of Economy, Family and Youth and by the State of Styria. COMET is managed by the Austrian Research Promotion Agency FFG.

References

Bannert, M., Molenar, I., Azevedo, R., Järvelä, S., Gasevic, D. (2017). Relevance of learning analytics to measure and support students' learning in adaptive educational technologies. LAK.

Boctor, L. (2013). Active-learning strategies: The use of a game to reinforce learning in nursing education. A case study. *Nurse Education in Practice, 13*(2), 96–100.

Bontchev, B., Vassileva, D. (2010). Educational quiz board games for adaptive e-learning. In *Proceedings of International Conference*. ICTE.

Boud, D., Keogh, R., Walker, D. (2013). Reflection: Turning experience into learning. Routledge.

Breuer, J. S., & Bente, G. (2010). Why so serious? On the relation of serious games and learning. *Eludamos. Journal for Computer Game Culture, 4*(1), 7–24.

Carroll, M., Curtis, L., Higgins, A., Nicholl, H., Redmond, R., & Timmins, F. (2002). Is there a place for reflective practice in the nursing curriculum? *Nurse Education in Practice, 2*(1), 13–20.

Davis, E. A. (2000). Scaffolding students' knowledge integration: Prompts for reflection in KIE. *International Journal of Science Education, 22*(8), 819–837.

Davis, E. A. (2003). Prompting middle school science students for productive reflection: Generic and directed prompts. *The Journal of the Learning Sciences, 12*(1), 91–142.

Deterding, S., Dixon, D., Khaled, R., Nacke, L. (2011). From game design elements to gamefulness: Defining gamification. In *Proceedings of the 15th international academic MindTrek conference: Envisioning Future Media Environments*, ACM.

Dit Dariel, O. J. P., Raby, T., Ravaut, F., Rothan-Tondeur, M. (2013). Developing the serious games potential in nursing education. *Nurse Education Today 33*(12), 1569–1575.

Fessl, A., Bratic, M., & Pammer, V. (2014). Continuous learning with a quiz for stroke nurses. *International Journal of Technology Enhanced Learning, 6*(3), 265–275.

Fessl, A., Wesiak, G., Rivera-Pelayo, V., Feyertag, S., & Pammer, V. (2015). *In-app reflection guidance for workplace learning* (pp. 85–99). Design for Teaching and Learning in a Networked World: Springer.

Hendricks, J., Mooney, D., & Berry, C. (1996). A practical strategy approach to use of reflective practice in critical care nursing. *Intensive & Critical Care Nursing, 12*(2), 97–101.

Hudson, J. N., & Bristow, D. (2006). Formative assessment can be fun as well as educational. *Advances in Physiology Education, 30*(1), 33–37.

Ifenthaler, D. (2012). Determining the effectiveness of prompts for self-regulated learning in problem-solving scenarios. *Educational Technology & Society, 15*(1), 38–52.

Jensen, K., Lahn, L. C. Nerland, M. (2012). Professional learning in the knowledge society. Springer Science & Business Media.

Kibble, J. (2007). Use of unsupervised online quizzes as formative assessment in a medical physiology course: Effects of incentives on student participation and performance. *Advances in Physiology Education, 31*(3), 253–260.

Kirriemuir, J., Mcfarlane, A. (2004). Literature review in games and learning.

Koch, J., Andrew, S., Salamonson, Y., Everett, B., & Davidson, P. M. (2010). Nursing students' perception of a web-based intervention to support learning. *Nurse Education Today, 30*(6), 584–590.

Lavis, J. N., Robertson, D., Woodside, J. M., Mcleod, C. B., & Abelson, J. (2003). How can research organizations more effectively transfer research knowledge to decision makers? *The Milbank Quarterly, 81*(2), 221–248.

Mann, K., Gordon, J., & Macleod, A. (2009). Reflection and reflective practice in health professions education: A systematic review. *Advances in Health Sciences Education, 14*(4), 595.

Michael, D. R., Chen, S. L. (2005). Serious games. games that educate, train, and inform (Lernmaterialien): Games that educate, train, and info.

O'hanlon, N., Diaz, K. R. (2010). Techniques for enhancing reflection and learning in an online course. *Journal of Online Learning and Teaching 6*(1), 43.

Pivec, M., Dziabenko, O., & Schinnerl, I. (2004). Game-based learning in universities and lifelong learning: "UniGame: social skills and knowledge training" game concept. *Journal of Universal Computer Science, 10*(1), 14–26.

Pollard, J. K. (2006). Student reflection using a Web-based quiz. In *7th International IEEE Conference on Information Technology Based Higher Education and Training, 2006, ITHET'06*. IEEE.

Prensky, M. (2001). Fun, play and games: What makes games engaging. *Digital Game-Based Learning, 5,* 1–05.
Prilla, M. (2014). Collaborative reflection support at work: A socio-technical design task.
Schön, D. A. (1987). Educating the reflective practitioner: Toward a new design for teaching and learning in the professions. Jossey-Bass.
Stanley, D., & Latimer, K. (2011). 'The Ward': A simulation game for nursing students. *Nurse Education in Practice, 11*(1), 20–25.
Thompson, N., & Pascal, J. (2012). Developing critically reflective practice. *Reflective practice, 13* (2), 311–325.
Van Eck, R. (2006). Digital game-based learning: It's not just the digital natives who are restless. *Educause review, 41*(2), 16.
Verpoorten, D., Westera, W., & Specht, M. (2011). Reflection amplifiers in online courses: A classification framework. *Journal of Interactive Learning Research, 22*(2), 167–190190.

Author Biographies

Angela Fessl Received the MSc degree for computer science in the field of Telematik, and the Ph.D. (with distinction) degree in informatics from the Graz University of Technology, Austria. She is a Post-Doc researcher in the area of technology-enhanced learning with a focus on reflective learning at the workplace and has a proven track of research and development on document/ content management and eLearning systems.

Gudrun Wesiak received the M.Sc. and Ph.D. (distinction) degrees from the University of Graz. She was a Senior Researcher at the Know-Center and is a Lecturer at Graz University. Her research focuses on psychological aspects in technology enhanced learning and assessment, as well as on the evaluation of knowledge technologies in learning and workplace settings.

Viktoria Pammer-Schindler received the Ph.D. and M.Sc. degrees (both with distinction) from the Graz University of Technology. She is an Assistant Professor in the Institute of Interactive Systems and Design Science, Graz University of Technology and Area Manager at the Know-Center. She is responsible for teaching, research, and innovation in the field of computer-supported working, learning, and creativity, with an emphasis on knowledge work.

Part IV
New Forms of Knowledge-Intensive Digitally Enabled Value Creation

The Digital Transformation of Healthcare

Andréa Belliger and David J. Krieger

Abstract In all areas of society we are experiencing a paradigm shift from thinking in terms of closed systems to thinking in terms of open networks. We live in a "networked" world that is characterized by networks both online and offline. Networks are non-hierarchical, inclusive, connected, complex, and open. They are constructed out of both humans and nonhumans. Networks today have become a kind of blueprint for the way in which society is being organized, including healthcare. Healthcare is no longer primarily something that takes place in the intimacy and confines of the doctor-patient relationship. Instead, health care is distributed throughout a complex network of both human and nonhuman actors such as databases, hospital information systems, digital health records, electronic health cards, online patient communities, health related apps, smart homes with ambient assisted living technologies, etc. Networks operate most efficiently when they conform to norms such as connectivity, flow of information, communication, participation, transparency, and authenticity. These norms guide the production and uses of health related information and knowledge. They condition how health related knowledge can create value both with regard to efficiency and quality of care. In this article, we take a look at how the norms of digital transformation have changed managing knowledge in health care networks.

1 Towards Networked Health

The digital transformation of healthcare is a complex and multi-sided phenomenon that cannot be easily reduced to a few common characteristics. This is especially the case when attempting to understand such new developments as self-tracking, big data and predictive analytics, e-health, mobile health, health apps, participative medical research, e-patient communities, electronic medical records, and shared decision-making in diagnosis and therapy. Although there remain significant

A. Belliger (✉) · D. J. Krieger
IKF, Lucerne, Switzerland
e-mail: andrea.belliger@ikf.ch

K. North et al. (eds.), *Knowledge Management in Digital Change*, Progress in IS,
https://doi.org/10.1007/978-3-319-73546-7_19

privacy and data security issues, many of these problems have been both legally and technically resolved so that new forms of using the potential of medical data are being implemented in hospitals, doctor's offices, clinics, rehabilitation centres, by insurers, researchers, and among all participants in the primary healthcare market. Beyond the traditional system of healthcare providers, however, a parallel universe has appeared in the domain of e-patients, online health communities, and personal health tracking. This is the world of connected citizens and healthcare consumers, the world of self-trackers and e-patients. It has often been noted that the small "e" in front of the word "patient" does mean electronic alone, but primarily "educated," "engaged," "enabled," and "empowered." The world of educated and empowered health consumers is based on the new possibilities of connectivity created by the Internet and also by mobile devices, apps, and the unprecedented availability of information. It is a world in which new values and norms arising from the affordances of digital information and communication technologies create new practices and new expectations. Connected healthcare also implies the connection of the primary and the secondary healthcare markets. Healthcare is no longer confined to traditional providers such as doctors, hospitals, laboratories, insurers, and regulators, but new players are joining in from areas such as mobility, telecommunications, logistics, and retail. Health-related information and knowledge is no longer locked up in the silos of the traditional healthcare system. This can be seen as a response to the transformative forces in information and knowledge management that are changing our relation to health, disease, prevention, well-being, work-life balance, and also what it means to lead a responsible and fulfilling life in today's information society. The "digital transformation" has reached healthcare and is changing closed systems into open, flexible, participative, and innovative healthcare "networks" (Belliger and Krieger 2016).[1]

There are many different aspects of connected health and it is a challenge to attempt to bring them all into relation with each other and gain an overview of what healthcare has become in the digital age. North and Kumta (2014) and North (2016) have proposed the model of a ladder upon which one begins at the bottom with data and enabling technologies and moves up the ladder to information, knowledge, actions, and competence until one reaches the top where the effects of these technologies appear. From the perspective of a data driven society, one climbs up from ubiquitous access to information through social media on to human-machine collectives up to digitally enabled, knowledge-based services and business models. The digital transformation of healthcare can also be viewed as effects of enabling, but also disruptive, technologies. In order to obtain an overview of what these developments are and how they are connected we propose using the metaphor of a tree. We propose to visualize the digital transformation of healthcare by using the image of a "digital health tree." This metaphor is intended to illustrate how the

[1]See Belliger/Krieger (2016) for a discussion of the impact of the digital transformation on organizations in general and on management practices in healthcare, business, education, and civil society.

different aspects of connected health are indeed "connected" and how the various areas, topics, concepts, and trends involved in the digital transformation of healthcare make up a more or less unified network of branches and fruit arising from the affordances of digital technologies.

2 The e-Health Tree

As with any tree, our digital health tree consists of roots, trunk, branches, and fruit (Fig. 1). The roots of the digital health tree reach deep into the broader social changes that have been initiated by the advent of a global network society. This is a society based on digital information and communication technologies. The digital transformation has affected every domain of society and amounts to a revolution comparable to the industrial revolution, which also changed every aspect of life from work to education to politics and science. Healthcare is one domain of society alongside other domains. The revolutionary character of digital technologies makes it not only possible, but also necessary to place a small "e" in front of almost all social activities. We speak of e-commerce, e-learning, e-banking, e-government, and also e-health. It can be said that the roots of these changes are an almost universal *connectivity*. Connectivity is an expression of the inherent tendency of digital technologies to link up not only computers to computers, but also people to people, and finally, as the Internet of Things and the 4th industrial revolution illustrate, almost everything to everything. This radically new event in the history of society has created a situation in which it is no longer adequate to think and act in terms of closed systems, bounded organizations, and distinct domains. Linking everything to everything by means of digital technologies transforms closed systems into open, flexible *networks*.

3 Networks—The Roots of e-Health

Networks are not new. Indeed, networks are perhaps the oldest form of human organization. What makes networks especially important today is the digital information and communication technologies that allow large, complex networks to be effectively managed. The affordances of digital technologies have created a situation in which many-to-many communication is possible (Castells 2005; Shirky 2008). Before the advent of new media, communication could either take place face-to-face in small groups, or when the number of people involved in cooperative action became larger, one-to-many, or top-down communication was necessary. Digital communication technologies eliminate the age-old spatial and temporal limitations on communication and allow large groups of people to coordinate their activities directly without the need for top-down management. This dismantles hierarchies, delegitimizes bureaucracies, and makes the typical top-down command

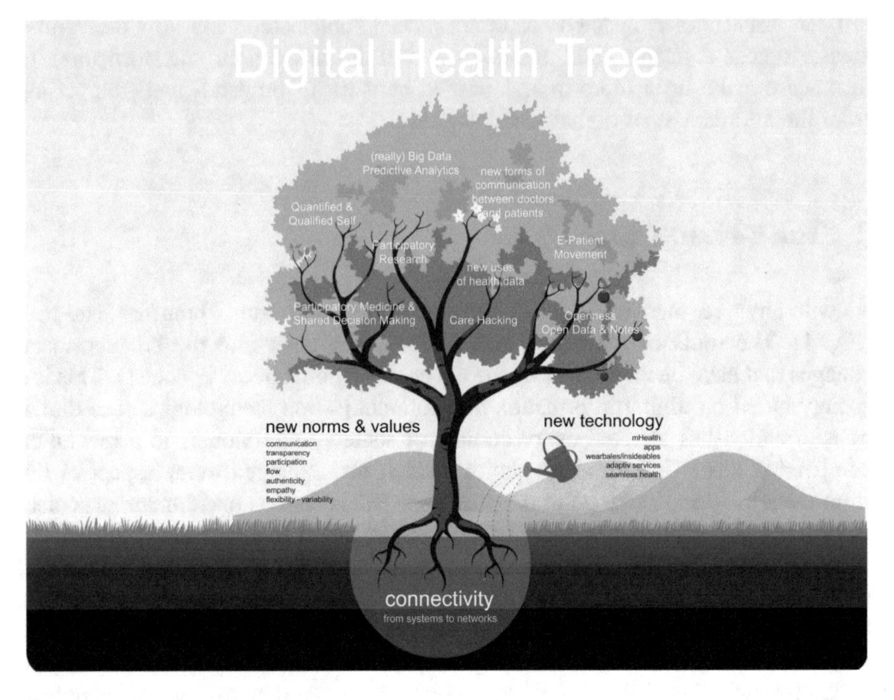

Fig. 1 The digital health tree

and control communication of the industrial age inefficient. This means that we entering into an information age and a global network society. As Castells (2005) points out, it is not that networks are new in human history, but "What is new is the microelectronics-based, networking technologies that provide new capabilities to an old form of social organization" (4). Castells (2005) goes on to point out that organizations in all areas of society are changing. He locates three characteristics of this new network society: (1) the "generation and diffusion of new microelectronics/ digital technologies of information and communication;" (2) the "transformation of labour that is able to innovate and adapt;" and (3) the "diffusion of a new form of organization around networking" (2005: 8). We might add a fourth important characteristic of today's world, namely, the omnipresent significance of information as central resource in all areas of society and a fundamental transformation in the ways in which information and knowledge are managed.

A network society based on information creates new ways in which knowledge and professional expertise are ordered. In the age of print media it was costly to produce and distribute information. The physical attributes of the medium sets limits on the amount of information that could be produced, stored, and distributed. This created an economy of scarcity with regard to information and knowledge that required the institution of a hierarchy of central authorities, experts, and gatekeepers who regulated how information was produced and distributed. Information and knowledge were organized in a kind of pyramid structure characterized by limitations, exclusions, and restricted access. Digital media changed this situation radically. Weinberger (2011) proposes replacing the traditional metaphor of hierarchical order, the pyramid, with a new symbol, the cloud. In the cloud, knowledge is non-hierarchical, unlimited, connected, inclusive, complex, and public. This is true for every kind of information and knowledge, including health-related information. To say that the digital health tree is rooted in connectivity is to assert that traditional hierarchies and pyramids in healthcare are breaking down and being replaced by more or less open and flexible networks. It is no longer surprising that informed patients have access to medical research about their condition, which even their doctors may not know of. More and more doctors are becoming willing to "let patients help" when it comes to offering the best options for therapies (Belliger and Krieger 2014).[2] We will return to these new developments below when describing the fruits of our digital health tree. First of all let's take a look at what is special about networks.

Networks are a unique form of social order that have their own typical characteristics as opposed to traditional ways of organizing in terms of either markets or hierarchies. As opposed to hierarchies or closed systems with centralized steering, networks are decentralized and do not have clear boundaries. They are constantly reconfiguring themselves by extending links and creating new hubs and are therefore flexible and can serve multiple purposes and have different identities, roles, and functions simultaneously. This means that networks cannot be effectively managed top-down, but require decentralized, collaborative forms of "governance" instead of bureaucratic command and control. Furthermore, and this is important for the trunk of our digital health tree, networks cannot easily control the flows of information that connectivity makes possible. If connectivity is the roots of our tree, then the free *flow of information* is the stem. The free flow of information not only allows many who previously did not have access to information the ability to use this information, but it creates a culture of *participation*. In the industrial age, many were excluded from access to information. They were also excluded from the means to produce, distribute, and use information. This had the effect of making people into passive consumers of products and services—"patient" comes from Latin *patiens*, i.e. the one who suffers or endures—and fostered a patriarchal healthcare system. Consumers have now become "prosumers" and are no longer passive.[3]

Ordinary people have become important participants in the information production and distribution value chain. Indeed, it has become almost a commonplace to speak of "participatory culture" in order to describe how consumers have become prosumers and information producers in their own right (Jenkins et al. 2009). Connectivity creates the conditions for the flow of information. Wherever information flows, it encourages and empowers participation. It produces change and

---

[2]See "Let Patients Help" by e-Patient Dave de Bronkart. http://www.epatientdave.com/let-patients-help/. On the e-patient movement in general see Belliger/Krieger (2014) and https://en.wikipedia.org/wiki/E-patient.

[3]See https://en.wikipedia.org/wiki/Prosumer.

opportunity in all areas of society, whether it be science, business, politics, education, or healthcare. Connectivity and flow lead to participation, which in turn demands *transparency* and *authenticity* from all involved, whether it be individuals, large corporations, or government and regulative agencies. These characteristics of networks, which can also be considered as specific "network norms" arising from the affordances of digital technologies can be said to constitute the trunk of the digital healthcare tree. They are the major drivers of change in social practices as well in technological innovation that are transforming healthcare today.

4 New Forms of Communication and Participation—The Fruits of e-Health

The digital health tree is of course not without its own fruit. Rooted in network connectivity, fed by the free flow of information and branching off into normative expectations of participation, transparency, and authenticity, these fruits are such things as the quantified self movement, health apps, big data and predictive analytics, personalized medicine, new forms of communication between doctors and patients, innovative ways of using health data, care hacking, crowd-sourced medical research, participative medicine, and shared decision making, to name only a few. These fruits can be seen as developments emerging from the new values and new technologies that make up the digital transformation of healthcare. They represent new ways in which health-related information and knowledge is being generated, distributed, and used to create value for all stakeholders. In the following, we take a closer look at some of these fruits that are currently ripening on the digital health tree.

4.1 Body Tracking and Quantified Self Movement

Apps, wearables, and cloud computing make it possible today to digitally track almost everything one does, including one's own fitness and health.[4] What was previously reserved for the chronically ill or professional athletes is now cheaply and easily available to anyone who has a mobile devices such as a smartphone or smartwatch. There are now hardware equipped with sensors, apps, and services for tracking the steps one takes, the distance one walks or runs, heart rate, calories burned, stress, blood pressure, sleep rhythms, and much more. It is possible to

[4]This is known as "personal informatics." "Personal informatics is a class of tools that help people collect personally relevant information for the purpose of self-reflection and self-monitoring. These tools help people gain self-knowledge about one's behaviors, habits, and thoughts." www.personalinformatics.org.

aggregate, evaluate, and visualize this data so that one has one's own long-term study of one's physical and mental condition. The medical laboratory as well as a doctor's advice now fit into one's pocket and are available anytime and anywhere. This has radically changed the way in which health-related data is generated and used. Doctors now prescribe apps instead of medication.[5] Patients come into the doctor's office with the laboratory results registered on their smartphones, as well as advice from their patient community or medical professionals. This deconstructs traditional hierarchies with regard to health information and democratizes healthcare in a similar way that access to the personal computer and the Internet put the power to create media in the hands of everyone and revolutionized the publishing industry. It is nonetheless still an open question just how this data can best be used. Many healthcare providers, including doctors, hospitals, laboratories, and patients as well are uncertain how this new connectivity in healthcare and the free flow of medical information will change their roles and relationships. There still remain many technical issues to be solved in making data from multiple different sources compatible and transferable so that the full potential of this information can be exploited at the point of care. An example of how this could be done is the platform Validic,[6] which promises to "simplify data access, integration, standardization, and storage" and thus bring data from different apps and wearables together and make it useful in day-to-day healthcare.

The Quantified Self movement networks individuals and organizations in new ways. Among the organizations involved there are medical and health service providers such as doctors, hospitals, laboratories, insurers, pharmaceutical companies, and medical technology companies. There are also many organizations from the area of prevention, for example, drugstores, pharmacies, telecommunications and IT companies, as well as large retailers who are interested in linking health data generated by self-tracking apps to their products and services. The increasing availability of high-quality devices, apps, and reliable tracking practices creates a situation in which patient monitoring is being integrated into diagnosis, therapy, and rehabilitation. Such devices and apps are increasingly being certified by government health agencies, such as the FDA in the USA or the NHS in the UK. Many new start-ups in the digital health sector are developing apps specifically for healthcare professionals instead of merely for people interested in self-tracking. In addition to this, public health agencies have become aware that self-tracking data can be easily aggregated into data on entire populations and thus become a valuable resource for prevention, epidemiology, and health improvement programs sponsored by governments, public health programs, and policy makers.[7] Important stakeholders in the networking of citizens and patients into the traditional healthcare

[5] See for example https://www.wsj.com/articles/doctors-prescribe-new-apps-to-manage-medical-conditions-1447094444; and https://www.newscientist.com/article/2121164-nhs-to-start-prescribing-health-apps-that-help-manage-conditions/.

[6] https://validic.com/.

[7] See for example the Elixir initiative of the European Union which attempts to bring together data from many different sources https://www.elixir-europe.org/about-us/what-we-do.

system are not only device producers, app developers, and service providers, but also governments and non-profit organizations. It has become apparent that these new forms of connected healthcare are not merely hype or isolated social movements among fitness enthusiasts. We are dealing with disruptive networks that are changing the traditional healthcare system.

4.2 Big Data and Predictive Analytics

Another important fruit on our digital health tree is big data and its analytics for health. Thanks to almost omnipresent connectivity and the enormous amount of data generated from apps, wearables, social media sites, online patient communities, consumer genomics, etc., it has become possible to do health data mining and apply the tools and techniques of big data and predictive analytics to this vast amount of information. Not only is there big data coming from self-tracking, but clinical research, electronic health records, as well as related sources such as life-style, hobbies, sport, dietary tracking, work-monitoring and so on. All these variegated data sources contribute to aggregating large data sets. These large data sets create a valuable pool for big data analytics in order to discover correlations that can only become visible when data sets are extremely large. There are, of course, challenges to make big data analytics in healthcare both technically and regulatively possible.

It is still not always easy to link this data, to manage, archive, share, and exploit it for health-related purposes. On the one hand, there are technical and regulative obstacles. On the other hand, however, it is in principle possible to link the complete genome data of a single person with the data in electronic health records, life-style data, and other self-reported health data in order create the foundation for truly personalized medicine. If aggregated data of this kind for many thousands of individuals, if not millions, can be gathered, aggregated, and provided to researchers, there is great potential to advance health care and discover cures for diseases that have long proved incurable. Beyond the promises of personalized medicine, there has arisen a new field of research known as "Health Data Mining and Predictive Analytics." It is an interdisciplinary endeavour situated at the interface of computer science, sociology, health research, medicine, statistics, data science, business intelligence, data visualization, machine learning, law, and other related areas. It promise to improve patient care, chronic disease management, hospital administration, and healthcare logistics by providing actionable knowledge on the basis of analysing large and variegated data sets.[8] International consortia and associations of stakeholders from all areas of medical research and

[8]See for example Health Catalyst https://www.healthcatalyst.com/catalyst-approach/.

healthcare are currently forming in order to make human health a "data driven" enterprise.[9]

4.3 *Participatory Research*

Participatory research involves ordinary people as "citizen scientists" in various stages of medical research. Almost everyone today carries a smartphone or other mobile devices that are fitted with sensors of many different kinds. If these devices have the appropriate apps installed, they can register vital data and transfer this data into the cloud or to a research platform so that it can become part of a clinical test or other research program. On the consumer genomics platform 23andMe, for example, circa 76% of the almost one million users donate their genome data for medical research. Those who participate in online patient communities for a specific disease, for example, Parkinson's disease, can donate their medical information, their self-tracking data, and their personal experiences with various forms of therapy and medication to the community and to professional researchers. The platform "antidote" goes a step further in that it creates the possibility for people who wish to participate in research to meet up with researchers who are looking for participants for clinical trials.[10]

There are also projects that enable patients to connect up with researchers who might be interested in investigating questions that come from the patients themselves. Not only do patients and concerned citizens donate their data, but they also participate in formulating research questions and research designs. This implies that patients are not only those who suffer from a disease, they are also in some cases those who know the most about their condition and most about what can help and thus are in a position to contribute in various ways to better health outcomes. This form of participation goes far beyond personal health management. Citizens and patients find themselves involved in innovation and a co-creation of value in the healthcare marketplace. Crowd power describes how medicine and healthcare can be advanced by connectivity, the free flow of information, participation, and cooperation in new networks involving patients, doctors, medical researchers, technology developers, and regulators.

---

[9]See for example the Global Alliance for Genomics and Health http://genomicsandhealth.org/; Elixir of the European Union https://www.elixir-europe.org/; and the Big Data to Knowledge BD2 K initiative of the National Institutes of Health in the USA https://commonfund.nih.gov/bd2k.

[10]See https://antidote.me/.

4.4 New Uses of Medical Data

It should be no surprise after what has already been said about self-tracking, big data analytics, and participatory research that people are more and more willing not only to generate their own data and manage their own health, but also to share this data with others, whether it be patient communities or research platforms. This does not imply, however, that patients are willing to relinquish all control over their data. On the contrary, they want to be able to decide themselves how their medical data are being used and by whom. Studies have shown that many people not only want to have access to their medical records, laboratory results, X-ray images, scans, etc., but would be willing to send these data to medical professionals or institutions. There is therefore good reason to assume that patients will increasingly demand to have access to their data and even to be able to take these data with them when they leave the hospital or doctor's office. Furthermore, they will want to be the ones who decide how this data is used and who can have access to it. This implies that patients will put pressure on healthcare providers to establish secure interfaces and interoperable platforms for the transfer of medical data in order to optimize communication and cooperation. It is no longer acceptable, that medical professionals and institutions along a treatment path do not communicate with each other and lay the burden of coordination on the patients. This is especially the case for the chronically ill and those with rare diseases, who often have to deal with many different healthcare providers.

In order to facilitate healthcare on all levels, data must be freed from confinement in silos and made accessible and transferable via platforms. Platforms are like markets in that they bring people together in order to share information, products, and services.[11] Platforms enable the co-creation of value. In order for all this to work and to create healthcare value, it must be clearly acknowledged that medical data belongs not only to those who produce it, but also to the patients and that patients are also entitled to make decisions about data distribution and use. To a great extent, patient ownership of medical data is today expressed by data protection regulations that require informed consent for any gathering, aggregating, and use of personal information. An example of how acknowledging patient ownership of data goes beyond informed consent is the "blue button" movement.[12] In the USA many government agencies, as well as healthcare providers allow patients not only to view their records online, but also to download their medical data. A "blue button" on the website of a clinic, doctor's office, or laboratory tells the patient that they have access to their medical data, that they can download these data and then decide themselves what to do with these data. For example, they can send them to other healthcare professionals, to their online patient community,

[11]See for example http://www.shareable.net/blog/11-platform-cooperatives-creating-a-real-sharing-economy.

[12]See https://en.wikipedia.org/wiki/Blue_Button; and https://www.healthit.gov/patients-families/about-blue-button-movement.

to a research platform, etc. Beyond merely being able to consent (or not) to use of data, patients have copies of their data and manage their medical information. The blue button illustrates more than technical interoperability, it also stands for a cultural interoperability that accepts patients as genuine stakeholders in healthcare and allows them to significantly participate in the entire healthcare process.

4.5 Open Notes and Open Data

Perhaps the first thing that comes to mind when speaking of health records and medical data is quite the opposite of what has been discussed above. One thinks almost inevitably of closed systems, weak interoperability, dominant players, no standardized and unified solutions, disinterested insurers, a focus on privacy, and informational asymmetries among stakeholders. It is not only e-patients, however, that are moving toward openness and transparency. Hospitals, doctors, and regulators are also being transformed by connectivity, flow, and participation. This is illustrated by the Open Notes initiative.[13] In 2010 several large hospitals conducted an experiment allowing 20,000 of their patients to read the notes that doctors and medical professionals take down during treatment. Contrary to expectations that this would cause many problems, the results, which were published in 2012,[14] showed that not only patients, but doctors as well were overwhelmingly satisfied with this new form of communication. In 2013 some agencies of the US government added the option of OpenNotes to the "blue button". Currently, more than 12 million patients in the USA have access to their notes and more institutions are joining in this movement toward transparency and open data.

The OpenNotes initiative corresponds to other open data initiatives in the healthcare sector. Recently, the US Dept. of Health and Human Services (HHS) started the website HealthData.gov (www.healthdata.gov.) with the intention of making public domain medical and healthcare information easily available to all for commercial as well as scientific uses. This information includes health service provider directories, clinical care provider quality information, databases on medical and scientific knowledge, community health performance data, consumer product information, information on government spending for healthcare, etc. What is different about this approach when compared to those discussed above are the explicit aims at furthering healthcare innovation in the private sector. "Our goal is

[13]https://www.opennotes.org/.

[14]http://annals.org/aim/article/1363511/inviting-patients-read-doctors-notes-quasi-experimental-study-look-ahead.

to unleash the power of private-sector innovators and entrepreneurs to utilize HHS data to create applications, products, services and features that help improve health and health care—while also helping to create jobs of the future at the same time."[15]

What these two examples of open data and open information illustrate is not only connectivity and the flow of information throughout large networks, but the value creating quality of transparency. It is not by locking knowledge away in silos that value in today's network society is created, but by sharing, which creates transparency and trust.

4.6 Care Hacking

Care hacking could be defined as any use of digital technologies, above all, the Internet, in order to take control of one's own health and use the healthcare system in new and unexpected ways. This can look a lot like real computer hacking, when Hugo Campus, for example, hacked into his implantable cardiac defibrillator in order gain access to the data that the manufacturer refused to give him.[16] Another well-known care hacker is Salvatore Iaconesi who used his computer skills to get access to his brain scans and medical records and to break the medical codes they were "encrypted" in.[17] He published this information on the website La Cura[18] and asked the online community for help. He received 500,000 responses of all kinds from around the world. After a successful surgery, he also implemented many of these suggestions. But the effect of the tremendous support of people throughout the world should be considered of great importance in itself, for it changes the way that medicine works. It gets people involved. It brings new, unforeseen, and unusual information into the healthcare system and disrupts traditional processes and protocols. A further important example of care hacking is the "openAPS" artificial pancreas system that was developed by concerned patients suffering from Type 1 diabetes and is offered free of charge as an open source product.[19]

4.7 Participative Medicine and Shared Decision-Making

Care hacking shows that the digital transformation of healthcare has empowered patients to use the Internet, computer skills, mobile devices, apps, etc. in order become active participants in managing their own health. This goes beyond

[15]https://www.healthdata.gov/content/about.

[16]https://www.youtube.com/watch?v=oro19-l5M8k.

[17]http://blog.ted.com/why-i-opensourced-cures-for-my-cancer-salvatore-iaconesi-at-tedglobal-2013/.

[18]http://opensourcecureforcancer.com/.

[19]See https://openaps.org/.

traditional prevention as well as traditional attempts to involve people in public health campaigns. Not only do people generate their own health-related data, they also share this data with friends, relatives, other patients, concerned medical professionals, and their healthcare providers. This situation changes the roles in healthcare. Patients are no longer passive receivers of diagnosis and therapy. They see themselves as active, informed, and self-determined partners in healthcare. They initiate preventive measures on the basis of their own health monitoring. They take responsibility to inform themselves about their condition, about alternative diagnoses and therapies, and about experiences of others affected by a certain condition. Today, patients have access to the latest research published about their diseases, about the reactions that others being treated for a disease have to medications, and much more. They use this information not to replace doctors, but to make the doctor patient relationship more a relation among partners, who are attempting together to attain a certain goal. An example of this situation is the Society for Participatory Medicine (participatorymedicine.org) which is a "not-for-profit organization devoted to promoting the concept of participatory medicine, a movement in which networked patients shift from being mere passengers to responsible drivers of their health, and in which providers encourage and value them as full partners."[20] A recent study that the Society for Participatory Medicine together with ORC Research conducted showed that 88% of those questioned believe that working with their healthcare providers as a partner will help their outcomes, and 84% believe that self-tracking and sharing of data with their health team would improve managing their health. A similar interest on sharing information and including patients in the healthcare process on the side of providers is attested by the Mayo Clinic, which engages with patients via social media. The Mayo Clinic also provides patients with information and opportunities for communication by means of a blogs, podcasts, discussion forums, and an extensive website.

This information and the many opportunities for patients to use this information in discussions with providers transform the traditional division of labour between patients and healthcare professionals, between lay person and expert. "Shared decision-making" formalizes participatory medicine into clear rules and processes for how doctor and patient share information, discuss options with regard to diagnosis and therapy, and reach collaborative decisions in situations of uncertainty. There has recently been much research and practical work done on how shared decision-making can be best implemented.[21] The Mayo Clinic maintains a national resource centre on shared decision-making.[22] In the UK, one speaks of "patient-centred care"[23] and the European Union has officially proclaimed that it is

[20]http://participatorymedicine.org/.

[21]See http://www.patient-als-partner.de/index.php?article_id=1&clang=2.

[22]http://shareddecisions.mayoclinic.org/.

[23]See Barry and Edgman-Levitan (2012). http://www.nejm.org/doi/full/10.1056/NEJMp1109283.

time to "put the patient in the driver's seat."[24] This is not only to be understood as a way of improving health outcomes and empowering citizens to take more responsibility for their own health, but also as the only promising solution to the rising costs of healthcare and the inefficiencies of large, top-down, bureaucratic institutions.

4.8 e-Patient Movement

The digital transformation of healthcare means that paternalistic medicine in which doctors are "gods in white" and patients have nothing to say is becoming increasingly dysfunctional. Returning to the metaphor of our digital health tree, connectivity, flow, transparency, and participation have brought forth many fruits which all have to do with how normal people access, understand, and use health-related information in new and unforeseen ways. Patients can now constructively participate in prevention, diagnosis, therapy, and rehabilitation. Perhaps nowhere is this new awareness of healthcare more appropriately expressed as in the slogans of the e-patient movement, "Gimme me my damn data" and "Let patients help."[25] We mentioned the Society for Participatory Medicine above. E-patients are not merely self-trackers. The Quantified Self movement operates primarily in the secondary healthcare sector of fitness and prevention, whereas the e-patient movement is mostly concerned with diagnosis, therapy, and rehabilitation. E-patients are creating new roles and relationships between patients and healthcare providers such as doctors, clinics, hospitals, insurers, regulative agencies, pharmaceutical companies, and the medical technology industry. The explicit goals of the Society for Participatory Medicine are: (1) "to guide patients and caregivers to be actively engaged in their health and health care experiences;" (2) "to guide health professional practices where patient experience and contribution is an integral goal of excellence;" and (3) "to encourage mutual collaboration among patients, health professionals, caregivers and others allowing them to partner in determining care."[26] As ePatient Dave de Bronkart puts it, "We perform better when we are informed better."[27]

Examples of how these goals are being realized are found in online communities such as Patients Like Me,[28] CureTogether,[29] and Acor.[30] What these knowledge networks have in common is that patients are connected to peers, to information

[24]https://ec.europa.eu/digital-single-market/en/news/putting-patients-driving-seat-digital-future-healthcare.

[25]http://www.epatientdave.com/let-patients-help/.

[26]http://participatorymedicine.org/about/.

[27]http://www.epatientdave.com/2012/10/21/we-perform-better-when-we%E2%80%99re-informed-better/.

[28]https://www.patientslikeme.com/.

[29]http://curetogether.com/.

[30]http://www.acor.org/.

throughout the Web, and to medical professionals. It has become possible that patients are sometimes better informed about their specific problems than their doctors are. With this information, e-patients are empowered to constructively contribute to diagnosis and therapy. With regard to participation in communities, they can provide information and support to other patients. In addition to this, they can donate their data to medical research and collaborate with scientists on agenda-setting for clinical studies and research programs. What we are witnessing is the emergence of a healthcare-related civil society in which patient advocacy organizations engage in promoting better health services by working with government agencies, healthcare providers, and health related industries such as pharmaceutical companies. Finally, the e-patient movement embodies the new transparency that the Web affords. Patients can rate the performance of medical service providers. CureTogether.com or in the German speaking world, Bertelsmann's "Weisse Liste"[31], for example, offer millions of ratings not only of doctors and hospitals, but also treatments and medications.

5 Implications for Managing Information and Knowledge

The digital transformation is a paradigm shift in how society organizes information and knowledge. Closed systems everywhere are being transformed into open, flexible networks in which top-down government is being replaced by decentralized, collaborative, multi-stakeholder governance. This is true for healthcare no less than for other areas of society. Major changes can already be seen in the merging of the primary and secondary healthcare markets and the new role of patients as partners in all aspects of healthcare from research to prevention, diagnosis, and therapy. The digital transformation is far from over. The e-health tree is still young and growing. There is every reason to believe that digital health networks involving not only the traditional healthcare providers, but also patients and citizens as well as responsible and innovative regulators will produce more and unexpected fruits in the future. Many new and "disruptive" technologies are promising even more radical changes in how health-related information and knowledge is managed. Leading thinkers are already speaking of a "blockchain revolution" that will link patients and health service providers into secure, immediate, and trusted networks of communication and cooperation (Tapscott and Tapscott 2016). Wearables are more and more becoming "implantables" that are directly integrated into the body allowing continuous steams of data to be fed into machine learning in order to create smart, personalized health services available everywhere and at all times.[32] The Internet of Things, artificial intelligence and chatbots will transform the home into a data-driven health supporting environment that proactively "nudges" us

[31]https://www.weisse-liste.de/de/.

[32]See https://www.forbes.com/sites/bijankhosravi/2015/07/31/forget-about-the-apple-watch-implantables-are-coming/#463824aa3b78.

toward a healthier life-style.[33] The future of healthcare, just as in other areas of society, is a future in which humans are so deeply and symbiotically integrated into socio-technical ensembles that health itself becomes a network attribute, or even a network effect. The challenge is to construct not only efficient networks, but also "healthy" networks in which many different kinds of health-related information and knowledge are merged and managed, such that digitally enabled and knowledge-based healthcare becomes an integral part of our lives.

References

Barry, M. J., & Edgman-Levitan, S. (2012). Shared decision making—The pinnacle of patient-centered care. *The New England Journal of Medicine 2012, 366*, 780–781.

Belliger, A., & Krieger, D. J. (2014). Gesundheit 2.0: Das ePatienten-Handbuch, transcript Verlag.

Belliger, A., & Krieger, D. J. (2016). Organizing Networks: An Actor-Network Theory of Organizations, transcript Verlag.

Castells, M. (2005). The network society: From knowledge to policy. In M. Castells, & G. Cardoso (eds.), *The network society from knowledge to policy.* Washington, DC: Center for Transatlantic Relations.

Jenkins, H., Purushotma, R., Weigel, M., Clinton, K., & Robison, A. J. (2009). Confronting the challenges of participatory culture: Media education for the 21st century. USA: Mit Press.

North, K. (2016). *Die wissenstreppe* (pp. 33–65). Wissensorientierte Unternehmensführung: Springer.

North K., & Kumta, G. (2014). Knowledge management: Value creation through organizational learning, Springer Science & Business Media.

Shirky, C. (2008). Here comes everybody: The power of organizing without organizations, Penguin.

Tapscott, D., & Tapscott, A. (2016). Blockchain revolution: How the technology behind Bitcoin is changing money, business, and the world, Penguin.

Weinberger, D. (2011). *Too big to know*. Nova Iorque: Basic Books.

Author Biographies

Andréa Belliger Prof. Dr., is Prorector of the University for Teacher's Training of Central Switzerland and Director at the Institute for Communication & Leadership in Lucerne, Switzerland. Her areas of specialization are network society, new media, e learning, knowledge management, social media and e-health, and digital transformation.

David J. Krieger Prof. Dr., is Director at the Institute for Communication & Leadership in Lucerne, Switzerland. His areas of specialization are systems theory, actor-network theory, new media, organization theory, and semiotics.

---

[33]See for example http://medicalfuturist.com/chatbots-health-assistants/; https://www.digitaltrends.com/cool-tech/artificial-intelligence-chatbots-are-revolutionizing-healthcare/.

Piloting Digitally Enabled Knowledge Management to Improve Health Programs in Rural Bangladesh

Piers J. W. Bocock, Tara M. Sullivan, Rebecca Arnold and Rupali J. Limaye

Abstract Until recently, digitally enabled Knowledge Management (KM) activities in developing countries have more often than not been dismissed as unrealistic given challenges with access to electricity and the internet. However, a number of recent examples of holistic KM activities, including digital elements, have demonstrated a measurable contribution to improved outcomes for some of the world's poorest people. This chapter focuses on such a case, looking at how a digitally enabled KM program was designed, piloted, and measured in two districts in Bangladesh. The program aimed to help rural community-based health workers be more informed about, and helpful in, providing health and nutrition guidance to some of the world's poorest people.

1 Introduction

Knowledge Management (KM) as a discipline has increasingly been accepted as an important approach worth including in international development activities, such as donors, foundations, non-governmental organizations (NGOs) and partners on the front lines in developing countries. However, KM activities that include digitally enabled approaches have more often than not been dismissed as unrealistic given challenges with access to electricity and the internet.

The United States Agency for International Development (USAID)[1] is the U.S. Government's main international development funder, with a mission to "partner to end extreme poverty and promote resilient, democratic societies while advancing our security and prosperity." Supporting foreign assistance efforts with thousands of

[1]http://www.usaid.gov.

P. J. W. Bocock (✉) · T. M. Sullivan · R. J. Limaye
Johns Hopkins Bloomberg School of Public Health, Baltimore, USA
e-mail: pbocock@learning4dev.org

T. M. Sullivan · R. Arnold
Johns Hopkins Center for Communication Programs (CCP), Baltimore, USA

K. North et al. (eds.), *Knowledge Management in Digital Change*, Progress in IS,
https://doi.org/10.1007/978-3-319-73546-7_20

organizations around the world—local, regional, national, and international—in poor and less-developed countries presents knowledge management challenges and opportunities. Over the past decade, USAID has funded a number of KM efforts with more regularity, across the Agency and in specific technical sectors. One of the best known is the Knowledge for Health (K4Health) project, which is the latest iteration of a KM effort that has been led by the Johns Hopkins Bloomberg School of Public Health's Center for Communication Programs (CCP) since 1978.

In 2011, USAID asked CCP to explore, with the Government of Bangladesh (GOB), ways in which an intentional KM strategy might improve collaboration and knowledge sharing across several government agencies, as well as the flow of information to frontline health workers, with the ultimate goal of improving health outcomes in rural communities. The resulting effort leveraged the "Knowledge Management Road Map" to create the Bangladesh Knowledge Management Initiative (BKMI).

2 Applying a Systematic Approach to Knowledge Management

One of the ever-present challenges of KM practitioners in the international development sector is making the case for investing the time, effort, and resources for intentional knowledge management as a key part of any intervention.

To help support this process, K4Health has been working with partners in the health sector, the international development community, as well as tapping the latest thinking and approaches to KM, to develop a systematic but flexible framework to help develop, pilot, and refine practical approaches to KM projects. The result is what is called the Knowledge Management Road Map, described below. (The Knowledge Management Road Map also notes the importance of social aspects of KM, including social systems, social capital, social networks, social software, and social benefit outcomes. Social aspects of KM are particularly important for facilitating knowledge exchange and for sharing tacit knowledge.)

The Knowledge Management Road Map (Fig. 1) outlines five steps that can help a user implement and systematize KM activity: (1) Assess needs; (2) Design strategy; (3) Create and iterate; (4) Mobilize and monitor; and (5) Evaluate and evolve. These five steps are detailed below.

Step 1. Assess Needs

All good KM initiatives start with getting an overall understanding of context and the health problems in a particular context and how KM tools and techniques might help. An effective assessment identifies needs, gaps, networks, stakeholders and resources based on a preliminary gathering of data and information and understanding of health and knowledge-related issues. Once existing and new data are analyzed, synthesize the key issues arising from the results, and make a set of recommendations for the appropriate KM solutions based on the findings.

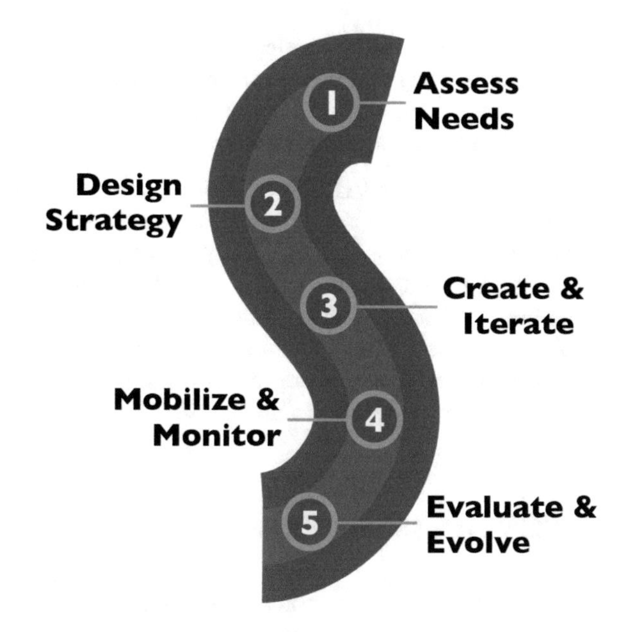

Fig. 1 The knowledge management road map for global health programs

Step 2. Design Strategy

An effective assessment is the foundation for identifying appropriate KM approaches. The assessment should ask what success could look like and what might be feasible depending on available resources, lessons from similar associated efforts and stakeholder input. Grounding KM interventions in a relevant theoretical framework, such as the stages of change, ideation, diffusion of innovations, or Bloom's taxonomy of learning, can improve likelihood of success. Once objectives and audiences are clearly defined, appropriate KM approaches can be designed.

Step 3. Create and Iterate

Creating, testing and iterating on the KM intervention will usually involve an interdisciplinary team, including subject matter experts, writers/editors, trainers/facilitators, and IT staff to name a few. It will also rely on understanding monitoring indicators that will help assess whether the KM initiative is meeting its goals. Gathering and incorporating through processes that encourage collaboration, tight feedback loops, and adaptation with stakeholders and intended audiences improves the quality of the products and approaches produced. Once all feedback is gathered and incorporated, final KM products and/or approaches can be produced, understanding full well that they will likely need to include moments to pause and reflect and, when necessary, adapt again.

Step 4. Mobilize and Monitor

After making adjustments during the create and iterate stage, if the approach appears to be working, then it can be scaled up according to needs and budget. Throughout implementation, it is important to review progress toward objectives and to make mid-course adjustments as necessary. Project monitoring includes reviewing progress toward indicators in your performance monitoring plan that may be measured using a variety of methods (online surveys, Google analytics, focus groups, interviews, etc.) to improve the quality of KM approaches and products through timely and appropriate adaptation.

Step 5. Evaluate and Evolve
To understand the impact of your KM work it is important to assess if you have achieved your KM objectives. Ideally, you should use rigorous evaluation designs that have before/after measures, a control or comparison group, and use random sampling methods to select participants. Strong evaluation designs can be cost prohibitive, so use the strongest design given your resources. Once data has been collected and analyzed, share results with key stakeholders and promote use of results in policy and practice moving forward.

3 The Bangladesh Knowledge Management Initiative

Bangladesh, one of the world's most populous countries (Fig. 2), has made significant progress in its development over the past few decades. But it is still fraught with poverty, flooding, and disparities in access to health, education, and social services. Significant gaps remain in neonatal, child, and maternal health indicators

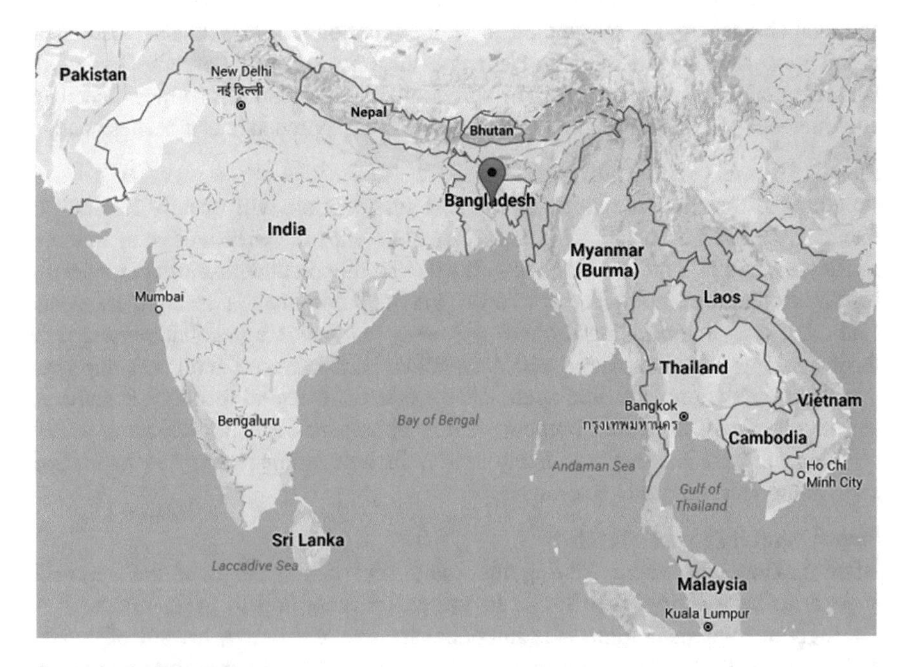

Fig. 2 Map of Southeast Asia

(Amin et al. 2001; Akhter 2004). Maternal mortality ratio is 194 deaths per 100,000 live births, and the under-5 mortality rate is 46 per 1000 live births (Chowdhury et al. 2013).

Malnutrition begins at birth for many infants, as nutritional deficiencies among pregnant women have substantially contributed to low birth weight rates in the country (Ahmed et al. 2012), and the majority of households (64%) lack a cleaning agent, such as soap.[2]

A key challenge to improving these indicators is that three-fourths of the total population of Bangladesh resides in rural areas. As a result, they experience difficulty accessing prompt, proper medical care when needed, as most trained medical personnel provide care in primarily urban areas (Mridha et al. 2009). This population relies on community health workers (known as fieldworkers) to provide accurate and often life-saving information to them in regular door-to-door visits and community-centered meetings. These fieldworkers, and what they know and can share, become a determining factor in the health of most of this country's population.

To address health challenges within the country, BKMI was designed to strengthen the capacity of the Government of Bangladesh, USAID implementing partners, and other stakeholders to develop and share effective and consistent social and behavior change communication (SBCC). To do so, coordination within the Ministry of Health and Family Welfare (MoHFW), and between the MoHFW and other stakeholders, was paramount. As such, BKMI sought to strengthen knowledge management around SBCC to ensure harmonization among programs and build the capacity of fieldworkers to deliver high-quality health information. The approach that BKMI took followed the Knowledge Management Road Map to pilot, test, and scale up a digital KM approach.

Step 1. KM Assessment in Collaboration with Government of Bangladesh
As described above, the first step in developing a knowledge management initiative is to conduct an assessment of the needs and the context. The KM assessment for BKMI was conducted with a broad range of stakeholders at both the national and community level. At the national level, where government policy and guidance related to health communication is developed, the assessment found there was very little coordination within the MoHFW around messaging that it wanted delivered by fieldworkers (whose job it was to communicate with local communities). Because of this lack of coordination, the assessment revealed frequent duplication of efforts, inconsistent or incorrect messaging, and—in some cases—no health messaging at all. What's more, the communication materials were often not aligned with current government policy.

[2]National Institute of Population Research and Training (NIPORT; 2014). *Mitra and Associates and ORC Macro.*

Step 2. Strategic Design
Working with MoHFW counterparts, the team identified an innovative, digital KM solution that would help address a number of identified challenges: consistency and accuracy of information; fieldworkers being up to date on the latest health issues; and the literal burden of carrying all of this information from door to door in massive binders. It would also strengthen community counseling, and connect to the efforts at the national level to harmonize health, family planning and nutrition messaging by government and non-government stakeholders.

The approach included an electronic toolkit (eToolkit) of counseling materials, and eight video- based eLearning courses. Both were provided to fieldworkers on "netbook" mini-computers.

The eToolkit (Fig. 3) is a digital library of select "gold standard" print and audiovisual SBCC materials in the Bangla language, presented in a simple graphic format, organized by topics and subtopics. It was designed for use as a counseling tool during home visits, courtyard meetings, and clinic-based counseling sessions. The eToolkit provides a full range of information on health, family planning and nutrition, which means that any fieldworker with access to the eToolkit can assist clients regardless of whether or not they have been trained on a particular topic.

The eLearning component was designed to supplement fieldworker training provided by the MoHFW. Fieldworkers usually received pre-service training,

Health, Population, Nutrition eToolkit for Field Workers

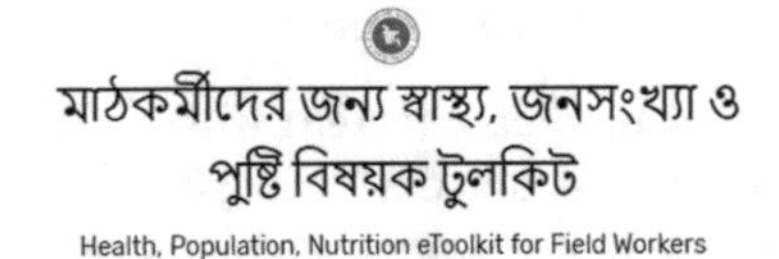

ইটুলকিট সম্পর্কিত তথ্য
Information about eToolkit

ইটুলকিট ব্যবহারবিধি
How to navigate eToolkit

স্বাস্থ্য
Health

পরিবার পরিকল্পনা
Family Planning

পুষ্টি
Nutrition

USAID

JOHNS HOPKINS Center for Communication Programs

BKMI

Fig. 3 eToolkit landing page

but very little in-service or refresher training. Eight courses were developed: two on family planning; two on maternal and child health; two on nutrition; one on interpersonal communication and counseling; and one on integrated messaging. The courses are video-based, to reflect fieldworkers' limited educational levels, and delivered in Bangla.

The netbooks—low-cost, rugged, and easily portable—housed both the eToolkit and the eLearning courses, designed in such a way as to be easily loadable onto the netbooks and require no internet connection; this reduced costs and improved accessibility (access to tools did not depend on connectivity), and also limited what fieldworkers (or their families) could do with the netbooks that was not related to their counseling work. (At the time, netbooks were the most appropriate technology; if the pilot were later, it's likely that tablets or mobile phones would have been chosen instead of netbooks).

Step 3. Create and Iterate

BKMI piloted the approach, with plans to iterate based on early results. One of the early steps was to establish the conditions and locations for the pilot (Fig. 4). Sylhet and Chittagong Districts were chosen because of their low performance on key health indicators. Only government fieldworkers were eligible to participate in the eHealth pilot. A total of 300 participants were selected by their supervisors. It was also important to include both types of fieldworkers (Health Assistants and Family Welfare Assistants), as integration of health, family planning and nutrition was an important objective of the eHealth pilot.

Fig. 4 Map of Bangladesh

Health Assistants were historically responsible for health, and Family Welfare Assistants were responsible for family planning. At the time, neither type of fieldworker was responsible for nutrition, though their job descriptions have since been updated.

In June 2012, three MoHFW Units issued a request for recent print and audiovisual SBCC materials from government and non-government organizations. Everything that was received was sorted, inventoried and tagged. In order to select the best ones to include in the eToolkit, all items were vetted on two levels. First, the materials were reviewed by national-level subject-matter experts from government and non-government agencies for accuracy, consistency with current government policies, and cultural acceptability. Items not approved during the first round of vetting were eliminated.

Secondly, fieldworkers themselves reviewed the items (Fig. 5). The most important criteria for the fieldworkers was whether or not they would use the particular item when counseling their clients. Some items were eliminated because they were too lengthy or text-heavy—even though fieldworkers mentioned that they would be good as reference materials for their own learning. The eToolkit was designed with the end users in mind: fieldworkers and their clients.

After vetting was complete, BKMI reviewed all of the items and placed them in categories and subcategories. Some items were removed because they were very similar to other items; it was important to include fewer items, so that fieldworkers could easily find what they need.

Prior to launching the eToolkit, BKMI tested its look and usability. The eToolkit was built on the standard K4Health toolkit platform; however, BKMI dramatically changed its format so that the interface would be graphic, colorful, and easy for a low-literate audience, and an audience that was not familiar with computers. Fieldworkers had a strong preference for colorful, realistic images rather than icons

Fig. 5 Fieldworker review

or illustrations. Lists of SBCC materials were accompanied by a thumbnail of each item, so fieldworkers could easily identify the items. No typing is required, and the search function was disabled. Rather, the design team limited the content in the eToolkit, organized it in a logical way, and pre-tested the organization and navigation with fieldworkers.

In addition to the eToolkit, eight video-based eLearning courses were loaded onto the netbooks. Each was around 12–15 minutes long, and included an assessment at the end. The two nutrition courses were adapted from Alive & Thrive training videos on breastfeeding and complementary feeding that had already been developed in Bangladesh. The six new courses were based on global best practices and current MoHFW policy. Outlines and scripts were reviewed by relevant USAID implementing partners and MoHFW colleagues, and were approved by USAID before beginning production.

MoHFW representatives were present during some of the filming of the eLearning videos, which helped ensure accuracy, relevance and appropriateness. For example, one MoHFW official noticed during filming that an actor was not dressed like a fieldworker; filming stopped while an appropriate apron was located and delivered to the filming location.

Prior to beginning the eHealth pilot, launches were held in the two districts to orient local government officials to the project and the pilot's objectives. Obtaining their approval and support were crucial for the smooth roll-out of the pilot. Following the launches, all participating fieldworkers received a two-day orientation in batches of 25. The orientation included hands-on orientation to the netbook, the eToolkit and the eLearning courses (Fig. 6).

Fig. 6 Fieldworker orientation

A minor implementation challenge was that fieldworkers' supervisors did not receive netbooks, nor an orientation to the eHealth pilot. While the fieldworkers were the primary actors in the eHealth pilot, supervisors have an important role to play in ensuring quality, accountability and performance. At the same time, it was clear that the netbooks were an attractive incentive, and that many other stakeholders wished to participate in the pilot because it gave them access to new technology.

Step 4. Monitor and Mobilize

To ensure proper implementation, eight Monitoring and Troubleshooting Officers (MTOs) circulated among the fieldworkers every two weeks to help them with technology-related issues, such as netbooks freezing, eToolkit icons deleted, or speakers not working properly. When netbooks became inoperable, MTOs provided replacement netbooks while the others were sent for repair. This minimized the amount of time that fieldworkers were without netbooks during the short piloting period. MTOs were also tasked to manually issue and collect eLearning course assessments to and from the fieldworkers to monitor their learning progress. The BKMI team, including colleagues from MoHFW, conducted three monitoring visits in Sylhet and Chittagong districts during June, July, and August 2013.

Monthly monitoring visits captured process indicators for the pilot. These visits also aimed to collect, via focus groups and interviews, some qualitative information from the fieldworkers, as well as from mothers who have at least one child under two years of age. Regular feedback from the field helped to ensure the smooth implementation of the pilot, and allowed the team to learn from challenges and successes. During the monitoring visits a number of focus group discussions took place with the fieldworkers, along with case studies and key informant interviews to collect their feedback and capture individual stories.

These focus groups and interviews revealed several important points: fieldworkers enjoyed using the netbooks and found them easy to use; the netbooks elevated the fieldworkers' status in their communities and led to the empowerment of fieldworkers; and both fieldworkers and mothers in the community like videos as a communication channel because of their entertainment and education value (Fig. 7).

Fig. 7 Fieldworker shares the eToolkit with clients

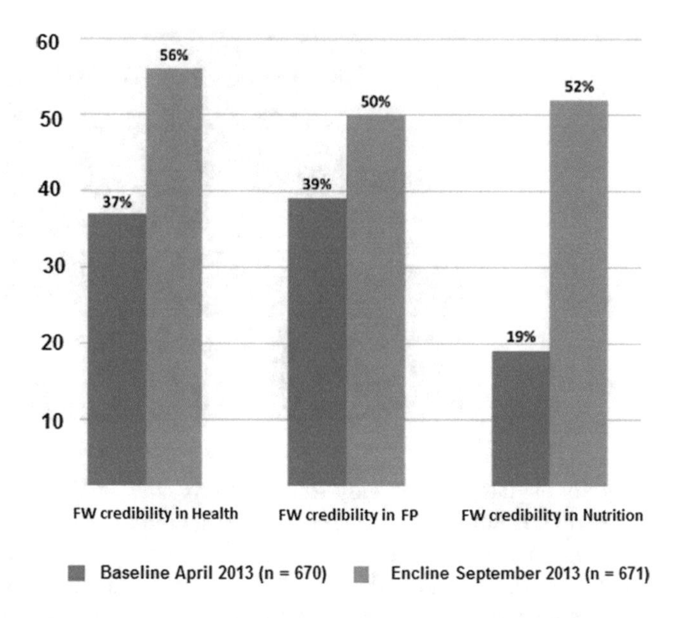

Fig. 8 Fieldworkers as trusted sources of information (pre-/post netbooks)

But even more importantly, perhaps, was the increase in perception by community members of fieldworkers as trusted sources of information (Fig. 8). Though the eLearning courses were designed for the fieldworkers' own use, fieldworkers also used them as a counseling tool with clients; fieldworkers became even more comfortable using the netbooks and the digital resources as time went on. Fieldworkers requested more videos on immunization, childhood illnesses, adolescent health, non-communicable diseases, side effects of family planning methods, food safety issues, healthy cooking and healthy eating.

Step 5. Evaluate and Evolve

An evaluation was conducted to assess the impact of the pilot on fieldworker behaviors and on client behaviors related to maternal, newborn, and child health (manuscript under review). A pre-post study measured fieldworkers' knowledge and behavior before and after the pilot. Comparing mean knowledge scores after and before the pilot, the difference in mean scores was significant ($p < 0.05$) related to knowledge of benefits of birth spacing, benefits of a small family, anemia prevention, recommended number of antenatal care visits, pregnancy and labor danger signs, proper attachment for breastfeeding, signs of adequate breast milk supply, newborn danger signs, preterm infant care, diarrhea prevention, and malnutrition symptoms. Post-intervention, fieldworkers were significantly more likely to counsel couples on all available contraceptive options and birth spacing benefits (Fig. 9).

Fig. 9 Fieldworker home visit

Related to family planning, a post-only study (Limaye et al. 2017) examined the effects of the intervention among fieldworkers' clients at different levels of exposure to the intervention: mothers with a child under the age of two who reported receiving a home visit from a fieldworker who had a netbook with the package (low exposure); mothers who reported receiving a home visit from a fieldworker who had a netbook with the package and were shown a digital resource (high exposure); and mothers who reported no home visit from a fieldworker who had a netbook (no exposure).

Both high and low levels of exposure were associated with higher odds of mothers reporting that the fieldworker discussed contraceptive choice, contraceptive side effects, and contraceptive side effects management compared to unexposed mothers. Mothers in the high exposure group had higher odds of contraceptive use compared to unexposed mothers. Results from both of these studies suggest that the knowledge a fieldworker gains through a digital health training package can be diffused to clients, positively affecting client knowledge and behaviors. Such interventions can empower fieldworkers by providing them relevant information at the point of care, which can enhance their credibility among the communities they serve.

Building on the success of the pilot, BKMI worked with MoHFW to scale up the eToolkit for fieldworkers in Bangladesh. The eToolkit is updated annually by a subgroup of the Bangladesh BCC Working Group (which includes representatives from both government and non-government organizations), with new content being added and outdated or redundant items being removed. The eToolkit is available online (on a website that is hosted by the MoHFW), offline via flash drive, and as an app for Android-based mobile devices. Nearly 24,000 Health Assistants have received Android-based tablets from the Directorate General of Health Services of the Ministry of Health and Family Welfare, and the project is advocating that the Directorate General of Family Planning also purchase tablets for its nearly 24,000 Family Welfare Assistants. The eLearning courses were updated in 2015 to reflect the latest GoB policies, and the courses are now available online, free of charge.

4 Conclusions

Low access to internet or electricity need not be a barrier to digitally enabled knowledge management efforts, provided the initiative is designed with the specific context in mind. Even without real-time access to the Web, vast amounts of content can be repacked in easily portable and sharable formats on digital tools that make a real improvement in the ability to communicate life-saving messages and lead to important changes in knowledge and behavior.

By ensuring that user needs are understood, content is appropriate, and the digital approach is designed and evolves in an iterative way, knowledge management tools can help improve access to information—and therefore improve conditions—for even the world's poorest people. And by regularly monitoring and learning from what the data is indicating, KM practitioners can learn about unintended but beneficial consequences, such as fieldworkers using content designed to improve their own education to help inform and communicate with their clients.

Perhaps the most important outcome of this pilot effort was the improvement in confidence of the fieldworkers themselves, as easy access to knowledge, tools, and content helped them feel empowered to make a difference in the lives of the rural communities they serve; as well as the increase in perception by those in the communities served by the fieldworkers that they were a trusted source of information and knowledge. This effort has proven to be a tangible example of how a clear KM framework can be used to develop a digitally enabled knowledge management initiative that lead to direct improvement of lives, in one of the poorest areas of the world.

There were important lessons learned by the BKMI team as well:

- Fieldworkers quickly learned how to use the netbooks, eToolkit and eLearning courses, even though most of them had not had much experience using computers prior to the eHealth pilot.
- Monitoring and Troubleshooting Officers were essential to the smooth implementation of the eHealth pilot. They ensured that fieldworkers were able to use the netbooks throughout the full pilot period, as they were able to quickly respond to any technological difficulties.
- Government cooperation and ownership of the pilot was essential, from the grassroots to the national levels. Most government colleagues were excited to be involved in such an innovative project, and appreciated that technological solutions were being offered.
- Future efforts to scale up the eHealth pilot should consider the importance of selecting the appropriate technology/device that include ease of use for health workers and appropriate features such as memory size. Providing upfront training, ongoing technical support, and routine updates to content and technology are also important for scale-up.
- To make scale-up of digital solutions truly sustainable, it will be important to support and strengthen capacity to make strategic investments in mobile devices; to define parameters for what government-issued devices will and will not

be used for; to coordinate the diverse stakeholders who have an interest in different uses of the tablets (such as data collection, counselling, job aids, surveillance, and real-time monitoring); to streamline and consolidate software whenever possible; and to design a comprehensive program for mobile devices and other technology-based solutions that includes plans for training, technical troubleshooting, supportive supervision, monitoring, maintenance and regular updating.

References

Ahmed, T., Mahfuz, M., Ireen, S., Ahmed, A. S., Rahman, S., & Islam, M. M. (2012). Nutrition of children and women in Bangladesh: Trends and directions for the future. *Journal of Health, Population, and Nutrition, 30*(1), 1.

Akhter, F. (2004). Decentralization and integration of health and family planning services in Bangladesh. *Development, 47*(2), 140–144.

Amin, R., Pierre, M. S., Ahmed, A., & Haq, R. (2001). Integration of an essential services package (ESP) in child and reproductive health and family planning with a micro-credit program for poor women: Experience from a pilot project in rural Bangladesh. *World Development, 29*(9), 1611–1621.

Chowdhury, A. M. R., Bhuiya, A., Chowdhury, M. E., Rasheed, S., Hussain, Z., & Chen, L. C. (2013). The Bangladesh paradox: Exceptional health achievement despite economic poverty. *The Lancet, 382*(9906), 1734–1745.

Limaye, R. J., Kapadia-Kundu, N., Arnold, R., Gergen, J., & Sullivan T. M. (2017). Utilizing digital health applications as a means to diffuse knowledge to improve family planning outcomes in Bangladesh. *Clinical Obstetrics, Gynecology, and Reproductive Medicine, 3*(2), 1–7.

Mridha, M. K., Anwar, I., & Koblinsky, M. (2009). Public-sector maternal health programmes and services for rural Bangladesh. *Journal of Health, Population, and Nutrition, 27*(2), 124.

Author Biographies

Piers J. W. Bocock, MBA has worked at the nexus of knowledge management, communications, information technology and organizational development for more than two decades, with a specific focus in international development. Currently leads USAID's Knowledge Management and Learning (LEARN) contract. He is an Associate on the faculty of the Johns Hopkins Bloomberg School of Public Health, where he teaches on knowledge management and learning.

Tara M. Sullivan, Ph.D., MPH has nearly 20 years of experience working in international health and development with a focus on program evaluation, knowledge management, family planning/ reproductive health, HIV & AIDS, and quality of care. She is the Director of the Knowledge for Health Project and is the Knowledge Management Director, Johns Hopkins Center for Communication Programs. She is also a faculty member in the Department of Health, Behavior, and Society, Johns Hopkins Bloomberg School of Public Health, where she teaches about knowledge management and learning.

Rebecca Arnold, MPH has 18 years of diverse social and behaviour change communication and knowledge management experience in a global setting, from grassroots to global levels. She works as a Senior Program Officer at Johns Hopkins Center for Communication Programs (CCP). Currently she is developing a dynamic learning and exchange platform to support The Challenge Initiative. Previously she was the Project Director and Technical Advisor for the Bangladesh Knowledge Management Initiative, based in Dhaka, Bangladesh.

Rupali J. Limaye, Ph.D., MPH, MA is a social and behavioral scientist, and has conducted research, managed programs, and taught classes related to public health and behaviour change for the last 15 years. She currently serves as a full-time faculty member at the Johns Hopkins Bloomberg School of Public Health within the departments of International Health and Health, Behavior & Society. She concurrently serves as a Research Director at the Johns Hopkins Center for Communication Programs.

Ubiquity and Industry 4.0

Fabricio Foresti and Gregorio Varvakis

Abstract The 4.0 industry is a new productive paradigm based on digitalization. The phenomenon is based, among other factors, in so-called cyber-physical systems—that allow absolute control of what takes place inside the factory, and even outside it, allowing full awareness of the entire process in the production chain. This awareness can be understood as ubiquity, that is, virtual presence in many places simultaneously. Thus, extensive bibliographic research—carried among articles published in the last five years—reveals that the new emerging business models with 4.0 Industry are essentially based on the ubiquity of information, products, and consumers. Therefore, ubiquity expresses new models of relationships with customers and suppliers, as well as innovative ways of producing and managing organizations.

Keywords Industry 4.0 · Ubiquity · Cyber-physical systems · Information
Knowledge

1 Introduction

Industry 4.0 (I4.0) is a new business paradigm that comes from digitization or virtualization, which tends to transform greatly the way goods and services are produced. The organizations, in turn, needs to precisely understand this new paradigm to survive in the digital economy. It is about the transformation that has been occurring and impacting the society and the economy since the beginning of 21st century, by means of the information and communication technologies (ICT), with significant impacts in both production of goods and provision of service (Blanchet et al. 2014, p. 7). This new industrial revolution will be characterized by "automation and total digitization of the processes" and intense use of ICT in the production of goods and provision of service (Roblek et al. 2016, p. 1). According to the authors, companies must find a way to intelligently use the connection of

F. Foresti · G. Varvakis (✉)
Universidade Federal de Santa Catarina, Florianopolis, Brazil
e-mail: g.varvakis@ufsc.br

K. North et al. (eds.), *Knowledge Management in Digital Change*, Progress in IS,
https://doi.org/10.1007/978-3-319-73546-7_21

consumers of goods and services (products) to meet opinions, social, demographic and even psychological information that influence the consumption of specific goods or services (products).

I4.0 emerges in a particular historical context. The so-called "post-industrial society [...] find its metaphor [...] in the net and in the virtual", according to De Masi et al. (2005, p. 135). One could say that the greatest metaphor of post-industrial society is I4.0 itself; an era focused on mass producing intangible goods (De Masi et al. 2005, p. 342). The term I4.0 is used to define the next industrial revolution (Hermann et al. 2016, p. 3929) and it was used for the first time in 2011 at the Hannover Messe, thus being associated with the German industry (Drath and Horch 2014, p. 56). Other nations use distinct terms, such as China, European Union, and the United States[1] (Kagermann et al. 2013, p. 67). We highlight the leading role of Germany, where competitiveness is due, among other factors, to the heavy use of ICT for many decades, so that today more than 90% of the German industry is backed heavily by ICT (Kagermann et al. 2013, p. 13).

The first industrial revolution happened in the 18th Century, a point in which manual production has been replaced by the use of steam engines, which allowed the mass production of material goods. The second industrial revolution occurred in the 20th Century, by means of electricity and production lines. Finally, the third industrial revolution happened in the late 20th Century, with the dawn of computing (Wahl 2015, p. 241). Those three revolutions that preceded I4.0 represent a "disruptive leaps in industrial processes resulting in significantly higher productivity", according to Blanchet et al. (2014, p. 7). Now, this economic trend linked to digitization promises many advances, just as the previous revolutions did. The term *revolution* itself is somewhat promising and a generator of high expectations. Among the term's promises are "gains in efficiency" and the emergence of "new business models", according to Roblek et al. (2016, p. 3).

It is worth highlighting that the first three revolutions took place over a period of only 200 years. The second revolution occurred 100 years ago, and the third one was in 1969, when the first program of automation systems was presented, a paradigm that continues to influence the present day (Drath and Horch 2014, p. 56). According to Drath and Horch (2014, pp. 56–57), "It is remarkable that Industry 4.0 announces an industrial revolution a priori." For the authors, it is still premature to determine if, in fact, this is a revolution, but they recognize that there is great potential for such, as most of the technology already exists, even if it is used for other purposes. However, according to the authors, the I4.0 is in the future.

The phenomenon of I4.0 is, to an extent, a consequence of the Information and Knowledge Society, and is closely related to the processing and use of information. I4.0 is the merge of the production with the ICT (Hermann et al. 2016, p. 3928). For

[1]According to Kagermann et al. (2013, p. 67): "'Smart Production', 'Smart Manufacturing' or 'Smart Factory' are used in Europe, China and the US to refer specifically to digital networking of production to create smart manufacturing systems, whereas the equally fashionable term "Advanced Manufacturing" embraces a broader spectrum of modernization trends in the manufacturing environment."

Batista et al. (2017, p. 16), the term is associated with a new evolution stage: "organization and management of the entire value chain in manufacturing industry process involved, sensor and actuator infrastructures." The term is associated with three interrelated factors: (a) the digitization of economic relations (from the simplest) supported by the net; (b) digitization of both products and services; (c) new business models. The activities will be linked in real-time to the communication complex through the net of things, people, and services, to the most promising technologies of I4.0 (Zezulka et al. 2016, p. 8). For (Blanchet et al. 2014, p. 7), I4.0 represents the "digitization and linking of all productive units in an economy." It is characterized, among other factors, by the advent of the cyber-physical systems and by a more advanced level of connection (Blanchet et al. 2014, pp. 8–9).

The I4.0 will require the establishment of a new types of relationship with the customer, based on the ubiquity of users and organizations. The ubiquity, understood as the permanent connection to the net, beyond the ubiquitous presence of technologies and sensors. Not only does it open space for new models of business, but also for a differentiated approach of the consumer, new forms of attendance, and interaction. The consumers exert the ubiquity through their personal Portable Devices (PD), i.e. smartphone. The smartphones tend to become a true virtual wallet, and they are the central infrastructure of the mobile money (Donovan 2012, p. 67). At the same time, they are also central for I4.0. The organizations, in turn, can exercise the ubiquity through the internet, uninterrupted services, full supervising (monitoring) of their customers/users via smart devices, embedded systems, and intelligent use of information. The ubiquity points out a trendy global connections between information systems, in addition to the cooperation between organizations and political state, regarding the sharing of information.

The mobile triad, made up of the PD, WiFi technology, and the Internet, stands out in the context of the I4.0. The PD performs the vital link of communication between humans, machines, and sensors. The "wireless communication technologies play a prominent role in the increasing interaction as they allow for ubiquitous internet access", according to (Hermann et al. 2016, p. 3932). The locative medias are also of great value and they tend to increasingly occupy space in products, services, and several environments. It was even through locative media and technologies such as RFID[2] tags that one of the fundamentals of I0.4, internet of things (or IoT), has emerged (Cooper and James 2009). Authors state that I4.0 is based on mobile computing, in cloud computing, and in big data (Roblek et al. 2016, p. 6). In other words, I4.0 is based on technology and networked information.

But the protagonist of I4.0 is the internet. According to Drath and Horch (2014, p. 57), "The major technical background of Industry 4.0 is the introduction of internet technologies into industry." The net is growing and getting specialized, and it can be subdivided into many levels, all connected to I4.0: the "Internet of Things" (IoT), the "Internet of Services" (IoS), the "Internet of People" (IoP), and the "Internet of Energy" (IoE), as point Lom et al. (2016, pp. 3–4). According to the

---

[2]Radio-frequency identication.

authors, IoT concerns electronic objects that collect and exchange data between themselves through the net, and it involves electronic devices, sensors, programs, connection. The IoS are devices connected to the net and that "expecting having set of smart services". The IoP is established by the merge of the IoT and IoS, and it originates the cyber-physical systems, with applications focused on the users, being thus a basic element when allowing the orientation of all the services offered to the users. It is the IoP that establishes the link between "environment and people, between the real world and cybernetic world, between reality and between virtual realities". Finally, the IoE is linked to the management of energy in the internet, via smart systems whose objective is the energy efficiency, with production under demand.

The IoT is one of the central components of the I4.0. According to Hermann et al. (2016, p. 3929), I4.0 has enabled the communication between "people, machines, and resources", characterized "by a paradigm shift from centrally controlled to decentralized production processes." The internet of things refers to objects that communicate via the internet (Cooper and James 2009). According to Lom et al. (2016, p. 1) the fourth industrial revolution is marked "by linking sub-components of the production process via Internet of Things". It is the IoT that will allow real-time communication between services (allocated in the clouds) and consumers (Batista et al. 2017, p. 17), i.e. the effectiveness of cyber-physical systems and the ubiquity of technology.

Authors point out some I4.0 principles, such as interoperability, virtualization, decentralization, real-time (or ubiquity), targeted services and modularity (Schlick et al.) (Lom et al. 2016). These are principles that synthesize quite well the promising scenario of this new production paradigm. Other authors indicate some requirements for the acceptance of the new paradigm by the industry, such as protection of the investments (in the already existing plans of the organizations), stability (to innovate without compromising the production), privacy of the data (to protect the organizational knowledge), and cybersecurity (to avoid not authorized accesses to prevent damages) (Drath and Horch 2014, p. 58).

As presented so far, it seems that with the advent of I4.0 opportunities and challenges emerge in two distinct senses. Firstly, new business models, in which services are a significant share in the added value of the products, will be needed to meet the demand of the fourth industrial revolution. Secondly, market opportunities arise related to the paradigm of ubiquity, the constant connection and consumer technologies. These will require adaptation from organizations and will involve new forms of relationship with the customer, for the survival in the new digital economy. Thus, it is a challenge to identify characteristics and elucidate management alternatives of the new business models based on ubiquity, which are emerging with the I4.0.

2 Ubiquity and I4.0

The essence of the Industry 4.0 is in the ubiquity and the exploration of information from the constant connection status of people, artifacts, and organizations. It is through ubiquity that cyber-physical systems (CPS) are established, expressing the need for uninterrupted communication, real-time, continuous flows and constant monitoring. It is through ubiquity that we establish cybersystems. The basis of the I4.0 is are CPS (Kagermann et al. 2013, p. 13), that is, the connection between the real universe and the virtual one (Roblek et al. 2016, p. 3). According to (Prause 2015, p. 160), I4.0 goes far beyond CPS and the "dynamic production networks". However, the cyber-physical systems or "the fusion of the virtual and the real world" stand out as a great promise. Authors show that CPS thematics, data, information and time, are among the most common when it comes to academic research related to I4.0. Among the clusters, that related to the information supply and real-time stands out (Hermann et al. 2016, p. 3930). The ubiquity is nothing but the interaction in real-time of people, objects and products (good and/or services) spatially distant, and real-time is one of the bases of I4.0 (Schlick et al.) (Lom et al. 2016).[3]

However, what it is ubiquity? According to the Merriam-Webster Dictionary (2017), ubiquity might mean "presence everywhere or in many places especially simultaneously", that is, widely spread. Ubiquity also might be understood as pervasiveness, telepresence, omnipresence, real-time, uninterrupted connection. Leite (2008, p. 106) distinguishes that the ubiquity in computing represents the power of numerous systems sharing the "same information".

Nowadays, in times of network and technological mobility, ubiquity shall be understood as the virtual presence (access) in several places at once, without constraints of time and space, and it is exercised through the network and other communication devices. The ubiquity's vectors are the ICT (Renau 2017, p. 808), and it is through PD that the users exert the ubiquity (SOURCES; GOMES, 2013, p. 72). That view comes from the highly technological context of the present day, and expresses "to be everywhere at the same time", or to be "omnipresent", according to Dias (2010, p. 56). In practice, it was the mobile phone that established the ubiquity (De Masi 2000, p. 198). It is worth noting that, currently, the condition of uninterrupted connectivity became extremely popular, with implications for the whole society (MADDOX;) (Mantovani and Moura 2012, p. 56). De Masi et al. (2000) p. 267 recognizes that the users have great "satisfaction" by the possibility of exerting the "ubiquity". Thus, the social interactions also become ubiquitous (Mantovani and Moura 2012, p. 68), either between users or between customers and organizations.

Consumers, suppliers, and organizations can exercise ubiquity trough the network, the technology, and the constant connection. The ubiquity exercised by organizations is a way to answer the ubiquitous client, besides to precisely control the production sites. Consumers and producers walk together in relation to the ubiquity. After all, the information and knowledge society constitutes a "ub-informational" true, according to (Godoy et al. 2015), where this context includes organizations. Although the I4.0 organizations are not ubiquitous in fact (present in every people, place, and thing, as the information is), they tend to be ubiquitous in the sense to exist in all or almost all places (virtually, and not

[3]Cyber-Physical System.

physically, in an informational level). The new paradigm expresses a ubiquitous organization (in which must be in many places at the same time and be able to adapt to the most diverse means), and having an "ubi" (or a place in the universe or global information system). However, all the organizations have a condition and particular situation (ubiquity) besides an ubi-quification (reason for being in a specific place). Exploring the organizations, the production of goods and services under the bias of ubiquity contributes to being better placed in the I4.0 context.

By understanding the meaning of the term ubiquity and the paradigm of I4.0, it is possible to explore the ubiquity in the context of organizations, in the offering of goods and services. We must ask ourselves: what is the reason and conditions of goods and services in the I4.0 context? What is changing is the proper production of goods and services time, which happens to be the immediate time or the so-called real-time; and the ubiquity transforms the time itself. With the advent of the internet, time became multiple, and, in fact, time itself might be changing (Elias 2010, p. 58). In the post-modernity, time is compressed to such point that it seems that there is only the present time (Harvey 2010, p. 219) or the real-time until it finally arrives at time zero.

Which are the impacts of that time zero? (Lévy and da Costa 1993, p. 114) asks "which would be the type of time secreted by computerization?" The author glimpses the phenomenon of the ubiquity and the proper I4.0 when affirming that the computation "[…] serves to the permanent mobilization of the men and the things that perhaps have started with the industrial revolution". This uninterrupted mobilization appears to be coming to zero time with the I4.0 and the CPS, and with the ubiquity of products, mainly through the services that are part of them, and with the consumers.

The writing was the echo […] of the socio-technical invention of the bounded time and the stock. Computing […] is part of the reabsorption work of a social, viscous space-time, of strong inertia, for the benefit of a permanent reorganization and in real-time of the social-technical arrangements: flexibility, tensioned flow, zero stock, zero deadlines (Lévy and da Costa 1993, p. 114).

The databases were the first computing instruments to promote ubiquity. Lévy and da Costa (1993, p. 115) points out the databases as true "mirrors" of reality, extremely *faithful* on the *current state of a specialty or a market*. According to the author, *the notion of real-time* was created by computing and it summarizes the *computing spirit,* which is the *condensation in the present time of the operation in progress*, a kind of *operational* knowledge, *chronological implosion*, *punctual time*, in counterpoint to the *circular time of the primary orality*, and to the *linear time of historical societies, a new pace*, which is no longer of the *history*, but of the *speed.* In fact, a new social time related to ubiquity has been established, which consequently accelerates the production cycle, and goods and services consumption (Lévy and da Costa 1993, pp. 117–118). Operationally, more than the maintenance of a condition, or far beyond the truth, what matters is the "speed and relevance of the implementation, and the speed and relevance of the operational modifications", according to Lévy and da Costa (1993, pp. 119–120).

In the I4.0 context, the concepts by Weiser of "ubiquitous computing" (1991) gain central importance, when naturally, simply and properly integrating the physical and the virtual world to users. This concept, coined in the early 1990s, argues that technologies tend to become "invisible" and fully adapted to the diverse human environments, and its essence resides to help users deal with the information overload. Although the author indicates a huge difference between ubiquitous computing and virtual reality, it is a concept of great value to the I4.0.

Through the information generated by technology, and by the constant connection of consumers and organizations, a new model of relationship based on ubiquity is established. Firstly, because of a cyber-physical system which aims to merge the real world with the virtual, this relationship can only be successful through the exercise of ubiquity. The continuous exchange of information between many systems (public and private) is going to enable a full and global cyber-physical system. Roblek et al. (2016, pp. 1–2) highlight the need to a "perpetual communication" among the several devices, a creation of an uninterrupted exchange of information channels in "real-time", organization reaction in "real-time" to the consumer in order to serve and to influence them. According to the authors, the companies must find ways to intelligently use the connection of consumers to know their opinion, besides taking advantages of the use of social, demographic and even psychological information that influence the consumption of a given product: connectivity is a central component of this new revolution. This feature is essential in the I4.0, and it is directly associated with the characteristic of simultaneity of the services, that is, a product without a service is only a good, it is not ubiquitous.

Ubiquity transforms both consumers and organizations. Now it is "anytime, anywhere, and through any medium", sustain Roblek et al. (2016, p. 6). That is the expression that best emphasizes the new paradigm of I4.0. The paradigm that presents opportunities to the organizations and researchers to influence the future and establish new business models, new products that are pure services or products, where services that demanded ubiquity have goods associated with them (Hermann et al. 2016, p. 3928). However, "[…] as the term itself is unclear […]", according to Hermann et al. (2016, p. 3928): "[…] companies are facing difficulties when it comes to identifying and implementing Industry 4.0 scenarios". The following section is an attempt to characterize the services in the context of I4.0 and to point a management alternative.

3 The Services, the Knowledge in the I4.0 Context

Before proceeding, it is necessary to define what services are and what difference they have in relation to material goods. Some authors point out that the term "service" is difficult to define because it contains much ambiguity, and there is no agreement about it, once some concepts are more extensive than others (Santos et al. 2015, p. 16). Still, according to Santos et al. (2015, pp. 16–17). *Among most of the definitions*, it is possible to remove central terms like *[...] experiences, interactions, activities, actions, processes, intangibility, performance, customer*. The authors highlight among the other terms, the *interactions* between the organization and the *customer*/user, therefore, it is closely linked to the *quality perceived by the customer*/user. According to the authors, if there were the possibility to conceptualize the term in a single sentence, it would be based *[...] in three characteristics that are frequently cited: intangibility, concurrency (customer involvement) and perishability* (Translated). Through that concept, the strong proximity of the definition of services and ubiquity is verified. In other words, the ubiquity clearly supports and characterizes the services. Also, we can emphasize that the concept of the product is directly associated with a production system in which a set of activities, parts of a process, transforms entries/inputs into exits/products. Therefore, the products are a combination of goods and/or services, and we identify the organizations incorporating services that might bring more value to their products. We observe this in service companies when they expand their services through the use of technologies, especially those ones with features that extend the simultaneity.

The I4.0 phenomenon indicates that new services must appear to meet the demands of this new production paradigm. However, will also a new paradigm in relation to the provision of service emerge with the advent of the I4.0? Hermann et al. (2016, p. 3932) maintain that the I4.0 is "[…] enabled through the communication between people, machines, and resources, the fourth industrial revolution is characterized by a paradigm shift from centrally controlled to decentralized production processes". In view of this central feature, one question emerges: is it possible to offer services in a decentralized manner with technological support?

We can say that the service sector will change in the same intensity as industrial production. This transformation is not only associated with pure services but to those ones linked with goods, forming, thus, a product. After all, when connecting "people, things, and data, new forms of organization emerge", according to Hermann et al. (2016, p. 3929) translated. The information has a central role in this new paradigm, as pointed out earlier. According to Blanchet et al. (2014, p. 7), the material objects tend to be naturally incorporated into the net. Thus, "the internet is combining with intelligent machines, systems production, and processes to form a sophisticated network. The real world is turning into a huge information system." Similarly, the necessary knowledge arises from many sources, resources, people, places, all combined with the existing knowledge in the organization and its responsibilities (Prause 2015, p. 164). As a result, it is essential the efficient connection and management of this informational flow, since services cannot be stocked due to its simultaneity once they are consumed and produced at the same time. Here, we would say that this is ubiquity in its edge of the production system.

Thus, the ubiquity of the technology and the users establishes new business models and customer relationship. The constant connection is a reaction (in real-time) of the organizations to meet the client 24/7, with no time-space limitations, and on a global scale, either in the production of goods or the provision of services or even in the consumption and production of this ubiquitous product (UP).

The banks and other financial organizations have been the first ones to act on the ubiquitous way. After that, the virtual stores started to sell their products *24/7* allowing the customer, from now on, to consume any product of the catalog from any place, at any hour. Nowadays, even physical stores are getting virtualized and they tend to become spaces of interaction with the customers, extending the informational flow in the productive system, and the consequential logistical complexity.

The ubiquity depends on the network and technologies. With the advent of the CPS, customers can, more than ever, experience, interact, allocate tasks and activities. Still, in the present time and through the internet, many tasks or services can be activated. Neither all the services are supported by the net, nor can be carried through long-distance. For example, a repair of a vehicle. However, with the advent of the CPS, a significant part of services that today are not possible will be in the future, particularly, the services that deal with objects of informational base, such as picture processing. The services of diagnosis of images stand out as an example (Dorow 2017). In the same way that there is nothing that hinders a product to be sold on the internet, there is nothing that hinders the offering of a service as well. Provision of services in this ubiquity era is giving a service without time-space restriction (to get degenerated), simultaneously, in real-time. This is the most striking feature of this new paradigm. Thus, we can affirm that the services will be offered on a global scale and they can be accessed at any time, from anywhere, by the increasing of the CPS. Even a touristic travel would be deeply influenced by the CPS if the most optimistic predictions are confirmed.

Here, creativity is the keyword, because the most creative ones will take advantages in the process. The digitization of production enables the emergence of new forms or models of the market (Zezulka et al. 2016, p. 8). Those who get the involved niches in the digitization of the economy will be successful. According to Prause (2015, p. 167), I4.0 express a "fragmentation, new structures, and new business models." According to the author, the very important value of information, in the I4.0 context open, for example, opportunities related to the "big data [...] but also in the establishment of new business models". The possibility to track a product throughout its life cycle, even after its purchase or consumption, opens unthinkable possibilities in the current business model, outstandingly under the protection of sustainability (Prause 2015, p. 167). Information and sustainability are important keywords in the I4.0.

The internet, the internet of things, big data, and cloud computing give rise to new flows of information, and new directions for the information management. The uninterrupted data entry throughout the production cycle (Zezulka et al. 2016, p. 9) composes the I4.0 phenomenon. According to Zezulka et al. (2016, p. 9), by means of the information, it is needed to "(ensure) data integrity, consistent integration of different data, obtaining new, higher quality data (data, information, knowledge) provision of structured data by means of service interfaces". Roblek et al. (2016, p. 1) point that it is needed to highlight that, mainly, the information exchange will happen between the machines themselves, in which "machines are streaming data via wireless sensors and sending these data to the smart service/product providers' centers, where large amounts of data are analyzed." There will be new flows of information that involves communication between human actors, non-human, knowledge of the context, and huge amounts of data. Therefore, the service sector needs to explore these new flows of information.

A piece of research shows the most recurrent terms in academic research related to I4.0. It verifies that data, control, and information, preceded of systems, processes, and technology are among the most usual terms (Hermann et al. 2016, pp. 3930–3931). The transparency of the information makes one of the related principles of I4.0, and it includes data analysis and provision of information (Hermann et al. 2016, p. 3932). Definitely the information has immense value in the I4.0 (Lom et al. 2016, p. 2). Authors point out that the new environment which comes from I4.0 does not state, necessarily, the creation of new technologies. What comes with I4.0 are new ways of using it, especially in relation to the use of the huge amount of available data, with the potential to originate new business-oriented models in this context (Drath and Horch 2014, p. 57).

Hermann et al. (2016, p. 3932) conducted a research to identify the main features of the I4.0, and as a result, they point four "design principles", as it follows: "interconnection, information transparency, decentralized decision, and technical assistance". According to the authors, the interconnection is about the establishment of standards, contribution and security, that will allow the communication between "machines, devices, sensors, and people with each other"; the transparency of the information in the analysis of data and in the provision of information; the technical assistance concerns for both physical and virtual assistance; and the decentralization comes from the possibility of interconnection between "objects and people", and of the "transparency of the information", enabled by the CPS. Service providers tend to more and more use sensors in their tasks. After all, it is through the sensors and its linking with digitized or virtual environments, that a true copy of the real or physical world establishes (Hermann et al. 2016, p. 3932). According to Prause (2015, p. 163), the business features of I4.0 are "openness, standardization, sustainability, cooperation and networking concepts as well the use of smart technologies comprising internet technologies".

New and sustainable business models have to ensure fairly shared business benefits among all stakeholders in the value chain and might be more complex, open, collective and evolutionary than the existing ones. Furthermore, they have to facilitate innovation, product development, financing, reliability, risk, intellectual property and know-how protection in a network environment. These considerations lead to different business areas for new business models (Prause 2015, p. 164).

Prause (2015, p. 161) points out the concept of fractal structures as adequate to approach new business models of the I4.0; a process of construction based "[…] on relations between material, personal and information", intensively. According to the author, this fractal approach "can be considered as the new structural and organizational building blocks of Industry 4.0, where the different fractals are connected by related information flows, which control the processes inside and between the networks of fractals." But what changes for organizations? According to Blanchet et al. (2014, pp. 9–13), many things change for them (Table 1).

Table 1 What changes in the organizations with the I4.0 (own elaboration based on Blanchet et al. 2014, pp. 9–13)

Output Personalization, local production and mass customization	With more freedom and flexibility in the production, it will be possible to meet the individual needs of the consumers, with lesser expenditures as in the previous model. In other words, the customization will be possible with a low production cost. It makes the distribution of parts easier, among others things. After all, only the data might be transferred, whereas the physical production might be done locally, through the 3D printers, for instance
Process Networked manufacturing and cluster dynamics	The organizations will operate in several places around the globe. In the same way, the competencies will be spread over many places. Suppliers based in small towns should allow a faster innovation flux. With the decline of boundaries between the information and the physical world, a kind of industrial democracy will emerge. There is a new distribution of power when performance barriers for smaller organizations are reduced. At the same time, it widens the complexity of production/value chain. The so-called "mobile manufacturing units" may emerge. They are factories or independent and reduced "production cells". It must transform the way emerging markets receive investments, and the localization needs will be reviewed
Business Models fragmentation of the value chain	The value chain will be rethought and restructured, fragmented, with the participation of many stakeholders. Thus, there are new challenges linked to the costs and profits, and a question about where the big profits in the future will take place raises: "*In the design, in process handling or in customer data expertise?*"; That is the way the new business model may emerge
Competition Converging frontiers	The classic industrial limits have declined, as well as the boundaries between applications that are industrial and the ones that are not. The methods, the forms of production, the ability of reproduction of products and services will be fundamental in the coming years. The services, as well as the products, can also be done on a large scale. Among the services, the digital ones with great infrastructure and efficiency become vital to the success of the I4.0. More and more the computing organizations and the telecommunications must come closer to the industry and perhaps become great players of the it (Google and Facebook, for instance). In a cyber-physical world, the new suppliers are the computing organizations (sensors, programs, computers) that might occupy the place of the traditional suppliers (of physical machines and equipment)

(continued)

Table 1 (continued)

Skills Interdisciplinary thinking is the key	The required competencies in the I4.0 has changed, and they are much more dynamic than the traditional ones. The core technologies will be those related to computing, robotics, biotechnology, and nanotechnology. Advanced competencies will be required, as social as techniques. The projects start to guide the thought, more than the products did. The organizational culture heads for a continues formation. There is more collaboration at the environmental work
Globalization Light footprint	Organizations will be focused on "selected hotspots" rather than a global physical presence; There will be places of open production ("makerspaces"), and "clusters"; There will be no need to maintain large production sites; it often will be more appropriate to transfer data and produce locally in small scales. Thus, the organizations will be more decentralized and flexible

The ubiquity also expresses the need for new techniques of information and knowledge management. New organizational expertise to deal with the constant analysis of information and the uninterrupted flow of data. New interdisciplinary teams connected by the net, without time-space limitation. The knowledge management in I4.0 becomes a function of the great amount of data arising from machines, users, sensors. The authors point out some differences in this context. Roblek et al. (2016, pp. 6–7) point out an emergence of a *knowledge management 4.0*, influenced by the internet of things, *that is arising from the phase of integration between people and people with documents, and passes to the phase of connecting between devices. KM processes are also located between the consumer and the manufacturer or service provider*. According to the authors, *products integrated with cloud computing in the field can provide data that enable a predictive maintenance and provide information about optimization possibilities in production*. Furthermore, North and Kumta (2014) present the impacts of ICT and consequent ubiquity on knowledge workers.

Considering the impact of ubiquity on the production system, the importance of knowledge as a critical resource is identified given the characteristics of simultaneity (user-production process, user knowledge), intangibility (no ownership of the product or part of it), perishability (simultaneous production and consumption—knowledge in immediate use). As knowledge management (KM) has significant importance in the production process of the ubiquitous product (UP), it is also impacted by the characteristics of the ubiquity.

4 The Impacts of the Ubiquity

We observer that a new revolution in opportunities, products, and demands is occurring. We also must pay attention to the characteristics of this productive system that has been emerging. It is observed that the ubiquity and the opportunities that are brought together in the happening of the Industry 4.0 are associated with the simultaneous production and consumption, inherent perishability and intangibility of a producing system of services. It brings implications that must be observed regarding the features of the "new" production system. Jardim-Goncalves et al. (2017) present an interesting view and the possibilities regarding implications for 4.0.

Initially we must consider that it is not possible to test, to check or even to inspect the quality of a ubiquitous product (UP). We can observe that the different stages of the process are coursing as planned/designed, although the ability to verify the quality only happens at the end of the production/consumption. Some component/stages can be verified but they characterize themselves neither for the ubiquity nor for the concurrence or contact with the customer. They are stages of production system that are distant from the user, and they are not ubiquitous.

The user cannot return a UP or fix it, as it happens to goods. Let us take as an example the automotive industry. It can stop the production line and fix a flaw of process or input, and even it can retain their products at the courtyard. Additionally, in an undesirable and costly scenario, the company can make a recall. But, when we look at a provision of service company, the same cannot occur because there is the need for another service with significant impacts on costs and image. As an example, we have Delta Airlines, that paralyzed its operations, suspending the flights in 2015, with a significant loss. In 2017, the same went to British Airways. Their passengers lost meetings (family or business), vacation, and other activities that they had planned. Anything that they did to compensate the consumers for the impact on non-provision of the service would be exactly a mere compensation because, when it happened, and if it happened, it was another service. The faults occurred in their systems, according to what was informed by those companies. It means that the informational flow was interrupted and the impacts were significant. That is an example of the need for observing the robustness of our processes. There is no foolproof process, no matter how much one invests in it as we noted in the failures that have resulted in accidents of Three Islands (USA), Chernobyl (Russia) and Fukushima (Japan). Therefore, we must be vigilant in our UP project in order to enlarge its strength or minimize the impact of failures, compensating and regaining our customers.

Similarly, we manage our employees being aware of their performance in the production, so we must be mindful and manage our customers to let them effectively and efficiently do the activities they set, once their failures can lead to errors or significant damage. Independently of the security systems that banks have developed in their virtual environments, it is still much associated with how the client keeps "their side of the operation" safe. If the client is not careful or if the bank extends the "security systems", making them complicated, the service quality diminishes, as well its use.

Another aspect to be considered is the simplicity of the process and the interfaces with the user, adding that the client, when part of the production process, has a perception of the necessary time for making it happen, as well they must have competencies to play "their part". Consequently, the more complex (interactions/steps and necessary competencies) or time demanding is the "consumption" of the product, the lesser will be the perceived quality, and the greater the likelihood of failures.

Thus, we can say that the UP must be created, designed, built, and operated as a significant attention to the characteristics that surround it, which are the immateriality, the simultaneity, and the perishability. Such features, in a production system with its production capacity significantly expanded in industry 4.0, must be correctly managed, under the risk of having failures and losses, rather than profit and success.

The possibilities arising from the ubiquity are many. In essence, it is the "total" transparency of the processes and information, even what one cannot see, like the inside of a machine. The exercise of ubiquity through the establishment of the CPS allows many possible roles for industry, in all the stages of the value chain. The real-time exploration allows to establish a new model of relationship with the customers and suppliers (customers take part of the production and influence it); precise control of the production, visualization of the emergence of given input, modification of the production in real-time according to the involved variables; to carry out forecasts, to identify trends; to investigate in a more precise way; to identify errors in real-time, and to reduce the need of recalls; to better know the collaborators and their moods, which allows a series of actions before the worker, allocations, and deployments; to better use the knowledge and information from the workers with more precision. In other words, the ubiquity will provide the total control, a kind of omniscience of what occurs in the factory, in all the value chain, and also in the society. According to Kagermann et al. (2013, p. 5), *In the manufacturing environment, these Cyber-Physical Systems comprise smart machines, storage systems and production facilities capable of autonomously exchanging information, triggering actions and controlling each other independently.* Yet according to Kagermann et al. (2013, p. 14), [...] *allows production to be configured more flexibly but also taps into the opportunities offered by much more differentiated management and control processes.* According to Blanchet et al. (2014, p. 7), *Using these technologies will make it possible to flexibly replace machines along the value chain. This enables highly efficient manufacturing in which production processes can be changed at short notice and downtime (e.g. at suppliers) can be offset.*

Numerous examples of the technology's impact and the potential for ubiquity can be observed in the 4.0 industry related literature, as mentioned by Thoben, Wiester and Wuest (2017, pp. 4–13) and CNI (2016). Rolls-Royce intending to use 3D printing for the production of turbine components, EMBRAER training its workers one year before the start of production through virtual environments, collaborative robots (cobots), interaction of sensing equipment and data processing systems among other possibilities. The possibilities and the changes are enormous in different industries and activities within.

The exercise of ubiquity through the establishment of the CPS allows many possible roles for industry, in all the stages of the value chain. Real-time exploration allows:

(a) a new model of relationship with customers and suppliers; customers take part of the production and influence it;
(b) precise control of the production, visualization of the emergence of given input, modification of the production in real-time according to the involved variables;

(c) to carry out forecasts, to identify trends; to investigate in a more precise way;
(d) to identify failures in real-time, and to reduce the need of recalls;
(e) better know the collaborators and their moods, which allows a series of actions before the worker, allocations, and deployments;
(f) real-time expresses new techniques of information and knowledge management;
(g) new organizational expertise to deal with the constant analysis of information and the uninterrupted flow of data;
(h) new interdisciplinary teams connected by the net, without time-space limitation;
(i) increasing of the security guards or surveillance, cybersecurity;
(j) total control, omniscience of what occurs in the factory and in society;
(k) increasing of the unpaid work or operation; customers work, operate;
(l) "total" transparency of processes and information.

References

Batista, N. C., Melício, R., & Mendes, V. M. F. (2017). Services enabler architecture for smart grid and smart living services providers under industry 4.0. *Energy and Buildings, 141,* 16–27. https://doi.org/10.1016/j.enbuild.2017.02.039.

Blanchet, M., Rinn, T., Von Thaden, G., & De Thieulloy, G. (2014). Industry 4.0: The new industrial revolution-How Europe will succeed. In *Hg. v. Roland Berger Strategy Consultants*. München: GmbH. Abgerufen am May 11 2014, unter http://www.rolandberger.com/media/pdf/Roland_Berger_TAB_Industry_4_0_20140403.pdf.

CNI. (2016). *Desafios para a indústria 4.0 no Brasil*, p. 34.

Cooper, J., & James, A. (2009). Challenges for database management in the internet of things. *IETE Technical Review*, *26*(5), 320–329.

De Masi, D. (2000). *O ócio criativo*, 7 (ed), p. 336. Rio de Janeiro: Sextante.

De Masi, D., Manzi L., Palieri, M. S. (2000). *O Ócio Criativo*. Rio de Janeiro: Sextante.

De Masi, D., Manzi, L., & Figueiredo, Y. (2005). *Criatividade e grupos criativos*. Rio de Janeiro: Sextante.

Dias, R.A. (2010). Tecnologias Digitais e Currículo: Possibilidades na Era da Ubiquidade. *Revista De Educação Do Cogeime*, *19*(36), 55–64. https://doi.org/10.15599/0104-4834/cogeime.v19n36p55-64.

Donovan, K. (2012). Mobile Money for Financial Inclusion. *WORLD BANK. Information And Communications For Development 2012: Maximizing Mobile*.

Dorow, P. F. (2017). *Compreensão do compartilhamento do conhecimento em atividades intensivas em conhecimento em organizações de diagnóstico por imagem*. Tese,PPG-EGC/UFSC.

Drath, R., & Horch, A. (2014). Industrie 4.0: Hit or hype? [Industry forum]. *IEEE Industrial Electronics Magazine, 8*(2), 56–58.

Elias, N. (2010). *Sobre o Tempo*. Jorge Zahar Editor.

Fontes, G. S., & de Lima, I. R. (2013). Cibercidades: as tecnologiasde comunicação e a reconfiguração de práticas sociais. *Informação e Informação*.

Godoy, V., Freddy, A., & Foresti, F. (2015). A ubiquidade proporcionada pelos dispositivos móveis e o fluxo da informação. *Datagramazero*.

Harvey, D. (2010). *Condição Pós-Moderna*. São Paulo: Loyola.

Hermann, M., Pentek, T., & Otto, B. (2016). Design principles for Industrie 4.0 scenarios. In *2016 49th Hawaii International Conference on System Sciences (HICSS)*. New York: IEEE, https://doi.org/10.1109/hicss.2016.488.

Jardim-Goncalves, R., Romero, D., & Grilo, A. (2017). *Factories of the future: Challenges and leading innovations in intelligent manufacturing. International Journal of Computer Integrated Manufacturing*. https://doi.org/10.1080/0951192X.2016.1258120.

Kagermann, H., Wolfgang, W., & Johannes, H. (2013). Recommendations for Implementing the Strategic Initiative Industry 4.0. *National Academy Of Science And Engineering*, Berlin/Frankfurt, p. 82.

Leite, J. (2008). *Os Três Tempos do Espírito: A Oralidade Primária, a Escrita e a Informática*. Logos.

Lévy, P., & Da Costa, C. I. (1993). *As Tecnologias da Inteligência*. Rio de Janeiro: Editora 34.

Lom, M., Pribyl, O., & Svitek, M. (2016). Industry 4.0 as a part of smart cities. In *Smart Cities Symposium Prague (SCSP), 2016*. https://doi.org/10.1109/scsp.2016.7501015.

Luciano, C., Varvakis, G., & Gohr, C. F. (2015). *Sistemas De Operações De Serviços. João Pessoa: UFPB*.

Mantovani, C. M. C. A., & Moura, M. A. (2012). Informação, interação e mobilidade. *Informação & Informação, 17*(2), 55–76.

Merriam-Webster. (2017). Ubiquity. Merriam-Webster. https://www.merriamwebster.com/dictionary/ubiquity.

North, K., & Kumta, G. (2014). *Knowledge management: Value creation through organizational learning*. Berlin: Springer Science & Business Media.

Prause, G. (2015). Sustainable business models and structures for INDUSTRY 4.0. *Journal of Security & Sustainability Issues*, *5*(2), 159–169. https://doi.org/10.9770/jssi.2015.5.2(3).

Richert, A. (2016). Educating Engineers for Industry 4.0: Virtual Worlds and Human-Robot-TeamsEmpirical Studies Towards a New Educational Age. In *Global Engineering Education Conference(EDUCON)*. Abu Dhabi, UAE.

Renau, J. (2017). Internet and mobile phone addiction. In Z. Yan (Ed.), *Encyclopedia of mobile phone behavior*.

Roblek, V., Meško, M., & Krapež, A. (2016). A complex view of industry 4.0. *SAGE Open, 6*(2), https://doi.org/10.1177/2158244016653987.

Santos, L. C., Varvakis, G., & Gohr, C. F. (2015). Um modelo para análise do alinhamento conceito-pacote: melhorando o projeto do sistema de operações de serviços. *Espacios (Caracas), 36*, 1–8.

Wahl, M. (2015). Strategic factor analysis for Industry 4.0. *Journal of security and sustainability issues*, 2029–2702.

Weizer, M. (1991). The Computer For The 21St Century. *Scientific American*.

Zezulka, F., Marcon, P., Vesely, I., & Sajdl, O. (2016). Industry 4.0—An Introduction in the phenomenon. *IFAC-PapersOnLine, 49*(25), 8–12. https://doi.org/10.1016/j.ifacol.2016.12.002.

Author Biographies

Fabricio Foresti is Information Science Ph.D. candidate and holds an MSc in Information Science from the Federal University of Santa Catarina—UFSC (2016), a BA in Library Science from UFSC (2006). He is Head of Technical Archives of the Engineering Council in Santa Catarina, Brazil and Web designer.

Gregorio Varvakis, Dr is Professor at the Federal University of Santa Catarina in the doctoral programs of Engineering and Knowledge Management and Information Science. He has a Ph.D. in Manufacturing Engineering (Loughborough University, 1991), a MEng Production Engineering (Federal University of Santa Catarina, 1982) and Mechanical Engineer (Federal University of Rio Grande do Sul, 1979). He is author of several technical books.

The DAO Case—Block Chain Technology Based Knowledge Intensive Business Models

Patrick Hofer

Abstract The DAO, the world's first "distributed autonomous organization" was founded on May 15, 2016. In just a few weeks, the investment fund managed to bring in USD 119.5 million from more than 50,000 investors. The DAO was not only the biggest crowdfunding campaign of all times, it was also ground zero for the biggest cybercrime in IT history. Just a month after its launch, hackers succeeded in siphoning USD 50 million out of the fund.

1 Introduction

In an age when "software is eating the world", the factors that determine business success or failure are changing. Disruptive technologies are breaking down entry barriers that had safeguarded profits in the past, thus radically transforming entire industries. Success now depends on how quickly an organization is able to develop and internalize new knowledge.

Blockchain is one of these revolutionary technologies that is included alongside artificial intelligence, virtual reality, the Internet of Things, self-driving cars and networked homes. A blockchain is a worldwide, decentralized accounting system invented in late 2008 by a mysterious creator known by the pseudonym Satoshi Nakamoto, but whose true identity remains unknown. Nakamoto published a nine-page PDF detailing the principles of Bitcoin—and was never heard of again.

Bitcoin was conceived of as an attack on the financial system, so its introduction by Nakamoto at the beginning of the financial crisis was hardly a coincidence. The Bitcoin community began to explode and not only attracted idealists, libertarians and anarchists who dreamed of a new monetary system, it also caught the attention of criminals, speculators and hackers.

And a 17-year-old by the name Vitalik Buterin, that like many others in his generation, was playing "World of Warcraft", an online game in which players

P. Hofer (✉)
Media Interface GmbH, Zurich, Switzerland
e-mail: patrick.hofer@media-interface.ch

K. North et al. (eds.), *Knowledge Management in Digital Change*, Progress in IS,
https://doi.org/10.1007/978-3-319-73546-7_22

purchase weapons and armor. But there was a problem for him and plenty of other online gamers. How could he get his hands on the weapons he needed to succeed in the game? He had the hard cash, but no way for him to make online payments. No credit card? Game over. This, of course, creates a barrier that shuts out a large segment of the world's population.

The online payment systems currently available are highly inefficient, and money transfers are not very secure. And endless information—card numbers, names, addresses, etc.—has to be revealed to unknown parties. But is all this information really necessary? After all, you do not have to provide your name and address when you make a payment with a banknote.

Buterin not only thought this absurd, it annoyed him that in addition we are expected to pay a fee for this inefficient service. Each time a payment is made by credit card, an array of companies are working in the background that require our security codes and verification data—and then they charge us for this "service" that does anything but inspire trust. The 17-year-old Buterin saw the massive lack of trust online as the reason for the complex processes and all the resultant fees—a billion-dollar business that forms the foundation of companies such as Visa, Mastercard and Paypal.

Buterin told the «*Der Stern*», "The day I took a closer look at Bitcoin for the first time, I understood that payments are possible without the middleman." He suddenly had a vision of a different world. Bitcoin money transfers are almost free. The fees are minuscule, no matter how large the payment or where it is being sent. And it's necessary only to enter the anonymous recipient address consisting of numbers and letters, along with the sender—practically like sending an email.

Soon it becomes clear to Buterin that blockchain technology could be used to do more than transferring virtual currency. Until this point, blockchain had been used as a digital vault for storing Bitcoins. In theory, however, the system could be used for all sorts of value, be it cars, houses or shares: a decentralized directory for all assets with a secure way to transfer these assets. Might it not be possible to eliminate bureaucracy entirely for these kinds of transactions? Landlords, notaries, entire government agencies?

That very same summer, shock waves were felt around the world when Edward Snowden reveals how the NSA has been infiltrating the internet. Buterin feels betrayed by his best friend, the internet, and the struggle for net neutrality becomes a personal mission. Not much later, he drops out of his computer science program at the prestigious University of Waterloo in Canada. Now 19, Buterin uses his savings to make a six-month world trip to help bring about the revolution.

2 Ethereum

In Tel Aviv, he meets software developers who are already working on making his dream a reality with the development of self-executing contracts; i.e. "smart contracts" that function without legal departments, since their execution is controlled by the blockchain. Thus, Ethereum is born.

But there's just one problem: Buterin notices that highly talented programmers are struggling to code even the simplest of contracts with Nakamoto's blockchain. He realizes that it's time to do something sacrilegious: a new blockchain must be created—a user-friendly, decentralized registry for everything. And, above all, a more powerful one—just as the Windows operating system replaced DOS a long time ago.

Buterin writes a concept paper that is a mix of politics, game theory and mathematics. He calls it "Ethereum" according to Aristotle's idea of ether as an omnipresent fifth element. His blockchain is also designed to be ubiquitous and run on all participating computers—a global computer, a gigantic ledger that is installed decentrally on all its users' computers. As with Bitcoin, money can be transferred—in this case with the Ether currency—in a way that is transparent to everyone. Ethereum also lets users create smart contracts; in other words, agreements that can be enforced without lawyers or courts.

He tells those in his inner circle about the idea, and in just a few weeks he puts together a nerdy army of high caliber programmers. Several of his team members already have impressive international careers. Although many in the Bitcoin scene see Buterin as a renegade, he soon catches the attention of Silicon Valley. Investors begin to regard him as a prodigy. Buterin announces his plans to launch a crowdfunding campaign to collect money "for a decentralized publishing platform with user-generated digital contracts and a Turing-complete programming language". The pitch manages to break a world record in crowdfunding—USD 18 million in four weeks.

3 The DAO

The story of The DAO begins in May 2016 with three men: Christoph Jentzsch, the former head tester at Ethereum, his brother Simon, a former manager of software projects for large corporations, and Stephan Tual, a former informatics specialist for companies such as Visa and BP and CTO of Ethereum. The roots of their project go back to another internet phenomenon: crowdfunding.

The 32-year-old Jentzsch, who also holds a physics degree, presents his idea to the world at an Ethereum Conference in London in November 2015. "How can we set up a company using blockchain?" he asks the audience. "Of course it has to be a DAO." DAO stands for "distributed autonomous organization". How it works: Ether, a virtual currency like Bitcoin, is used to fund projects for applications built with Ethereum blockchain technology. Investors buy shares, or tokens, with their Ether funds, enabling them to vote for or against projects. If the applications that are outsourced and developed make money, the holders of the tokens receive a portion of the proceeds.

The idea impresses tens of thousands of people. In the course of just a few weeks, shares in the form of tokens valued at USD 150 million are sold.

By comparison, the most successful project up to this point was a Kickstarter campaign that brought in USD 20 million. The Jentzsch brothers become stars of the tech world practically overnight. Next they take a proposal to The DAO to court investors for Slock.it—the company they created but do not own. Brilliant. But then they run into a problem.

4 Making off with the Loot

It is Friday morning and software developers in the western hemisphere wake up to the news that The DAO has been hacked. Christoph Jentzsch is lying on the floor of his home office, taking deep breaths, trying not to panic. Meanwhile, the community wonders what is going on when the first message appears: "I think The DAO is being drained right now; unfortunately, I am on a train to work, so cannot investigate, but looks like recursive call exploit of some kind."

An unknown hacker has discovered an error in the smart contract, or the underlying rules of The DAO. "We had hired a security company in Seattle specifically for this purpose, and even they didn't notice," Simon Jentzsch tells «Wired». Tual was furious and was convinced that it was an attack on his team's idea. It must be an insider who is very familiar with the programming language of the Ethereum blockchain.

The attacker exploits a function actually designed to protect investors: If a person rejects an investment on The DAO because they do not agree with the majority decision, the individual can withdraw their shares, which are then transferred to a sub-account, or "child DAO". The attackers repeat this operation over and over again before the system notices that the shares have long been parked elsewhere. In this way, hackers amass DAO tokens valued at USD 53 million—a disaster that spreads panic in the community. But not all is lost—there's an additional security function.

The system freezes child DAOs for 28 days before the shares can be permanently withdrawn. "The deadline is July 16," says Christoph Jentzsch in June. In the meantime, there's still time to stop the hacker. Except that no one knows who should take on the task. Should it be Slock.it, the developer of The DAO? Or perhaps the Ethereum Foundation as an organization of the underlying blockchain? Or possibly The DAO investor crowd itself?

In any case, Buterin is not interested in helping as one of his tweets makes clear: "Reminder: the Ethereum Foundation has no involvement in the DAO." The technology is supposed to remain neutral, and the coding does not get involved with the problems of its users. But this detachment causes an uproar in the community. The futuristic wars that unfold in the following weeks add a new chapter to the history of the internet.

5 DAO Wars

Then a letter claiming responsibility surfaces. The author claims simply to have used the technology in a smart way. He does not see himself as a hacker—and many agree. Blockchain forums, Slack channels and Reddit debates boil over with no end to the finger-pointing. Then another several million disappear. It's a second hack, a counter-attack by a self-proclaimed Robin Hood group. "Dao is being securely drained. Do not Panic," writes Alex Van de Sande, head designer at Ethereum.

To prevent the attackers from withdrawing the tokens, the solution of a "fork" is discussed. This involves modification of the basic code of the Ethereum blockchain with a fork programmed into the chain that enables all DAO investors- to cash in their tokens at a fixed price, as in a currency reform. This requires all parties to update their software. The old DAO continues to exist on the old Ethereum-blockchain, but without any investors will become extinct.

The shares belonging to the hacker and to the vigilante group will become void —much to the chagrin of the attackers, of course: "A soft or hard fork would amount to seizure of my legitimate and rightful ether, claimed legally through the terms of a smart contract. Such fork would permanently and irrevocably ruin all confidence in not only Ethereum but also the in the field of smart contracts and blockchain technology. Many large Ethereum holders will dump their ether, and developers, researchers and companies will leave Ethereum. Make no mistake: any fork, soft or hard, will further damage Ethereum and destroy its reputation and appeal."

6 The Fork

Just before the 28-day deadline expires, Buterin intervenes. It becomes clear that Ethereum's movers and shakers are becoming nervous about the survival of The DAO. It makes up 17% of the blockchain—in other words, "too big to fail" (and we all know where that can lead). On July 20, 2016, Buterin announces that the key miners, which provide the processing power for the technology, have accepted the fork and have installed the new version. One group protests, however. Wasn't the original vision to create precisely the kind of system that is safe from the interference of dishonest people such as corrupt officials, politicians, administrators, CEOs and lawyers? The code itself was supposed to be the law. If you failed to see the weakness in the software, that was your own fault, since the software code was publicly available.

In protest, they stick with the old blockchain and christen it Ethereum Classic—a coup d'etat of sorts. And soon speculators and competitors in the blockchain scene begin buying into the blockchain. Instead of losing value, The DAO begins making gains again. Now the "evil twin" is functioning as a parallel currency and starts trading on the exchanges, which allows the attackers to move a portion of their purloined 3.6 million in Ether Classic to safe shores.

Now it's August, and there's chatter among the founders in the forums about how the hackers want to gain access to their account and why no one is really pursuing the perpetrators. The investors in The DAO get back all their original stake, but the project itself crashes and burns spectacularly. "It all simply grew too fast," Simon Jentzsch explains in a YouTube video.

7 New Knowledge Creation

The story of The DAO is highly pertinent to knowledge management. The dynamic way in which technology develops means that knowledge can no longer be communicated using the traditional models. Learning by experimentation is the rule, not the exception.

Knowledge can no longer be developed by individual learning or companies; instead, it demands cross-company partnerships. Or, as Tual writes on the Slock.it blog: "There are some things which one can only learn through experience, either one's own, or that of others."

What happened with The DAO serves as a cautionary tale for entrepreneurs whose task it is to develop and collect new disruptive knowledge within their organization.

Lesson 1: Experience improves security over time
Security is paramount when it comes to developing technologies. The most effective way to minimize risk is to set a cap on investment and transaction volumes.

Lesson 2: Stay aware of "unknown unknowns"
It's impossible to completely rule out attacks on new technologies. Despite comprehensive security audits and open source community reviews, the vulnerability will not be found because no one knows where to look. Projects that are based on disruptive technologies will never be entirely secure. It is essential to be agile when responding to incidents.

Lesson 3: Tooling is immature, but things are improving
Wait for the right time, when the technology is mature. The software tools available at the time of The DAO project were not yet ready.

Lesson 4: It is imperative to develop governance and voting mechanisms adapted to decentralized systems
Many people in the community wanted to exercise leadership and exert power over how the decentralized organization should take shape. How to conduct votes or deal with crises is another issue, leading to the question of who should be in charge.

Was it the responsibility of the founders or the creators of the underlying technology? Of course, the investor community itself was responsible for this type of organization.

Yet the instruments used to organize the community fell short. Forums such as Reddit or Slack were inadequate, because anyone could put in their two cents—regardless of whether they were actual token holders. Public forums were easy prey for social engineering attacks.

In times of crises, authority is needed, which was sorely lacking in the case of The DAO. This is the nature of decentralized systems, which is both a blessing and a curse. It turned out that a little tweet by Buterin was all it took and the decision was made.

Lesson 5: Launch gradually
One important lesson from Slack.it was that much more consideration needs to be given to the issue of complete decentralization. The lesson is that a DAO needs to be introduced gradually. A proposal was made to incorporate a sort of sandbox for future products, so that participants can first test out the new organizational form in small increments before making any real decisions.

Lesson 6: Minimal complexity
The less code, the better. A rule of thumb is that 15–50 errors sneak into every 1000 characters of code. For this reason, smart contracts should be kept as simple as possible.

8 Bottom Line

Some important insights were gained during the project within just a few months. The failure itself will go down in history and will also have an impact on the future of blockchain technology. For the many people involved in this project, the disappointment still had a silver lining, and the vision of the world computer lives on.

Since the theft, the value of the Ethereum currency multiplied by the factor 40. The Slock.it founders have found a first investor to invest USD 2 million in the company. What happened with the 3.6 million stolen by the attackers in Ether Classic currency remains to be seen. The value as of June 14, 2017 is USD 68.7 million. Had The DAO not been attacked, the world's first distributed autonomous organization would have a market capitalization of more than USD 4 billion.

But the story does not end there. In an interview with leaders of a trading platform for virtual currencies, «Bloomberg» had discovered that the attackers were a group of people in Switzerland. This information was passed on to the Boston office of the FBI. Further agents and New York's justice department are expected to pursue the case further.

However, before this thriller finds its way to the big screen, we are likely to see a few more episodes. Will the attackers be found or will they reveal their identities themselves? How will law enforcement authorities and courts assess this case? What is the legal jurisdiction of a DAO? Who is liable for what? And was any crime actually committed in the legal and ethical sense?

Author Biography

Patrick Hofer is a Swiss Transformation Advisor with Media Interface, a digital business development consultancy based in Zurich, Switzerland. Media Interface is helping organisations transition to the experience economy. From small retailers to global playing brands Mr. Hofer has over 20 years of experience in Tech. Besides his career in professional services, Patrick Hofer was part of the AI startup Artificial Life and founder of the largest Swiss platform for Media Arts and Digital Entertainment. Patrick holds an Executive MBA in Business Engineering from the University of St. Gallen.

Startup and Technology Hubs

Christian Kreutz

Abstract Digital transformation is affecting the economy. Startups act as role models for new software-driven business solutions by benefiting from a global innovation eco-system. To assure constant innovation, they are run on flat organizational models. A key success factor lies on their collaboration around open source software, and on how it sets a standard for the future of open innovation.

1 Introduction

Every day, more and more promising new startups with competing business models are founded; something that would be impossible without the internet. For most companies, the New Product and Service Development Process is changing drastically in terms of speed, location (ir-)relevance and sophisticated international collaboration opportunities. Thanks to their openness and flexible organization models, startups are being able to include faster digital technologies to challenge large companies.

Established companies, which have tight communication policies and rely on internal innovating capacity, can face various challenges such as sustaining their products and services. "Unless you have the capacity to innovate alternative business models, someone else will likely get a crack at it first." (Baur 2017). To achieve this in times of ongoing innovation, companies have to transform themselves from different angles. They have to absorb digital changes, and find new ways to be able to participate in a global and highly dynamic open innovation eco-system.

C. Kreutz (✉)
University of Applied Sciences, Darmstadt, Germany
e-mail: ck@crisscrossed.de

K. North et al. (eds.), *Knowledge Management in Digital Change*, Progress in IS,
https://doi.org/10.1007/978-3-319-73546-7_23

Despite the controversy about its business practices, Uber shows that a transformation in building a business without owning its main assets—a fleet—can be done. Uber found a way to connect different existing services (e.g. location based intelligence) to form a new business model. It does not even have to provide all the software required by its services, as they rely on high quality technology such as Google Maps. On top of that, Uber benefits from worldwide open source software and can tap into existing mobile platforms to offer its app. In the end, Uber is challenging the existing taxi market because it is able to leverage software to a better user experience and to use data for innovation (e.g. automated driving).

But, what makes startups so special? And, can they act as role models for innovation and business sustainability in the digital age? What defines a startup? Initially, startups are loose ideas that are incrementally developed to a potential business model, for example, through business incubators. Startups are here meant as newly established companies or organizations, developing new or improving existing business models, that rely mainly on digital channels for their products and services, and that exploit all potential digital technologies to their advantage. So, while they are mostly legally founded as companies with a classic structure, they still differ greatly in their management and on how they are built through various creative financing forms, such as business angels and crowdfunding.

Startups rely on the internet and digital technologies in at least three ways. First, the digital space is their main approach to market and have access to clients. Second, they use digital technologies to run most of their operations, from accounting to support. Third, their own products and services are mostly software-driven.

Software is the foundation of all the startups' operations. Still, they would never be able to write all the required software by themselves if they were independent from a global innovation network for open source software and software services.

That's why open source software plays a central role in their success; thanks to it, they are able to incorporate cutting edge technological innovation. But in order to benefit from open source software, a fruitful collaboration is conducive, where a group of startups provide better software than one single organization. Open knowledge sharing and constant learning is a key requirement for startups to either be able to constantly improve their service with clients and consumers directly, or to be able to use and improve cutting innovative software solutions for all their operations.

Startups have found a formula to efficiently tap into this large open source eco-system to innovate faster:

- They experiment with new flat managing models for higher productivity, ongoing innovation and to give employees more responsibilities.
- They experiment with agile processes for flexible product and service development in dynamic markets that are much closer to consumers.
- They run a lot of their operations openly by default, for a more effective knowledge sharing and faster learning.

If startups act here as role models and tell us about the future of the economy, we can work with the following assumptions:

- Future successful companies are those, which form or become active players in global open innovation networks and master digital collaboration.
- Software collaboration is only the beginning of a much larger change entailing all types of products and services that will be developed in an open source approach.
- These open networks challenge traditional organizational management models and require them to adopt their business and management styles.
- In terms of knowledge, it means that highly innovative companies will only sustain their services if they transform into open learning organizations.

2 The Case of Open Source Software for Global Innovation

Global open source software collaboration stands out as one of the most successful examples for digitally enabled knowledge creation. It has led to a global competition for the best software solutions and a division of labour to develop high quality solutions, where startups are so far benefit the most.

But what makes software in general, and particularly open source software, such an important innovation driver? The internet infrastructure, along with most web services, relies on worldwide open source software collaboration. There is hardly any company or government that does not rely on open source software. For instance, more than 2/3 of web server traffic runs over open source software server.[1] According to a study by Linåker, Runeson and Regnell analyzing the role of open source software at Sony, "Today, co-created, open system software dominates." (Munir et al. 2017). The market for software products is fast-moving and highly innovative; a great share of software development is completely open and happening around-the-clock. Software is a great example for digitally enabled knowledge creation:

- Software is represented as codified knowledge and can take full advantage of the internet and its information sharing capability.

[1] W3techs https://w3techs.com/technologies/overview/web_server/all.

- Software code collaboration, and obtaining a potential end product, can be easily done in a decentralized fashion.
- Software developing or programming is knowledge-intensive and requires profound competencies.
- The output and performance of software development is a decisive factor for the competitiveness of businesses.

Hence, digitally enabled knowledge creation through software, gives us great insights about the future of online collaboration.

But what precisely is open source software and why has it become such a critical asset?

Open source software is first of all, as its name indicates, software that is developed openly, meaning the programming code behind it is publicly available, for example through websites, and can be downloaded at any time. Furthermore, open source software is freely accessible to enable collaboration in a decentralized fashion and (anonymously) programming code contribution. Depending on the software license, the code can be changed and adapted and used for private or commercial purposes. Some benefits of open source software are:

- Making software open accessible discloses weaknesses and offers opportunities to improve through the help of programmers and end-users.
- It allows to join synergies by talented programmers and companies to develop together a better-quality end-product.
- The software can be used for many different purposes and alongside being developed in different ways.

It can attract a community of developers and end-users that potentially require additional services, leading to new business models.

3 Github—A Global Open Software Collaboration Network

A great way to analyze open source software development from a bird's eye view are software project portals such as Github.com. Over the past years Github has become the focal point for global software collaboration with 5.8 million active users, 331 thousand organizations and almost 20 million software projects.[2]

Github can be described as a social network for programmers, with social media features akin to those from Facebook, but it focuses on working efficiently on software programming code. Software code is usually organized in a software repository with a version control system, similar to a Wikipedia article that has been written and changed over time. The history feature allows to see how the code

[2]The state of the Octoverse 2016 https://octoverse.github.com/.

developed from the start to the present state. Github offers an easy interface to such repositories, where programmers can discuss about code problems, such as bugs or changes, or simply offer improved code. Code repositories facilitate cloning code to develop in different directions and, most importantly, to divide software programming into different tasks for an easy decentralized collaboration.

If developers want an additional feature, they can simply copy the public available code, add own programming logic and use the software for another purpose. This technique is called forking—a central efficient mechanism for digitally enabled knowledge creation. Take for instance, Tensorflow, a popular machine intelligence library by Google, that has as of June 2017 over 19 thousand contributions by almost one thousand programmers, not all employees of Google.

Currently, 731 errors and potential improvements are in the queue; in the past 2 years close to seven thousand such issues were solved. Only a smaller group of software repositories have this strong participation, but it highlights how even firms with great capabilities for software development such as Google, strategically decide to work in a larger open eco-system. Lastly, Tensorflow software has led to over 30 thousand software forkings, which are parallel developed by programmers for additional purposes or to improve the software. Moreover, it is important to understand that documenting the whole software development code makes it incredibly transparent and promotes learning from failure.

"Designing and managing these kinds of innovation communities is going to become increasingly important to the future of open innovation, and innovation in general." (Chesbrough 2017). Frequently, such open source software projects are managed or hosted by an organization willing to make such an investment, because as much as you get back from the community, it also requires quite some resources depending on the size of the project. Google makes this investment with a business goal in mind: to create a community of potential clients for additional cloud services to use Tensorflow for data analytics. A case study at Sony mobile shows "assets not seen as competitive advantage nor a source of revenue are made open to OSS [open source software] communities, and gradually, the organization turns more open." (Munir et al. 2017). This open community approach is attracting every year more and more developers, who see this kind of approach as essential in their work.

That's why large software firms are participating on Github also to attract talented programmers such as Microsoft, Google and Facebook with between 200 and 1300 software repositories.

Still, the majority of contributions on Github do not come from big software firms, rather from individuals and startups, who create their own projects. For example, the software library Vue.js, used on many websites, was initially programmed by a single independent programmer, and is now successfully challenging rival software libraries from Facebook and Google.[3]

Looking at the distribution of contribution geographically, the most contributions come from the USA and Europe, but also countries such as India or Brazil are

[3]Vue.js: https://github.com/vuejs/vue.

under the top ten contributing nations.[4] Lastly, it is not only companies, but also governments, such as the different ministries and authorities in the United States under code.gov, who are releasing their software investments as open source.

Startups would not be able to offer their services, if they were to develop the software by themselves and provide all the competence in-house. That's why they actively engaging in an open innovation network to learn from each other, including competitors. It is for example for most startups a standard to have an engineering blog, where they share a lot of their inside programming work with the world. Many technology firms encourage their employees to share their knowledge platforms such as Stackoverflow, which is a large Q&A portal for programmers with great quality contributions.[5] It has over a 22 million voluntarily provided answers to 14 million programmers' questions.

This openness by most startups is conducive for organizational learning and a must in a highly innovative software market. Learning is a center part of this process, where products and services are improved in fast and short intervals. Developing products the agile way means involving customers and releasing products early on, and developing a product in collaboration. "One strength of this startup innovation model is the constant, direct feedback that new products and sections of code are exposed to both internally from colleagues and externally from clients." (Richter 2015).

The open source software collaboration has also led to a range of new business models. Mapbox is a digital cartography and geo-data company that successfully challenges geo-data services from Google and Microsoft. Founded in 2010, Mapbox mainly offers custom digital maps for different purposes. On the one hand, it depends on the large open data project called Openstreetmaps, where volunteers worldwide created a high quality global map with valuable geo-data. And on the other, it supports this community with a range of open source mapping libraries and applications, which are used by clients from Mapbox, but also by other companies to offer other services, e.g. mobile apps. And lastly, Mapbox also relies heavily on the work of open source software from others.

The over 200 employees of Mapbox engage themselves in various projects outside their main work and affiliate themselves with the topic. These software projects have a strong vision to promote open geo-data worldwide, to be used by anyone without license costs and other charges. In this way, Mapbox's approach attracts also much talent, who want to work particularly for a company that fosters such an open source environment approach.

Altogether, Mapbox has found a profitable business model relying entirely on open source software and open data through Opentreetmaps, which makes them very resilient in a highly competitive software market. It also underlines that competition and openness can work together for the benefit of all participating companies.

Although particularly startups are benefiting a lot from this global division of software development, only a fraction of startups are able to become profitable

[4]https://medium.com/@hoffa/github-top-countries-201608-13f642493773.

[5]Stackoverflow 2016 Survey: https://insights.stackoverflow.com/survey/2016.

businesses. And despite relying heavily on software in their services, open source software is only one puzzle piece in their pursuit for a successful business.

4 The Case of Sensorica: Open Source Innovation Beyond Software

Another interesting case of future open innovation is the open value network Sensorica, which focuses on open source hardware development such as machines. Founded in 2011, Sensorica provides innovative solutions for companies and governments through open and decentralized innovation processes, without any traditional central management. In other words, instead of relying on its own limited research and development departments, companies can here tap into a large network of experts who try to find best solutions for the clients' problems. Sensorica is a decentralized collaborative organization that consists of partnerships between innovation labs or individual experts that like to cooperate. Such physical hubs are critical for hardware development as they provide the required infrastructure to build prototypes, and to test and optimize these. The solutions range from robotics to scientific instruments.[6]

"The whole idea of Sensorica starts with the observation that open-source communities have demonstrated that they can innovate very effectively." (Kreutz 2017). Sensorica picks up on the successes of open source development and builds a predictable and reliable business model behind it. Companies and governments approach Sensorica for specific product solutions, for example, the health sector. These solutions are then developed through the participation of various network partners in direct or remote collaboration. Sensorica has developed an efficient managing process to guide participating experts from idea to solution and to allocate resources and finally manage payments. Sensorica has learned and proved that competitive innovation, in form of idea challenges, does not necessarily lead to the best solutions; on contrary, they might hinder best solutions. To prevent this, competing experts need to start collaborating for the best-fit end product.

Let's suppose that this project is about building a bridge, and you choose the best design from ten designs of the bridge, and you discard the other nine. Now there's a high probability that the nine others that you have discarded have some good features that the best design you choose doesn't have. Imagine you can put all the best features from all these nine propositions together, you get the best possible design, but in competitive crowdsourcing you end up with one best out of ten instead of the best possible one.

Potentially, every person can be affiliated to the network and can contribute at any of the different steps of a process to achieve an end-product (e.g. project management, testing, planning). The more people are available, the more stable the

[6]Sensoria Projects: http://www.sensorica.co/home/what-we-do/projects.

process becomes; therefore, a certain critical mass of contributors is required, which could be challenging. The key to this is to set the right incentives in monetary form or recognition.

"There are different types of roles and some activity can be broken down into very small tasks that are taken sporadically by individuals. Other roles require more core involvement and sustained involvement by individuals. [...] if you're like a core member of the project, maybe we carve out a bigger portion of the budget for you." With this approach, Sensorica was able to accomplish dozens of projects in the past years, and moved the merits of open source software development to manufacturing hardware. However, according to Tiberius Brastaviceanu, one large challenge remains: "Networks don't know yet how to manufacture." Sensorica is great at finding great solutions, but is not able to provide the whole supply chain necessary to manufacture complex products.

5 Transforming into an Agile Organizations

As demonstrated by diverse open innovation examples, open source collaboration entails a lot of potential and promises, but also face various challenges. Established companies have reacted to this rising competition and to challenges for digital transformation in various ways. The Dutch banking group ING went recently through such a change process. They realized that they needed to become a "technology company operating in the financial-services business." And they were resolved to achieve that through more agility: "Agility is about flexibility and the ability of an organization to rapidly adapt and steer itself in a new direction. It's about minimizing handovers and bureaucracy, and empowering people." Their objectives were to execute quicker to market, increase employee engagement across departments for more collaboration, and to improve client experience particularly through their digital channels. They gave up on traditional hierarchy, formal meetings and detailed planning. Not all personnel was willing to adapt to the new 'structure' and that's why ING concluded from their experience that culture was the most important element of that change. "You need to look beyond your own industry and allow yourself to make mistakes and learn. The prize will be an organization ready to face any challenge."

This example shows the burden established organizations have to take to (a) adjust their operations to the increasing digitization of services and (b) to be able to innovate quicker to dynamic markets. Organizations that want to benefit from this open source innovation, need to adapt their strategy and organizational model, and work long-term on culture change and openness from within and to the outside.

That is why many organizations prefer less radical steps and approach the challenge more slowly. A popular way is the innovation lab or technology hub, often founded by companies in a special location, in order to create a more open sphere of innovation. These hubs are meant to provide the startup type of open collaboration, often incorporating different projects and interdisciplinary teams with enough time and resources to think about new ideas instead of incrementing the old ones. These hubs often hold event incubators and organize challenges such as Hackathons to attract outside talent. Successful challenges attract a diverse group of people necessary for breakthrough ideas. Typical incentives are incubators, so that good ideas find their required support. This way, many companies hope to benefit from such ideas, which could later be incorporated into the company systems. But here lies a great challenge: Collaboration culture, flat organizations, and the self-understanding of the team players are not compatible with the old hierarchical organizational style of companies.

Whatever way one chooses to tap into this global innovation network, it requires a new type of openness within the organization and to the outside world. It starts by asking what knowledge needs to be kept inside and confidential? Or can it be open by default in order to build a greater eco-system around products and services? This transformation requires encouraging employees to build relationships and practice constant learning. Because future successful companies are those ones that overcome the traditional closed innovation model and actively participate in a global innovation eco-system with great opportunities.

References

Baur, P. (2017). *The advance of open innovation interview with Henry Chesbrough*. Retrieved from http://news.sap.com/henry-chesbrough-interview-advance-of-open-innovation/.

Chesbrough, H. (2017). The Future of Open Innovation: The future of open innovation is more extensive, more collaborative, and more engaged with a wider variety of participants. *Research-Technology Management, 60*(1), 35–38.

Kreutz, C. (2017). *Sensorica: Self managed open network innovation. Interview with Tiberius Brastaviceanu*. Retrieved from https://www.crisscrossed.net/2017/06/23/Sensorica-Self-Managed-Open-Network-Innovation/.

Munir, H., Linåker, J., Wnuk, K., Runeson, P., & Regnell B. (2017). Open innovation using open source tools: A case study at Sony Mobile. *Empirical Software Engineering*, 1–38.

Richter, F. (2015). *Startup innovation*. Retrieved from http://10innovations.alumniportal.com/startup-innovation.html.

Author Biography

Christian Kreutz is a Political Scientist and a Programmer by passion. He is a Digital Innovation Consultant, who has worked with various organizations, such as the United Nations, World Bank, GIZ and the German government. He is also the Founder of WeThinq, an open innovation platform service, which supports organizations and companies to solve challenges through collective intelligence. He writes on crisscrossed.net and teaches at the University of Applied Sciences in Darmstadt.

Digital Science: Cyberinfrastructure, e-Science and Citizen Science

Roberto C. S. Pacheco, Everton R. Nascimento and Rosina O. Weber

Abstract Digital change and scientific development have mutual implications. On one hand, science and technology development has been a major factor to digital change. On the other hand, the digital era has brought major changes to scientific knowledge production. First, there is a *cyberinfrastructure*—not only infrastructure for computing, but a major virtual lab where all professionals in science and technology (e.g., researchers, engineers, technicians) can collaborate and exchange data, information, and knowledge. In Europe, this new infrastructure is referred to as *e-science*. Second, the digital era has increased coproduction beyond frontiers of traditional players, bringing other participants to scientific development. Such kind of co-work is central to both *citizen science* and *transdisciplinary* knowledge coproduction, where non-academic players engage in activities such as planning, data gathering, and impact assessment of science. In this chapter, we define digital science as a convergent phenomenon of cyberinfrastructure, e-science, citizen science and transdisciplinarity. We examine how digital science has been a disruptive factor to traditional scientific development, changing productivity, expanding frontiers and challenging traditional processes in science, such as planning and assessment.

1 Introduction

Digital change and scientific development have mutual implications. On one hand, science and technology development has been a major factor to digital change. On the other hand, the digital era has brought major changes to scientific knowledge production.

R. C. S. Pacheco (✉) · E. R. Nascimento
Federal University of Santa Catarina, Florianópolis, Brazil
e-mail: pacheco@egc.ufsc.br

R. O. Weber
Drexel University, Philadelphia, USA

K. North et al. (eds.), *Knowledge Management in Digital Change*, Progress in IS,
https://doi.org/10.1007/978-3-319-73546-7_24

The digital era has already fostered scientific development, offering technologies to support online learning, virtual labs, global research networks and worldwide sharing of computer facilities. Nevertheless, as we can learn from other digital changes, technological applications are only one of many impacts of the digital era. In science, digital technologies are changing scientific knowledge production not only in scale and speed, but also in its own nature, bringing new stakeholders and challenging traditional structures.

This new state of affairs in science has been studied in different fields, including *cyberinfrastructure* (CI), *e-science* (eS), *citizen science* (CS) and transdisciplinarity. In this chapter, we review these fields as a convergent concept, named *digital science*, a system that yields common (digital) spaces for knowledge coproduction based on open access and connectivity. We begin by addressing the conceptual backgrounds of CI, eS, CS, and transdisciplinarity. We then propose the convergent notion of *digital science*, discussing its challenges, trends and impacts.

2 Conceptual Background

Cyberinfrastructure (CI) (Atkins 2003) was first proposed as a vision of how computing technologies can be used to efficiently and effectively support scholarly research or *researchers* (i.e., scientists, engineers, humanists). According to the National Science Foundation (NSF) such vision is a combination of technology, interdisciplinary teams and new educational and workforce initiatives. It is basically "computing systems, data, information resources, networking, digitally enabled-sensors, instruments, virtual organizations, and observatories, along with an interoperable suite of software services and tools" shared by interdisciplinary teams of professionals who develop, deploy and use "transformative approaches to scientific and engineering discovery and learning" (Council 2007, p. 2).

In the last decade, several authors have proposed alternative definitions to CI (e.g. Stewart 2007; Bietz et al. 2010; Marshall et al. 2011; Young and Lutters 2015). These authors seem to agree that CI is not only a vision, but rather a complex arrangement of technological, human and organizational factors that foster research endeavors that would not be possible without such infrastructure. CI applies cutting-edge information technologies enabling a new paradigm of data-driven scientific research, where investigation is collectively conducted with access to large volumes of complex data (Shahand et al. 2015). Most importantly, this new paradigm is not only computationally driven, but it is an end-to-end integration performed by cross-disciplinary collaboration with impact to the way science is conducted (Ribes and Lee 2010). One of the main goals is to foster efficiency through leveraging and reusing research efforts to encourage reproducibility (Gil et al. 2007).

In Europe, CI refers to the concepts of *e-science* (eS) and *e-infrastructure*. The term e-science was defined by John Taylor, the director of Research Councils in the UK Office of Science and Technology (OST), as a "global collaboration in key areas of science and the next generation of infrastructure that will enable it." (Hey and Trefethen 2002). The term e-infrastructure is used either as a synonym of CI (e.g. Almes et al. 2004) or as the infrastructure component of CI (in this case, e-science can be seen as a scientific method which foresees the adoption of cutting-edge digital platforms known as e-infrastructures, Foster 2003). Thus, although e-science and e-infrastructure have been used interchangeably across Europe (Jirotka et al. 2013), sometimes they are perceived either as technological and/or a conceptual paradigm to a new way of pursuing scientific development.

Similarly as to what happened to CI, e-science has also been studied in the last decade with alternative definitions. Some authors have neglected the notion of a regional analogy of CI (e.g. Taylor 2001), considering e-science a paradigm to global collaboration based on computationally intensive scientific research (Cho 2007). In this view, the ultimate goal of e-science is to help increase the rate of scientific discoveries by empowering scientists with CI in collaborative research conducted in sustainable global networks (Alvarez et al. 2007).

In this chapter, we consider this last notion of CI as the technological, organizational and cultural infrastructure that enable e-science, a global community of cross-disciplinary researchers investigating complex phenomena as a global network, working with large amounts of data using cutting-edge information technologies. In this notion, CI is a sociotechnical structure to e-science that can also be applied to other global collective actions, such as the ones studied in this book.

Most studies consider that the term "citizen science" (CS) was introduced in 1989, in the United States, when 225 volunteers collected, tested and reported results about rain samples (e.g. Bonney 1996; Gharesifard et al. 2017). Six years later, Alan Irwin referred to *citizen science* as a "science developed and enacted by citizens themselves" (Irwin 1995 p. xi). As pointed out by Gharesifard et al. (2017) p. 381, only recently the term was added to the Oxford English Dictionary as "the collection and analysis of data relating to the natural word by members of general public, typically as part of a collaborative project with professional scientists" (Dictionary 2007).

In sum, CI and eS enable scientific development by crowdsourcing in global research networks (Law et al. 2017) while CS opens space to general public to engage in investigative endeavors. This last phenomenon has also been referred to as a particular kind of transdisciplinarity, where scientific and non-academic players join efforts in coproduction of knowledge (Frodeman 2013; Klein 2017).

The interest in CI, eS, CS and transdisciplinarity has been growing over the years, as shown in Fig. 1.

As can be seen in Fig. 1, in the last 14 years, 621 articles were published in CI, eS, CS and Transdisciplinarity. The search was performed in different databases, so the results included redundancies and articles not precisely related to the original terms. We then analyzed the full documents, checking abstracts, keywords and goals. Consequently, content analysis was applied only to 116 articles. The most frequent and relevant terms are presented in Fig. 2.

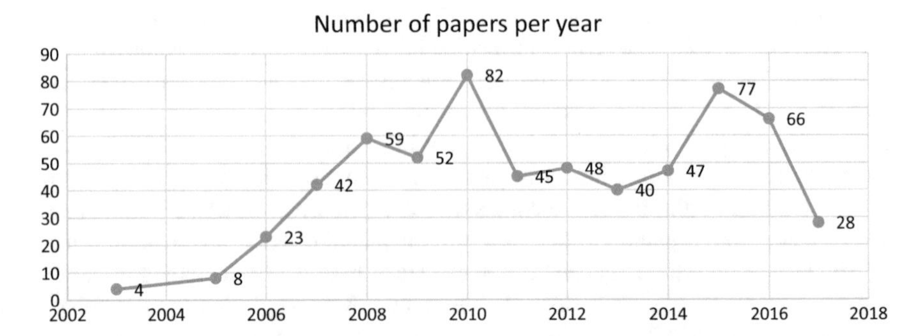

Fig. 1 Total of articles published in CI, eS e CS in the last 15 years (Sources: Scopus, Web of Knowledge, Emerald Insight and EBSCO host). Papers with "Transdisciplinary" AND "Knowledge" AND ("co-production" OR "coproduction"); "Citizen science" AND "Cyberinfrastructure"; "e-Science" AND "Cyberinfrastructure"; "Transdisciplina*" AND ("Cyberinfrastructure" OR "e-Science" OR "Citizen Science")

Fig. 2 Word cloud of related terms to CI, eS, and CS

As can be seen in Fig. 2, the content analysis revealed methods, technologies and fields of applications related to CI, eS and CS. In Fig. 3, we detached the most important terms and their corresponding definitions.

As it can be seen in Fig. 3, conceptually CI, eS and CS offer a multilayer system to contemporaneous science, perceived as a joint effort of scientific and non-academic players with open access to their resulting knowledge. It is a co-working process based on data analysis and community monitoring. These worldviews are methodologically supported by new workflow systems and knowledge engineering methods (i.e., semantic web applied to science, linked data, network ontology) and related fields. Such methods are applied by multidisciplinary teams with new technologies (cloud computing, on-demand access grid computing and connectivity technologies) that allow co-working and sharing of huge amounts of data and applications in service-oriented computing.

Concepts

Crowdsourcing: "a new paradigm for utilizing the power of "crowds" of people to facilitate large scale tasks that are costly or time consuming with traditional methods." (Yan et al., 2009, p. 347).

Collaboratories: "virtual entities that allow scientists to collaborate with each other across organizations and physical locations" (Gil et al., 2007, p. 25)

Open access: is "an alternative to the traditional subscription-based publishing model made possible by new digital technologies and networked communications (...) with no expectation of direct monetary return and made available at no cost" (McLellan, 2003, p.52).

Data science: "the application of quantitative and qualitative methods to solve relevant problems and predict outcomes". (Waller and Fawcett, 2013, p. 78).

Methods

Community Based Monitoring (CMB): "a process where concerned citizens, government agencies, industry, academia, community groups and local institutions collaborate to monitor, track and respond to issues of common community concern" (Whitelaw et al., 2003, p. 410).

Scientific workflow systems: is "a system that orchestrates and manages virtual experiments for scientists" (Rygg, Sumitomo and Roe, 2006, p. 2).

Semantic e-Science: "is an approach supporting research collaboration in which all the services of data access, integration, provenance, and data processing need semantic representation". (Le Dinh, Nomo, and Ayayi, 2015, p. 38).

Linked data: was first proposed by Tim Berners-Lee (2006) to indicate how data should be published on the web (i.e., as a network of machine readable, connected data disposed on a non-proprietary format according to RDF standards).

Network ontology: "is a formal specification that describes the capabilities of the network" (Koderswaran and Joshi, 2009, p. 4). It is also a "a meta-ontology that draws on established ontologies and controlled vocabularies" (Srinivasan et al., 2007, p. 325).

Technologies

Service-oriented computing: "is a paradigm that utilizes services as fundamental elements for application design" (Escoffier, Hall, and Lalanda, 2007, p. 474), a distributed computing and e-business processing that changed the software applications design, architecture, delivery and use (Amir and Zeid, 2004, p. 192).

Cloud computing: "a model for enabling ubiquitous, convenient, on-demand network access to a shared pool of configurable computing resources (e.g., networks, servers, storage, applications, and services)". (Mell and Grance, 2011, p. 3).

Grid computing: "refers to the large-scale integration of computer systems (via high-speed networks) to provide on-demand access to data-crunching capabilities and functions not available to one individual or group machines". (Foster, 2003, p. 81).

Connectivity technologies: "Connectivity technologies are those that provide communications and connectivity between systems, including enterprise network management, videoconferencing systems (e.g. routers, VoIP, Ethernet)" (Sethi, Larson and Tafti, 2014, p. 6), mobile, IoT, and others.

Fig. 3 CI, eS and CS concepts, methods and technologies

3 Digital Science

The three layers described in Fig. 2 are foundational backgrounds to science in the digital era. Although recent studies have acknowledged transformational digital changes to science, when it comes to the notion of "digital science" most authors detach mainly the technological impacts of the digital era over traditional scientific development (e.g. Buchanan 2016).

The terms "open science" and "Science 2.0" have been used to acknowledge the impact of digital technologies to scientific communication and access to non-scientists (Curtis 2015). The digital era is though, a multidimensional phenomenon characterized by convergence and integration with impacts over institutional, political, technological, economic, social and cultural aspects of every collective activity. The digital era has brought a much broader change than new ways of communicating and giving access to science. It is challenging stakeholders' roles and limits, inviting them new worldviews of science, where all players have responsibilities and a common goal to share in sustainable development.

In this chapter, we define *digital science* as a system shared by scientific and social communities engaged in solving complex problems based on common good and sharing a set of methods, data, information, technological and methodological infrastructure. In Fig. 4, we present a general view of this digital science system.

As represented in Fig. 4, digital science stakeholders include researchers, citizen scientists, students, professors, policy makers, business people, and social workers. They work in *collaboratories* sharing service oriented, highly connected technologies; particularly grid and cloud computing. They create a common knowledge space following a scientific workflow system and working as a transdisciplinary team. Data is, at the same time, an online income and an open outcome in

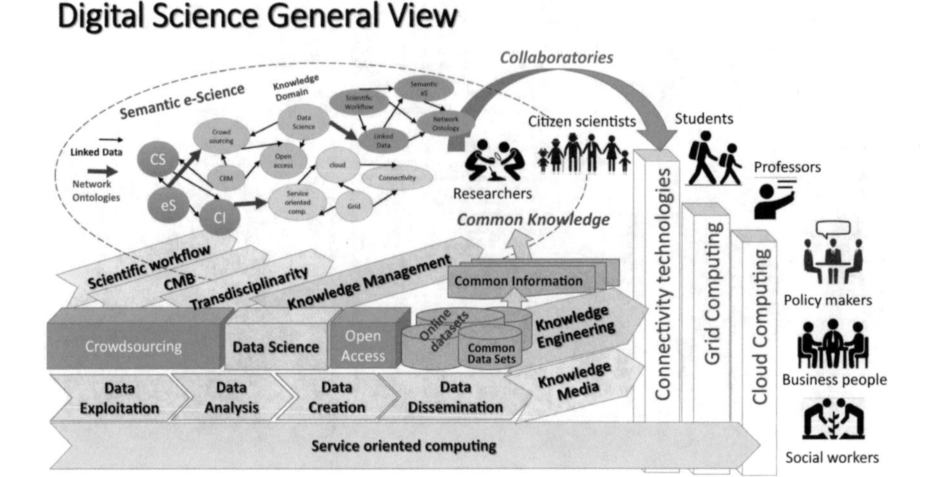

Fig. 4 Digital Science general view

crowdsourcing projects monitored by interested communities. Knowledge engineering and data science methods (e.g. Waller and Fawcett 2013) and techniques allow the creation of a semantic space to science, where data and ontologies are linked according to their meaning and purpose.[1] Knowledge media helps to make the outcomes accessible and socially acknowledged and knowledge management offers a set of practices and guidelines to make digital science teams work as sustainable virtual organizations.

A literature review will probably not find a single case that covers all elements in the complex digital science system represented in Fig. 4. However, there are already thousands of digital science projects (Bonney et al. 2014). CI, eS and, mainly, CS have several fields of application, including biology (Wei et al. 2016), environmental science (Dickinson et al. 2012; Donnelly et al. 2014), astronomy (Raddick et al. 2009), volunteered geographic information (Goodchild 2007), public governance (Georgiadou et al. 2014), games (Newman et al. 2012), health and education (Toerpe 2013).

The variety and amount of cases and its multidimensional nature make digital science a particular instance of knowledge society. As such, two factors are particularly suitable to a further analysis in this chapter: the competences required from digital science players and the challenges and trends to digital science future.

As a complex system, digital science requires several competences from its components. A digital scientist needs computational skills to use word processors, spreadsheets, internet, scientific databases, communication systems and social

[1]Both data and ontologies are domain specific (e.g., health, law, etc.). In Fig. 4 we use "Digital Science" as a domain, with it concepts, methods and technologies represented as liked data and network ontologies (i.e., the *knowledge domain* here is what we know about digital science).

networks, besides the ability to manage large amounts of information, visit and possibly create websites. There is also a need for working attitudes such as work either alone or collectively, discern accessible from acceptable, and treat personal matters with sensitivity.

These skills allow digital scientists to co-work in computational platforms. Nevertheless, when it comes to citizen science and transdisciplinary knowledge, another set of competences are required. Digital scientists need openness (to learn from different sources, including non-academic collaborators), humbleness (to acknowledge his/her knowledge limits), empathy (to maintain shared spaces of learning and dialog), reciprocity (to offer help and knowledge), conflict management and diversity fondness (to deal with inescapable different worldviews).

3.1 Digital Science Challenges and Trends

In Fig. 5 we present a schematic view of major trends and challenges to digital science, according to three major classifications—sociocultural, technological or political/institutional dimensions.

Citizen science and transdisciplinarity have common concerns regarding citizen engagement in scientific projects, including motivation, training, project management, data governance, and conflict management (particularly when it comes to agreement on the relevance, nature, causes and consequences of a problem). These and other related issues (such as ethics and equity) are mainly social and cultural factors in digital science.

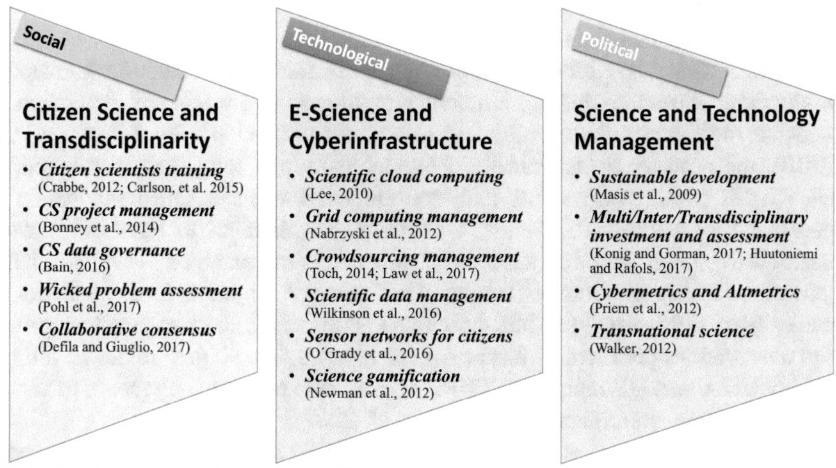

Fig. 5 Digital Science challenges and trends

The second set of subjects in Fig. 5 refers mainly to technological challenges and trends in designing, developing and maintaining cloud and grid computing as suitable infrastructures, so science can deal effectively with crowdsourcing demands and large amounts of data. Since mainly technological in nature, we refer to eS and CI as the major ground to such concerns, where other studies have pointed out several trends, such as the use of sensors and other IoT outcomes by citizen scientists and new practices such as gamefication and Massive Open Online Courses (MOOCs).

The third set in Fig. 5 refers to political and institutional elements of digital science. There is a need to a clear and collective view of the role of science in sustainable development (i.e., grow socially, environmentally and economically fair), with current and effective methods and techniques to plan, assess, control and disseminate scientific knowledge. These demands have implications in fields such as science assessment, science and technology indicators, public policies and science and technology management. Digital science makes processes such as planning, funding, hiring, controlling and communicating scientific endeavors a complex and multi-institutional system, with several impacts in traditional science and technology management.

4 Digital Science Impacts

Conceived as the convergence between CS, CI, eS and transdisciplinarity, digital science has impacted several players, factors and dimensions. However, since it is a new phenomenon, it might be too soon to completely foresee how digital science will change scientific development. Here we mention some innovations to traditional scientific knowledge production, new trends in science and technology management and stakeholders.

Digital science has brought major changes in traditional scientific knowledge production. Current computing infrastructure allows large groups of researchers to work as multidisciplinary worldwide teams. As pointed out by Ribes and Lee (2010), the impacts are technically, geographically and long term distributed. In some fields such as astronomy and environmental sciences, the participation in worldwide networks is becoming a necessary requirement to have one´s work acknowledged. On the other hand, even researchers not engaged in digital science networks can benefit from its results. They can use common large databases in studies (e.g., eBird—a global bird-watching community with more than five million bird observations per month—has been used in more than 90 peer-reviewed articles and book chapters—Bonney et al. 2014) or even in experiments by means of virtual labs and remote experimentation (Alves et al. 2007).

By including citizens as active players in scientific and technological knowledge production, digital science calls for inclusive public management. Public administration proposals such as *New Public Service* (NPS) (Denhardt and Denhardt 2007) can foster coproduction between citizens, academics and government.

This can change traditional science and technology planning and governance, which has so far been conducted exclusively by academics (peers) and public administrators. Citizen participation demands normative principles such as transparency, accountability, and coproduction of public policies and projects (Salm Jr. and Pacheco). Traditional science and technology management is challenged to review its processes to include more players. For instance, digital science challenges the traditional notion of "peer review", since knowledge production is no longer exclusive to the academic community. Interdisciplinary and transdisciplinary projects require new methods of assessment (Huutoniemi and Rafols 2017). Grant policies and funding criteria are also challenged (Koenig and Gorman 2016) and will no longer be exclusively a public administrators and academy concern. As citizens become to acknowledge the impact of government decisions to scientific development society becomes more demanding of participation in science and technology management.

As with other contemporaneous phenomena based on worldwide connectivity, digital science challenges traditional institutional systems. These include not only public funding agencies but also universities, research institutes, firms and social organizations. All these systems can make digital science an instrument to fulfill their mission, but this requires cultural and organizational transformative changes. Universities have to consider multidisciplinary and interdisciplinary curricula, flexible structures [(Hoover and Harder 2015), and agile trans-institutional relationships (especially with firms in innovation)]. Firms should create organizational practices and processes to benefit from digital science in areas such as open innovation (Seltzer and Mahmoudi 2012), crowd capital creation (Lenart-Gansiniec 2016) and brand control (Bal et al. 2017).

Despite being a recent phenomenon, digital science has already promoted changes and opened opportunities to everyone involved or affected by scientific development. However, these seem to be initial states of a new era for science. The impact of digital science in the progress of science and technology opens interesting questions for investigation in subjects such as organizational change, assessment and evaluation, funding, planning, equity and productivity.

5 Final Remarks

In the last decades cyberinfrastructure, e-science, citizen science and transdisciplinarity have not only been evolutive but have also been disruptive factors to traditional scientific knowledge production. In this chapter, we referred to these four approaches as digital science, conceived as a complex system where large transdisciplinary teams work in scientific coproduction, sharing common goals, open access, service-oriented and highly connected computing. Digital science results from the combination of technological, methodological, institutional, economic, social and cultural elements.

Digital Science is not only changing traditional scientific knowledge production and science management, but it is also encouraging all citizens to become both beneficiaries as well as co-producers of science. Citizen scientists not only comprehend the benefits of science to sustainable development, but join scientists to produce knowledge. This is probably the most significant change in the way future generations will realize science's place in society.

Naturally, this can only be done in educated and conscientious societies. Digital Science demands stakeholders such as policy makers, researchers, educators, and business people to think and act conscientiously in the role of education, culture and values in common good and sustainable development.

References

Almes, G., Cummings, J., Birnholtz, J. P., Foster, I., Hey, T., & Spencer, B. (2004). CSCW and cyberinfrastructure: opportunities and challenges. In *Proceedings of the 2004 ACM conference on Computer supported cooperative work*. ACM.

Alvarez, H. L., Chatfield, D., Cox, D. A., Crumpler, E., D'cunha, C., Gutierrez, R., Ibarra, J., Johnson, E., Kumar, K., & Milledge, T. (2007). CyberBridges a model collaboration infrastructure for e-Science. In *Seventh IEEE International Symposium on Cluster Computing and the Grid, 2007. CCGRID 2007*. Rio De Janeiro: IEEE.

Alves, G. R., Gericota, M. G., Silva, J. B., & Alves, J. B. (2007). Large and small scale networks of remote labs: A survey. *Advances on remote laboratories and e-learning experiences* **15**.

Atkins, D. (2003). Revolutionizing science and engineering through cyberinfrastructure: Report of the National Science Foundation blue-ribbon advisory panel on cyberinfrastructure.

Bal, A. S., Weidner, K., Hanna, R., & Mills, A. J. (2017). Crowdsourcing and brand control. *Business Horizons, 60*(2), 219–228.

Bietz, M. J., Baumer, E. P., & Lee, C. P. (2010). Synergizing in cyberinfrastructure development. *Computer Supported Cooperative Work (CSCW), 19*(3–4), 245–281.

Bonney, R. (1996). Citizen science: A lab tradition. *Living Bird, 15*(4), 7–15.

Bonney, R., Shirk, J. L., Phillips, T. B., Wiggins, A., Ballard, H. L., Miller-Rushing, A. J., et al. (2014). Next steps for citizen science. *Science, 343*(6178), 1436–1437.

Buchanan, M. (2016). Digital science. *Nature Physics, 12,* 630.

Cho, K. (2007). A test of the interoperability of grid middleware for the Korean High Energy Physics Data Grid system. *International Journal of Computer Science and Network Security, 7,* 49–54.

Council, C. (2007). *Cyberinfrastructure vision for 21st century discovery*. Arlington, VA: National Science Foundation.

Curtis, V. (2015). *Online citizen science projects: an exploration of motivation, contribution and participation*. The Open University.

Denhardt, J. V., & Denhardt, R. B. (2007). *The new public service: Serving, not steering*. New York: ME Sharpe.

Dickinson, J. L., Shirk, J., Bonter, D., Bonney, R., Crain, R. L., Martin, J., et al. (2012). The current state of citizen science as a tool for ecological research and public engagement. *Frontiers in Ecology and the Environment, 10*(6), 291–297.

Dictionary, O. E. (2007). Oxford English dictionary online. *JSTOR*.

Donnelly, A., Crowe, O., Regan, E., Begley, S., & Caffarra, A. (2014). The role of citizen science in monitoring biodiversity in Ireland. *International Journal of Biometeorology, 58*(6), 1237–1249.

Foster, I. (2003). The grid: Computing without bounds. *Scientific American, 288*(4), 78–85.

Frodeman, R. (2013). *Sustainable knowledge: A theory of interdisciplinarity*. Berlin: Springer.

Georgiadou, Y., Lungo, J. H., & Richter, C. (2014). Citizen sensors or extreme publics? Transparency and accountability interventions on the mobile geoweb. *International Journal of Digital Earth, 7*(7), 516–533.

Gharesifard, M., Wehn, U., & Van Der Zaag, P. (2017). Towards benchmarking citizen observatories: Features and functioning of online amateur weather networks. *Journal of Environmental Management, 193,* 381–393.

Gil, Y., Deelman, E., Ellisman, M., Fahringer, T., Fox, G., Gannon, D., et al. (2007). Examining the challenges of scientific workflows. *Computer,* **40**(12).

Goodchild, M. F. (2007). Citizens as sensors: The world of volunteered geography. *GeoJournal, 69*(4), 211–221.

Hey, T., & Trefethen, A. E. (2002). The UK e-science core programme and the grid. *Future Generation Computer Systems, 18*(8), 1017–1031.

Hoover, E., & Harder, M. K. (2015). What lies beneath the surface? The hidden complexities of organizational change for sustainability in higher education. *Journal of Cleaner Production, 106,* 175–188.

Huutoniemi, K., & Rafols, I. (2017). Interdisciplinary in research evaluation. In J. T. Klein, R. C. S. Pacheco, & R. Frodeman (Eds.), *The Oxford handbook of interdisciplinarity*. New York: Oxford University Press.

Irwin, A. (1995). *Citizen science: A study of people, expertise and sustainable development*. Psychology Press.

Jirotka, M., Lee, C. P., & Olson, G. M. (2013). Supporting scientific collaboration: Methods, tools and concepts. *Computer Supported Cooperative Work (CSCW), 22*(4–6), 667–715.

Klein, J. T. (2017). A Taxonomy of interdisciplinarity. In J. T. Klein, R. C. S. Pacheco, & R. Frodeman (Eds.), *The Oxford handbook of interdisciplinarity*. New York: Oxford University Press.

Koenig, T., & Gorman, M. E. (2016). The challenge of funding interdisciplinary research: A look inside public research funding agencies.

Law, E., Gajos, K. Z., Wiggins, A., Gray, M. L., & Williams, A. C. (2017). *Crowdsourcing as a Tool for Research: Implications of Uncertainty*. In *CSCW*.

Lenart-Gansiniec, R. (2016). Crowd Capital-Conceptualisation Attempt. *International Journal of Contemporary Management, 2016*(Numer 15 (2)), 29–57.

Marshall, J. J., Downs, R. R., & Mattmann, C. A. (2011). Software reuse methods to improve technological infrastructure for e-Science. In *2011 IEEE International Conference on Information Reuse and Integration (IRI)*. IEEE.

Newman, G., Wiggins, A., Crall, A., Graham, E., Newman, S., & Crowston, K. (2012). The future of citizen science: Emerging technologies and shifting paradigms. *Frontiers in Ecology and the Environment, 10*(6), 298–304.

Raddick, M. J., Bracey, G., Gay, P. L., Lintott, C. J., Murray, P., Schawinski, K., et al. (2009). Galaxy zoo: Exploring the motivations of citizen science volunteers. arXiv preprint arXiv:0909.2925.

Ribes, D., & Lee, C. P. (2010). Sociotechnical studies of cyberinfrastructure and e-research: Current themes and future trajectories. *Computer Supported Cooperative Work (CSCW), 19*(3–4), 231–244.

Salm, J. F., Jr., & Pacheco, R. C. New Public Service through Coproduction. *The Oxford Handbook of Interdisciplinarity*.

Seltzer, E., & Mahmoudi, D. (2012). *Planning in public: Citizen involvement, open innovation, and crowdsourcing*.

Shahand, S., Van Kampen, A. H., & Olabarriaga, S. D. (2015). Science gateway canvas: A business reference model for science gateways. In *Proceedings of the 1st Workshop on The Science of Cyberinfrastructure: Research, Experience, Applications and Models*. ACM.

Stewart, C. (2007). Indiana university cyberinfrastructure newsletter. *Retrieved November, 1*(2008), 2007–2003.

Taylor, J. (2001, July). Talk given at UK e-Science Town Meeting. JM Taylor.

Toerpe, K. (2013). The rise of citizen science. *The Futurist, 47*(4), 25.

Waller, M. A., & Fawcett, S. E. (2013). Data science, predictive analytics, and big data: A revolution that will transform supply chain design and management. *Journal of Business Logistics, 34*(2), 77–84.

Wei, J. W., Lee, B. P. Y., & Wen, L. B. (2016). Citizen science and the urban ecology of birds and butterflies—A systematic review. *PLoS ONE, 11*(6), e0156425.

Young, A. L., & Lutters, W. G. (2015). (Re)defining land change science through synthetic research practices. In *Proceedings of the 18th ACM Conference on Computer Supported Cooperative Work & Social Computing*. ACM.

Author Biographies

Roberto C. S. Pacheco, Dr., works at the Federal University of Santa Catarina (Brazil). His main research interests are in knowledge engineering, coproduction, innovation and interdisciplinarity. He is the Founder of a graduate program in Knowledge and Management Engineering (EGC/UFSC) and Founder of Instituto Stela, a knowledge engineering research institute in Brazil. He has coordinated several national and international projects in current information scientific systems (including Lattes Platform in Brazil). He is a co-editor of the second edition of Oxford Handbook of Interdisciplinarity.

Everton R. Nascimento, M.Sc., works at the University of Mato Grosso State (Brazil). His main research interests are in intellectual capital, knowledge management, coproduction, complex systems and innovation. He is a Doctoral Student of the Graduate Program in Knowledge and Management Engineering (EGC/UFSC) and Science Computation Teacher in University of Mato Grosso State (UNEMAT). His background is in artificial intelligence, graphic computation and software engineering, and his professional experience includes analysis in complex and adaptive systems and perception analysis of complex systems.

Rosina O. Weber, Dr., works at Drexel University (United States). Her main research interests are in engineering knowledge from text, context, metadata, and humans for knowledge identification, extraction, sharing, reuse, adaptation and learning and applications for cyberinfrastructure; also e-Science through intelligent methods such as case-based reasoning, neural networks, information extraction, natural language understanding, and genetic algorithms. Some of the domains in which she has worked are software assurance, finance, military, law, microbiology, and health sciences.

Glossary

Adaptive learning (environment) Learning activity (or system), where content and its presentation are tailored to the individual needs and preferences of the learner.

Augmented intelligence The concept describes systems that enhance (rather than replace) human capabilities such as creativity and interpretation by integrating knowledge from diverse sources including current state and past experiences and by generating and evaluating hypotheses.

Behaviouristic learner model Black Box learner model, where only the proper response to stimuli is required. Not really outdated, because it may be applied to a variety of informal learning processes.

Bitcoin A worldwide cryptocurrency and digital payment system invented by an unknown programmer, or a group of programmers, under the name Satoshi Nakamoto. It was released as open-source software in 2009.

Blockchain A blockchain—originally block chain—is a continuously growing list of records, called blocks, which are linked and secured using cryptography. Each block contains typically a hash pointer as a link to a previous block, a timestamp and transaction data.

Care hacking Any use of digital technologies, above all, the Web in order to take control of one's own health and use the healthcare system in new and unexpected ways.

Cognitive computing Systems that learn at scale, reason with purpose and interact with humans "naturally".

Cognitive learner model Learner model addressing the internal processes of the learner.

K. North et al. (eds.), *Knowledge Management in Digital Change*, Progress in IS,
https://doi.org/10.1007/978-3-319-73546-7

Competences Abilities, commitments, knowledge, and skills that enable a person (or an organization) to act effectively in a job or situation.

Computer-based scaffolding Is technology-based guidance and assistance in order to support learners in achieving their learning goals.

Connectivist learner model Learner model emphasizing the connections among learners as wells as learning items. It is a specialization of the constructivist model.

Connectivity The affordances of digital information and communication technologies, which potentially connect everything creating the possibility of global, real-time communication and data-exchange.

Constructivist learner model Learner model emphasizing the learner‘s task of meaningful extension of his existing knowledge base by constructing ontology extensions.

Continuous process improvement A basic mind-set that aims to increase process stability in small steps. Every step includes the four phases, (1) Plan, (2) Do, (3) Check and (4) Act. Only stable processes can be controlled and lead to high quality products or services.

Crowdfunding Is the practice of funding a project or venture by raising monetary contributions from a large number of people. Crowdfunding is a form of crowdsourcing and of alternative finance.

Cyber-physical system (CPS) The term refers to the tight conjoining of and coordination between computational and physical resources.

Decentralized autonomous organization (DAO) A DAO, sometimes labelled a decentralized autonomous corporation (DAC), is an organization that is run through rules encoded as computer programs called smart contracts. A DAO's financial transaction record and program rules are maintained on a blockchain.

Data driven knowledge discovery Based on business intelligence methods and clear goals for data analysis, the data driven knowledge discovery enables new opportunities for gaining knowledge by decreasing workload for domain experts.

Digital assessment Is a framework based on information and communication technologies in order to implement different forms of assessment.

Digital transformation The changes associated with the application of digital technology in all aspects of human society.

Domain expert A person who has long experience and proper education in a specific domain. Many of them can be seen as knowledge workers. Domain experts are not necessarily analysis experts, but often they are specific nominated process experts.

DOS MS-DOS; acronym for Microsoft Disk Operating System is a discontinued operating system for x86-based personal computers mostly developed by Microsoft.

E-Patients Patients who use the Internet to become educated, engaged, and empowered with regard to their own health and demand to be accepted as partners in healthcare.

Ethereum Is an open-source, public, blockchain-based distributed computing platform featuring smart contract (scripting) functionality. It provides a decentralized Turing-complete virtual machine, the Ethereum Virtual Machine (EVM), which can execute scripts using an international network of public nodes.

Explicit knowledge Employees' explicit knowledge refers to work-related knowledge that can be translated into formal, systematic language–manuals and guidelines for instance–that is effortlessly accessible and usable.

Feedback Feedback is the possibility to trigger an already finished run again—to learn based on the first run. The new knowledge can be used for an improved second run. Validation plays a big role in the feedback concept.

Fieldworker Frontline government employees who provide health information and referral services to families in their homes, in the community and at healthcare facilities.

Flow The uncontrollable und unexpected movement of information through digital networks.

Intelligent tutoring system Is a technology-based system simulating human tutors based on artificial intelligence.

Kickstarter Is an American public-benefit corporation based in Brooklyn, New York, that maintains a global crowdfunding platform focused on creativity.

Knowledge Refers to the tacit or explicit understanding of people about relationships among phenomena. It is embodied in routines for the performance of activities, in organisational structures and processes and in embedded beliefs and behaviour. Knowledge implies an ability to relate inputs to outputs, to observe regularities in information, to codify, explain and ultimately to predict.

Knowledge discovery Once knowledge is created, it exists within a company. Knowledge workers have to manually find the "right" artefacts searching in different data stores. Knowledge is often hard to find and elaborate—especially if it is tacit knowledge.

Knowledge graph A knowledge graphs is a graph-based knowledge representation. Knowledge graphs contain concepts associated via directed and labelled edges.

Knowledge management enables individuals, teams and entire organisations as well as networks, regions and nations to collectively and systematically create, share and apply knowledge to achieve their strategic and operational objectives.

Knowledge organization systems Expression of semantic meaning through classification logic.

Knowledge protection We define knowledge protection as a set of capabilities comprising and enforcing technical, organizational, and legal mechanisms to protect tacit and explicit knowledge necessary to generate or adopt innovations.

Knowledge work is an activity based on cognitive skills that has an intangible result and whose value added relies on information processing and creativity, and consequently on the creation and communication of knowledge.

Knowledge worker is a person who primarily engages in knowledge work. Also called "Creative Class" (Florida) or "white collar", "gold collar" workers.

Learning analytics Collection of technologies to accumulate meta data on the learning process, preferably with from a large number of learners, followed by conclusions on and readjustments of the learning process.

Learning goal Boundary object for integrating emerging ideas and professional endeavours with more mature forms of knowledge.

Learning management system Digital learning environment which facilitates the delivery of learning objects, their presentation to learners and the organisation of learning processes.

Learning oriented architecture Business architecture from the perspective of connecting just in time learning to business demands for evolved organizational competencies.

Learning pathway Sequence of knowledge objects that is traversed by a learner through a learning content store.

Localized learning The organizational learning in which knowledge from outside the company is adapted to the specifics of the company, successfully assimilated and finally successful applied in the organizational context.

Mining To form a distributed timestamp server as a peer-to-peer network, bitcoin uses a proof-of-work system. The work in this system is what is often referred to as bitcoin mining. The signature is discovered rather than provided by knowledge.

Networks Open, flexible, decentralized, collaborative, multipurpose associations of both humans and nonhumans (computers).

Ontology Collection of terms and relations among them, usually from a special field of knowledge (=domain).

Participative medicine Healthcare which encourages and enables patients to help in diagnosis and therapy.

Patient community Online community of e-patients for purposes of support, participating in medical research, sharing knowledge, etc.

Proximity A measure to describe the closeness between two or more partners in relation to a certain characteristic. For example, a high professional proximity between two partners indicates that they have a very similar professional background, that they share similar beliefs and have similar experiences.

Quality The term can be interpreted as product quality/service quality as well as process quality. To reach high quality generally, the most important factor is to define the exact goal. For example: manufacturing high quality products is only possible if a detailed specification is given and the specification includes all necessary parameters and the use case.

Quantified self People who use hardware with sensors, for example smartphones or smartwatches, in order to register their bodily and mental states for purposes of self-improvement, sport, life-style, and healthcare.

RDF The Resource Description Framework provides a graph-based data model to represent semi-structured data on the web. The RDF data model allows for expressing statements or facts in the forms of triples composed of a subject, predicate, and object. A set of triples is denominated an RDF graph.

Reddit Is an American social news aggregation, web content rating, and discussion website. Reddit's registered community members can submit content such as text posts or direct links. Registered users can then vote submissions up or down that determines their position on the page.

Satisficing Human strategy to copy with information overload.

Semantic annotation The manual or automated process of transforming unstructured content (like natural language text) into a structured representation (like linked data). A prominent example is entity linking which links phrases in a text document to the entity in a knowledge graph which the phrase refers to.

Shared decision making Formalized procedure for participative medicine, usually between a doctor and a patient, for the purpose of making consensual decisions about diagnosis and therapy.

Slack Slack is a cloud-based set of team collaboration tools and services, founded by Stewart Butterfield. Slack began as an internal tool used by their company, Tiny Speck, in the development of Glitch, a now defunct online game. The name is an acronym for "Searchable Log of All Conversation and Knowledge".

Smart contract Is a computer protocol intended to facilitate, verify, or enforce the negotiation or performance of a contract. Smart contracts were first proposed by Nick Szabo in 1996.

Social connectivity Patterns of recurring individual's behaviour in terms of availability and responsiveness using technical connectivity, such as (computer) networks, (mobile) devices and (social) applications.

Social learning Generally used to describe different methods of learning, from learning in social networks to collaborative learning. Specially used to describe a certain behaviouristic learner model.

SPARQL The SPARQL Protocol and RDF query language (SPARQL) defines the recommended language to query and manipulate data modelled in RDF.

System Closed, functionally determined, and hierarchical form of social order.

Tacit knowledge Employees' tacit knowledge refers to highly personal, work-related knowledge deeply rooted in both action and commitment in specific contexts–experience-based knowhow and skills, for instance–that is difficult to formalize and articulate.

Tacit knowledge externalization This process refers to employees' mechanisms–storytelling and collaboration, for instance–for articulating their tacit knowledge into explicit concepts usable and understandable for the organization with the chief target to increase an organisation's success.

Transparency Knowledge about where information comes from, what it is good for and what can be done with it.

Ubiquitous product (UP) A product with property of being ubiquitous, not subjected to a time frame for consumption or space constraint for delivery and production means that the interval between production and consumption is reduced to zero in any setting.

Ubiquity The quality of being present everywhere or in many places, especially simultaneously.

Virtual currency Virtual currency, also known as virtual money, is a type of unregulated, digital money, which is issued and usually controlled by its developers, and used and accepted among the members of a specific virtual community.

Printed in the United States
By Bookmasters